Biomedical Engineering: Integrating Medicine and Technology

Biomedical Engineering: Integrating Medicine and Technology

Edited by Luke Madison

New York

Hayle Medical,
750 Third Avenue, 9th Floor,
New York, NY 10017, USA

Visit us on the World Wide Web at:
www.haylemedical.com

ISBN: 978-1-63241-833-3

Cataloging-in-Publication Data

Biomedical engineering : integrating medicine and technology / edited by Luke Madison.
p. cm.
Includes bibliographical references and index.
ISBN 978-1-63241-833-3
1. Biomedical engineering. 2. Medicine. 3. Bioengineering. 4. Technology. I. Madison, Luke.
R856 .B56 2020
610.28--dc23

Table of Contents

Permissions

List of Contributors

Index

Preface

The application of engineering principles and design to biology and medicine for applications in healthcare is under the scope of biomedical engineering. The development and management of medical equipment within industry standards is also an important focus of this field. Research in biomedical engineering spans a number of significant fields, from bioinformatics, biomechanics and biomedical optics to genetic engineering, neural engineering and clinical engineering. Prominent biomedical engineering applications include among others, the development of biocompatible prostheses, regenerative tissue growth technology, imaging methodologies and therapeutic biologicals. Many medical devices have also been developed owing to research in biomedical engineering, such as dialysis machines, corrective lenses, ocular prosthetics, artificial organs, implants, pacemakers, etc. Medical imaging technologies such as MRI, PET, PET-CT, X-rays, CT scans, ultrasound, electron microscopy, etc. have been developed due to research in biomedical engineering. This book is a valuable compilation of topics, ranging from the basic to the most complex advancements in biomedical engineering. The various studies that are constantly contributing towards advancing technologies and evolution of this field are examined in detail. It is an essential guide for both academicians and those who wish to pursue this discipline further.

After months of intensive research and writing, this book is the end result of all who devoted their time and efforts in the initiation and progress of this book. It will surely be a source of reference in enhancing the required knowledge of the new developments in the area. During the course of developing this book, certain measures such as accuracy, authenticity and research focused analytical studies were given preference in order to produce a comprehensive book in the area of study.

This book would not have been possible without the efforts of the authors and the publisher. I extend my sincere thanks to them. Secondly, I express my gratitude to my family and well-wishers. And most importantly, I thank my students for constantly expressing their willingness and curiosity in enhancing their knowledge in the field, which encourages me to take up further research projects for the advancement of the area.

Editor

Fusing multi-scale information in convolution network for MR image super-resolution reconstruction

Chang Liu[1,2,3,4] iD, Xi Wu[5*], Xi Yu[1,4], YuanYan Tang[6], Jian Zhang[7] and JiLiu Zhou[5]

*Correspondence:
xi.wu@cuit.edu.cn
[5] Department of Computer Science, Chengdu University of Information Technology, Chengdu 610225, China
Full list of author information is available at the end of the article

Abstract

Background: Magnetic resonance (MR) images are usually limited by low spatial resolution, which leads to errors in post-processing procedures. Recently, learning-based super-resolution methods, such as sparse coding and super-resolution convolution neural network, have achieved promising reconstruction results in scene images. However, these methods remain insufficient for recovering detailed information from low-resolution MR images due to the limited size of training dataset.

Methods: To investigate the different edge responses using different convolution kernel sizes, this study employs a multi-scale fusion convolution network (MFCN) to perform super-resolution for MRI images. Unlike traditional convolution networks that simply stack several convolution layers, the proposed network is stacked by multi-scale fusion units (MFUs). Each MFU consists of a main path and some sub-paths and finally fuses all paths within the fusion layer.

Results: We discussed our experimental network parameters setting using simulated data to achieve trade-offs between the reconstruction performance and computational efficiency. We also conducted super-resolution reconstruction experiments using real datasets of MR brain images and demonstrated that the proposed MFCN has achieved a remarkable improvement in recovering detailed information from MR images and outperforms state-of-the-art methods.

Conclusions: We have proposed a multi-scale fusion convolution network based on MFUs which extracts different scales features to restore the detail information. The structure of the MFU is helpful for extracting multi-scale information and making full-use of prior knowledge from a few training samples to enhance the spatial resolution.

Keywords: Super-resolution reconstruction, Multi-scale information fusion, Convolution network, Magnetic resonance imaging

Background

A higher magnetic resonance image (MRI) resolution often results in fewer image artifacts, such as the partial volume effect (PVE), and a higher algorithm accuracy in the post-image processing steps (e.g., image registration and image segmentation). However, the MR resolution is affected by various physical, technological and economic limitations. Thus, increasing the spatial resolution is of considerable interest in the field of medical image processing. Conventional super-resolution (SR) methods using Bicubic and B-spline interpolation [1, 2] compute new voxel gray-values according to certain smoothness assumptions. However, these methods are not always valid in non-homogeneous areas and result in blurred images.

Super-resolution technologies have been implemented in the following two major categories. (1) During the acquisition stage, k-space data can be manipulated, and the parameters can be configured to improve the spatial resolution [3, 4]. (2) During the post-processing stage, conventional image super-resolution methods can be adapted and applied to MRI. Peled and Yeshurun [5] and Greenspan [6] applied an iterative back-projection method to 2D and 3D MRI super-resolution. The resolution enhancement [7] and non-local method [8] were also implemented and extended to reconstruct a high-resolution image from corresponding low-resolution image with inter-modality priors from another HR image [9].

Recently, sparse coding (SC)-based super-resolution approaches have been shown to have good performance and accuracy in several applications, including de-noising [10], and restoration [11]. Donoho [12] reconstructed MRI from a small subset of k-space samples to solve the super-resolution problem. Yang et al. [13] and Zeyed et al. [14] implemented sparse representations of natural images and successfully adapted these representations to MRI [15]. The sparse representation-based super-resolution method involves several steps. First, low-resolution and high-resolution dictionaries are trained by overlapping patches cropped from low- and high-resolution images, respectively. Based on this, the low-resolution images are considered sparse combinations of patches in the low-resolution dictionary space. Finally, the solved sparse coefficients are mapped onto a high-resolution dictionary space and used to reconstruct the high-resolution version. Since the conventional sparse representation method trains a dictionary based on a gradient or Sobel features, the reconstructed high-resolution images are not robust and are sensitive to noise. Additionally, the independent SC of the sequential patches cannot ensure the optimal reconstruction of entire dataset [16, 17].

Deep learning algorithms, such as the deep forward neural network or multiple layer perceptron, have recently regained their popularity [18–27] due to an improved computer infrastructure (i.e., software and hardware) and increased amount of available training data. Deep convolutional neural networks (CNN) are specialized deep forward neural networks that use a convolution operation on 1D, 2D and 3D grids (e.g., 1D time series, 2D and 3D images). Successful applications in computer vision date back several decades [28, 29]. Recently, CNN-based methods have resulted in a significantly reduced error rate that is comparable to or better than that achieved by humans in many computer vision applications, such as image classification [30], object detection [31], face recognition [32], and natural image super-resolution [26, 27]. CNN-based super-resolution learns image representations from training data similarly to all other deep learning approaches. Thus, it often produces better results than conventional feature-engineering-based methods, such as SC, when a large amount of training data is available [26], formulated the super-resolution problem into a function approximation problem. These authors have implemented a cascading convolution neural network to solve the problem of natural image super-resolution reconstruction. The end-to-end optimization of a large amount of training data produced a better result than the SC-based approach.

In the literature, a few studies have addressed the MRI super-resolution problem using the deep CNN approach. A higher dimension (3D) MRI is associated with a huge computational burden and complicates the training more than a 2D version. In addition, a large amount of training data is not always available. To overcome these challenges, we were inspired by studies using multi-scale analyses and residual networks [15, 33]; we fused multi-scale information and propagated this information along the convolution network. Unlike conventional deep CNN learning, we observed that fusing multi-scale information in a convolution network makes it easier to achieve 3D MRI super-resolution using a limited amount of training data. In addition, the experiments indicated that the multi-scale fusion convolution network (MFCN) preserved detailed image information during the reconstruction procedure, which is essential for medical image applications.

The contributions of our work include the following three aspects:

- We illustrated different convolution responses using different convolution kernel sizes experimentally and demonstrated that fusing different responses was beneficial for recovering detailed information from a low-resolution image. Conventional CNNs can learn different scale information from different convolution layers, but they are unable to integrate different scale information and decrease the error during the back-propagation procedure.

- To overcome the drawback of conventional CNNs and integrate multi-scale information induced by different convolution layers, we developed an MFCN. The proposed network, which is stacked by a multi-scale fusion unit (MFU), is a full convolution network that is capable of learning end-to-end mapping between low- and high-resolution images, makes full use of prior knowledge from high-resolution images, and uses multi-scale information to infer missed details in low-resolution images. This network exhibits an outstanding performance in MRI reconstruction. The proposed network also has a faster convergence speed than the traditional convolution network. This network is capable of learning feature maps and provides exact guidance for the design of network architecture.

- Contrary to the argument that "deeper is not better" [26], we found that a larger kernel size, an increased number of kernels, and a deeper structure are all beneficial for improving the reconstruction performance. However, these features increase the computational burden and converge more slowly. Considering the ideal trade-off between performance and speed, the adopted network structure has achieved a better performance with both simulated and real MRI data compared to some classical SR methods.

The remainder of this paper is organized as follows. "MRI super-resolution with deep learning" section presents detailed information regarding the implementation of MFCN for solving the super-resolution problem. "Experiments" section provides extensive validation using both simulated and real brain MRI datasets. A discussion and conclusion are presented in "Discussion" and "Conclusion" sections respectively.

MRI super-resolution with deep learning

Problem formulation

In the field of medical image analysis, a low-resolution MR image, L, can be presented as a blurred and down-sampled version of a high-resolution image, H, as follows:

$$L = DSH + e \tag{1}$$

where e is the noise, D is a down-sampling operator, and S is a blurring filter. The degradation procedure is shown in Fig. 1.

In Eq. (1), the high-resolution image can be estimated by minimizing the following cost function:

$$\tilde{H} = \arg \min \|DSH - L\|^2 \tag{2}$$

where $\tilde{H}$ is the reconstructed high-resolution image. However, the above problem is ill-posed, and it is difficult to find a perfect solution that satisfies Eq. (2). Normally, image patches are extracted to alleviate the ill-posed nature of the problem as follows:

$$\tilde{H}_i = \arg \min \sum_i^m \|DSH_i - L_i\|^2 \tag{3}$$

where H_i and L_i represent the i-th patch cropped from the high- and low-resolution images, respectively, $\tilde{H}_i$ is the i-th reconstructed high-resolution patch and m is the number of patches. Therefore, the key issue becomes identifying the mapping relationship, DS, in Eq. (3) that maps the high-resolution images onto the low-resolution images.

MFCN for achieving MRI super-resolution

Analysis of the network architecture

While implementing super-resolution reconstruction using deep learning, it is natural to acquire a mapping from the low- to high-resolution images. Generally, the low-resolution image is up-sampled to have the same size as the high-resolution image before SR. Previous studies [26] have successfully implemented natural image SR with convolution neural networks. The SR based on the deep convolutional network is easy to implement due to its end-to-end learning strategy. An overview of SR based on the deep convolutional network is shown in Fig. 2.

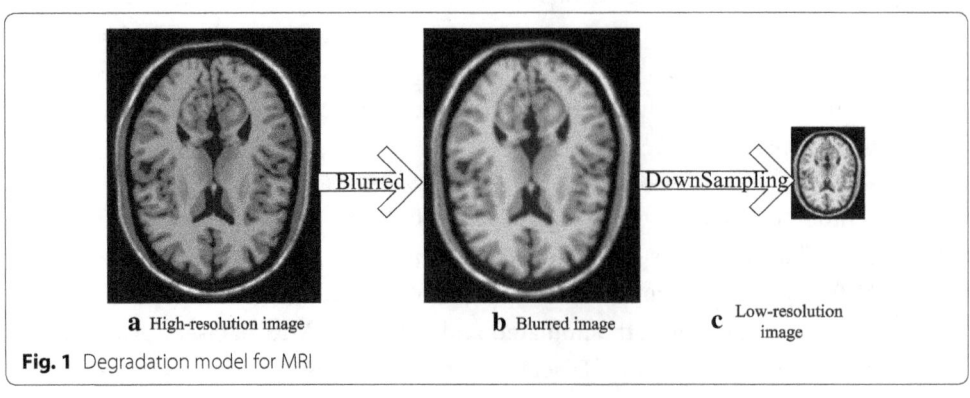

    a High-resolution image   **b** Blurred image   **c** Low-resolution image

Fig. 1 Degradation model for MRI

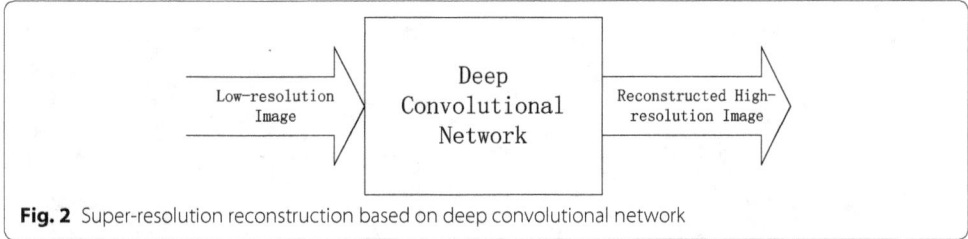

Fig. 2 Super-resolution reconstruction based on deep convolutional network

The success of convolution neural networks in SR mostly depends on the contribution of the learned convolution kernels from the training samples. To investigate the effects of different convolution kernels in SR tasks, we generated two distinct kernels with sizes of 3×3 and 15×15 for a better visual representation. Then, the two kernels were applied to a simple low-resolution image. The convolution results and the difference between the high-resolution and low-resolution images are shown in Fig. 3. As shown in the first row, the main difference between the high-resolution and low-resolution images is at the edges. Therefore, the task of SR is to recover detailed information, such as edges. Furthermore, the second and third rows in Fig. 3 show that convolution operations with different kernel sizes yield varying responses along the edges, and the strengths of the responses depend on the size of the convolution kernels. Due to the receptive field range of the convolution kernels with different sizes, the larger convolution kernels induce stronger responses along the edges. Consequently, these convolution responses are extracted as multi-scale information of the convolution kernels.

Design of multi-scale network architecture

Due to the forward and back propagation mechanisms in the convolution neural network, we constructed a simple convolution network stacked by two convolution layers as shown in Fig. 4. Both convolution layers have only one convolution kernel. In the convolution network, the input low-resolution images are submitted to the network and convoluted using the following convolution layers sequentially to obtain the feature maps. This procedure is called forward propagation. After the final convolution layer, the errors in the feature maps and high-resolution images, and the difference images, are computed based on the Euclidean distance of the loss layer. The difference images are very important for adjusting the kernel parameters of the final convolution layer. All parameters of each layer are adjusted using stochastic gradient descent.

Due to the multi-scale properties of different kernel sizes, fusing different scale convolution responses is assumed to accelerate the SR procedure. In the following study, we developed a simple MFCN as shown in Fig. 5. As depicted in Fig. 4, the MFCN has two convolution layers, and each layer has only one convolution kernel. We added a fusion layer to the network shown in Fig. 5. The function of the fusion layer is simply to add feature maps from (b) and (c). Initially, the fusion image had more details than the feature map in (c). Moreover, compared with the difference image I in Fig. 4, the difference image (f) in Fig. 5 is darker, which indicates less error between the recovered image and high-resolution image and is beneficial for accelerating the convergence in the training phase.

Therefore, it is desirable to design a convolution network that combines different scale information. Reconstructed images benefit from end-to-end learning of low/

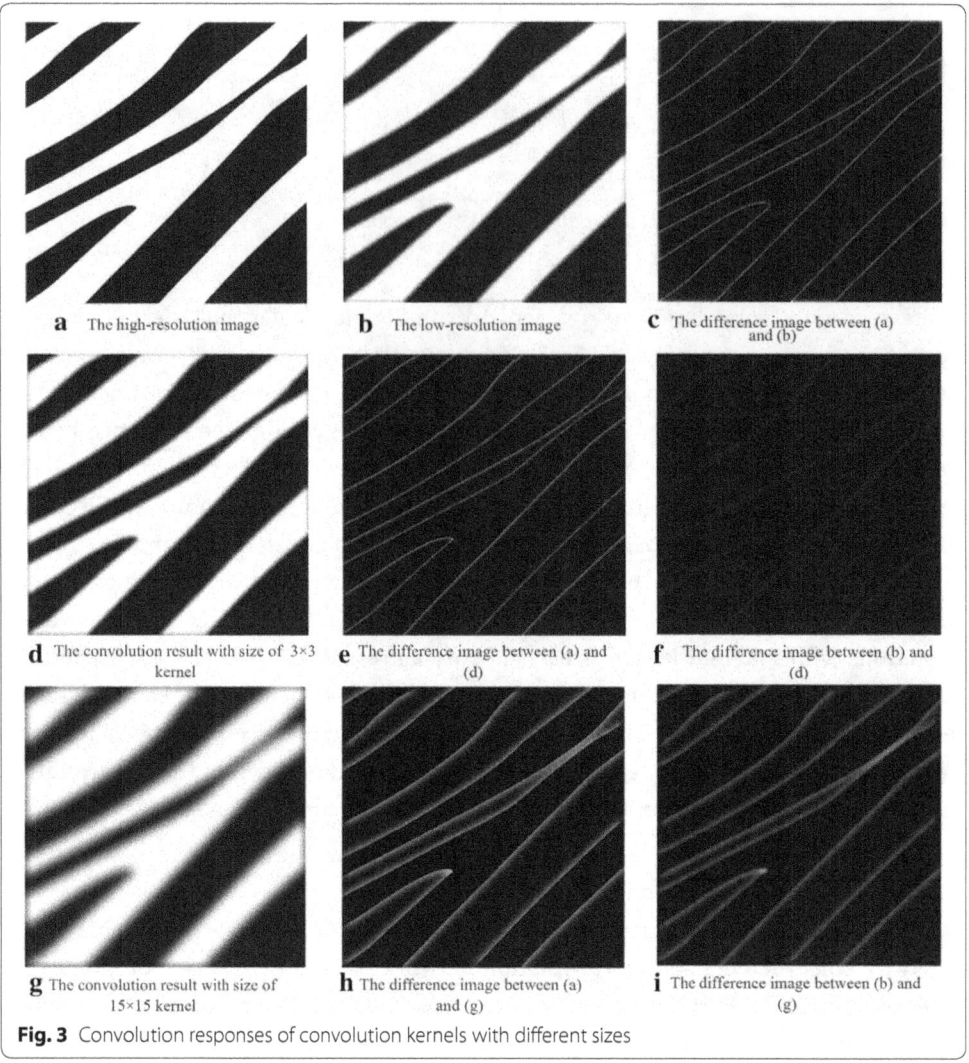

a The high-resolution image   **b** The low-resolution image   **c** The difference image between (a) and (b)

d The convolution result with size of 3×3 kernel   **e** The difference image between (a) and (d)   **f** The difference image between (b) and (d)

g The convolution result with size of 15×15 kernel   **h** The difference image between (a) and (g)   **i** The difference image between (b) and (g)

Fig. 3 Convolution responses of convolution kernels with different sizes

high-resolution images and multi-scale information propagation through the whole network structure. Inspired by residual networks [33], we defined the following structure, i.e., the MFU, to fuse different convolution paths as shown in Fig. 6:

$$x_{i+1} = f(x_i, W_0) + \sum_{j=1}^{J} SP_j(x_i, \bar{W}_i) \tag{4}$$

where f represents the convolution layer and ReLU. $SP_j(x_i, \bar{W}_i)$ denotes the j-th sub-path in which input x_i is convoluted by some convolution kernels. J is the number of sub-paths. According to the main path and several sub-paths, different scale information is extracted by various convolution kernels, and then, the multi-scale information is combined in the fusion layer based on additional operations. Output x_{i+1} retains more detailed information than the output from the traditional convolution network that is simply stacked by a convolution layer and helps accelerate the convergence. "Experiments" section provides a validation.

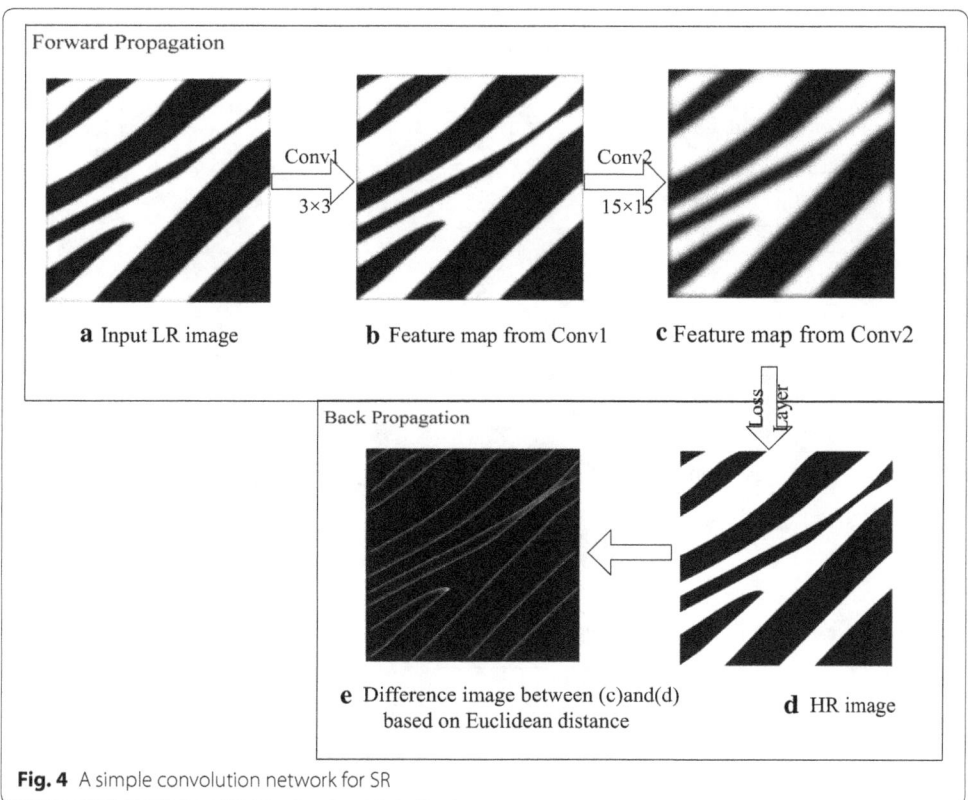

Fig. 4 A simple convolution network for SR

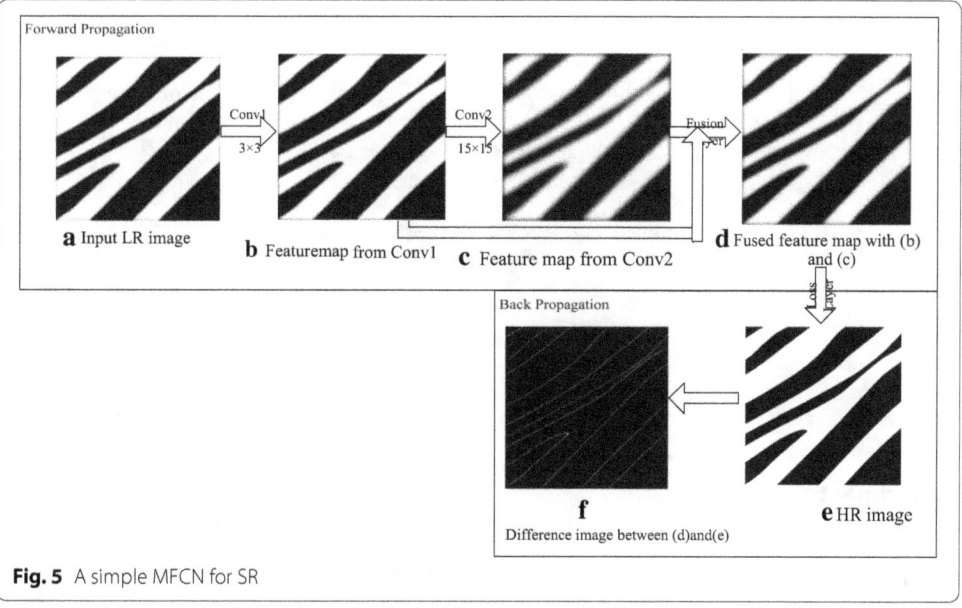

Fig. 5 A simple MFCN for SR

Based on the above-mentioned MFU, we developed the MFCN shown in Fig. 7. This network is stacked by a few MFUs and a reconstruction layer, in which the reconstruction layer is a convolution layer with one kernel.

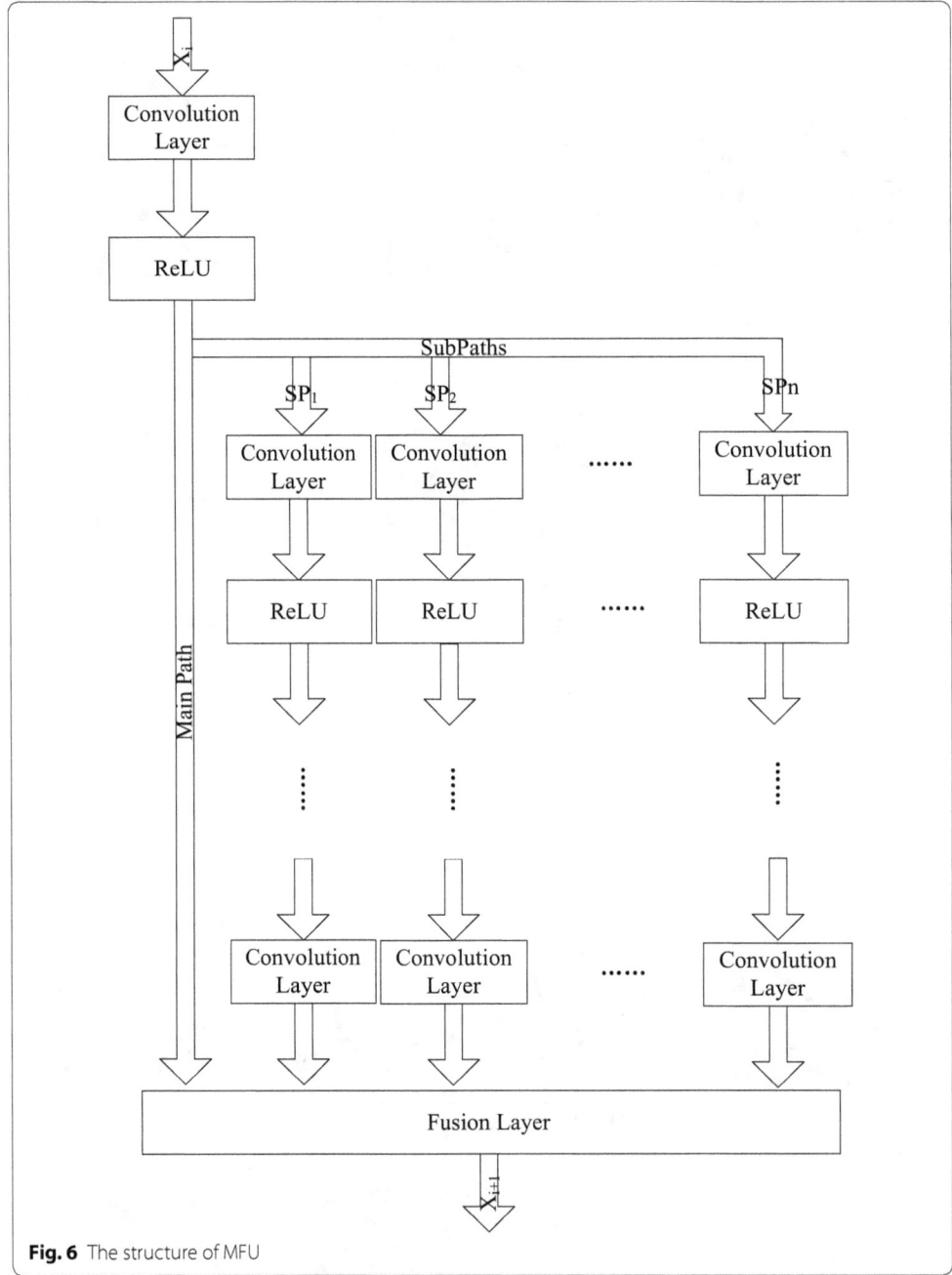

Fig. 6 The structure of MFU

Experiments

To evaluate the reconstruction performance of the proposed MFCN for structural MR images, we designed an extensive set of validation experiments using both simulated and real MR images. Furthermore, several methods were employed for comparison, including bicubic interpolation, non-local mean (NLM) [11], sparse coding [13], and super-resolution convolution neural network (SRCNN) [26].

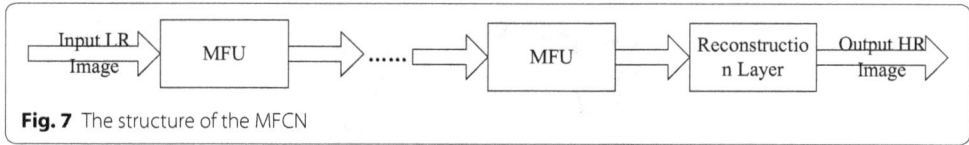

Fig. 7 The structure of the MFCN

Experimental settings

The proposed MFCN was run on an Ubuntu 14.04 with an Intel Xeon E5-2620 processor at 2.4 GHz, K80 GPU and the 96 GB of RAM based on the Caffe deep learning framework [34].

Brain MR image sets

In this paper, the proposed MFCN was tested using different MR image sets, including both simulated and real images.

- Simulated MR images were generated using an MRI simulator and obtained from the BrainWeb brain database [35]. The simulation provides volumes acquired in the axial plane with dimensions of $181 \times 217 \times 181$ pixels and 1 mm^3 resolution.
- Real T1-weighted brain MR images were obtained from thirty subjects and were acquired using a GE MR750 3.0T scanner with two different spatial resolutions of $1 \text{ mm} \times 1 \text{ mm} \times 1 \text{ mm}$ and $3 \text{ mm} \times 3 \text{ mm} \times 3 \text{ mm}$. For the high-resolution MR images, each anatomical scan had 156 axial slices with a size of 256×256 pixels. The low-resolution MR images only included 52 axial slices.

Similarly to the pre-processing step in [15], the skull and skin were removed from the MR images using a brain extraction tool (BET) [36] to eliminate the influence of the background. The resulting MR image is shown in Fig. 8. For the training set, high-resolution patches were extracted from each slice of a brain region with a size of 33×33 pixels. To obtain low-resolution patches, a blurring and down-sampling operation was applied to the extracted brain regions. Then, a bi-cubic interpolator was implemented. Finally, low-resolution image patches were acquired from the interpolated brain region.

For the learning-based method, we constructed the same training set to ensure consistency. The sparsity regularization parameter was set to 0.01 for the sparse coding-based reconstruction as reported in the literature [15]. The learning rate was set to 0.001. The network was trained using mini-batches of size 32.

Quantitative performance measures

To quantitatively evaluate the performance of the reconstruction of different MR image sets, three different metrics were used to compare the original high-resolution images (x) with the reconstructed images (y).

- The signal-to-noise ratio (SNR) was used to compare the level of the reconstructed image with the level of the background noise:

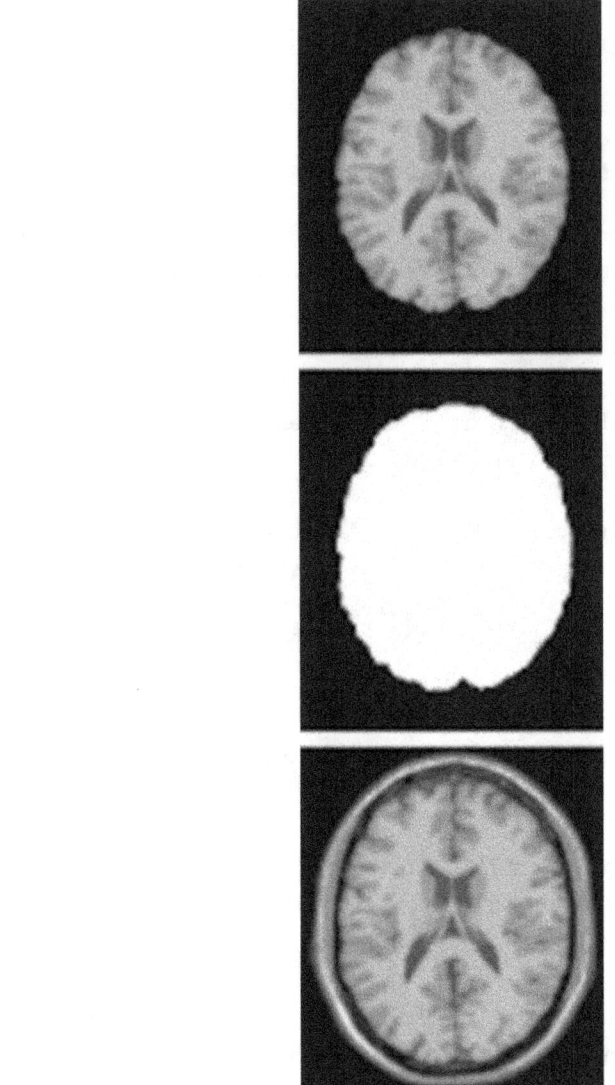

Fig. 8 The MR image (top), its binary mask (middle) and the extracted brain region (down)

$$SNR\,(x,y) = 10\log_{10}\left(\frac{\sum_k |x_k|^2}{\sum_k |x_k - y_k|^2}\right) \tag{5}$$

where x_k and y_k are the image intensities at position k.

- The peak SNR (PSNR) was used to measure the reconstruction accuracy between the reconstructed image and the original image:

$$PSNR\,(x,y) = 10\log_{10}\left(\frac{R}{\sqrt{\frac{1}{|\Omega|}\sum_{k\in\Omega}|x_k - y_k|^2}}\right) \tag{6}$$

where Ω is the brain region, and R is the maximum pixel value in the low-resolution image.

- The structural similarity index (SSIM) [37] was used to measure the similarity between the two images, which is more consistent with human visual systems and perception.

$$SSIM\,(x, y) = \frac{(2\mu_x\mu_y + c_1)(2\sigma_{xy} + c_2)}{(\mu_x^2 + \mu_y^2 + c_1)(\sigma_x^2 + \sigma_y^2 + c_2)} \tag{7}$$

where $c_1 = (k_1 L)^2$, $c_2 = (k_2 L)^2$, and L are the dynamic range, $k_1 = 0.01$, and $k_2 = 0.03$. The terms μ_x and μ_y are the mean values of images x and y, respectively; σ_x and σ_y are the standard noise variance in images x and y, respectively; and σ_{xy} is the covariance of x and y.

Network architecture analysis

To achieve deep learning, a large number of parameters must be tuned, which affected the reconstruction performance of the proposed network. In this section, we discuss these various factors and investigate the best trade-off between performance and speed in the simulated data. For the simulated data, the training set was constructed from the BrainWeb database using 30 real MR brain slices acquired by sampling 600 random image locations from each slice, and the test data were obtained from the slices excluded from the training set. For a baseline, the parameter configuration is listed in Tables 1, 2 and 3, where nMFU is the number of multi-scale fusion units, MFU_n is the n-th multi-scale fusion units, s_k is the size of convolution kernel, n_k is the number of convolution kernel, nSubPath is the number of sub-paths in each MFU, and nLayer is the number of convolution layers in each sub-path.

Table 1 The network configurations of the baseline in MFCN

Network	nMFU	s_k/n_k in reconstruction layer
$MFCN_{BL}$	2	5/1

Table 2 The MFU_1 configurations in $MFCN_{BL}$

s_k/n_k of main path in MFU	nSubPath	nLayer in each sub-path	s_k/n_k of in sub-path
9/32	1	1	1/32

Table 3 The MFU_2 configurations in $MFCN_{BL}$

s_k/n_k of main path in MFU	nSubPath	nLayer in each sub-path	s_k/n_k of in sub-path
3/64	1	1	1/64

Table 4 The different kernel size configurations

Network	s_k/n_k in nMFU	s_k/n_k in reconstruction layer
$MFCN_{BL}$	9/32, 3/64	5/1
$MFCN_{s713}$	7/32, 1/64	3/1
$MFCN_{s957}$	9/32, 5/64	7/1
$MFCN_{s1159}$	11/32, 5/64	9/1

Parameter discussion for main path

In this section, we develop several networks that have the same structure as $MFCN_{BL}$, except for a different kernel size and number in the main path in the MFU, and a final reconstruction layer to examine the reconstruction performance.

Kernel size Several image recognition and recognition experiments have demonstrated that if the number of kernels in each layer increases, the performance will improve. However, increasing the number of kernels also requires more time to train the network. Therefore, we compared the influence of the different kernel sizes on the reconstruction performance. The detailed parameter configuration is shown in Table 4.

The average PSNR with an upscaling factor of 3 is shown in Fig. 9. The proposed network with different kernel sizes always achieved a better performance than the bi-cubic interpolation and SC. Furthermore, the $MFCN_{BL}$ and $MFCN_{s957}$ networks had a comparable PSNR, while the $MFCN_{s713}$ network had a worse performance. One possible reason is that the $MFCN_{s713}$ network had limited descriptive power for the super-resolution reconstruction due to the fewer parameters. However, we also observed that although the PSNR of the $MFCN_{s1159}$ network increases as the iteration number increases, it always performs worse than the $MFCN_{BL}$ and $MFCN_{s957}$ networks within limited iterative numbers, which probably illustrated that the

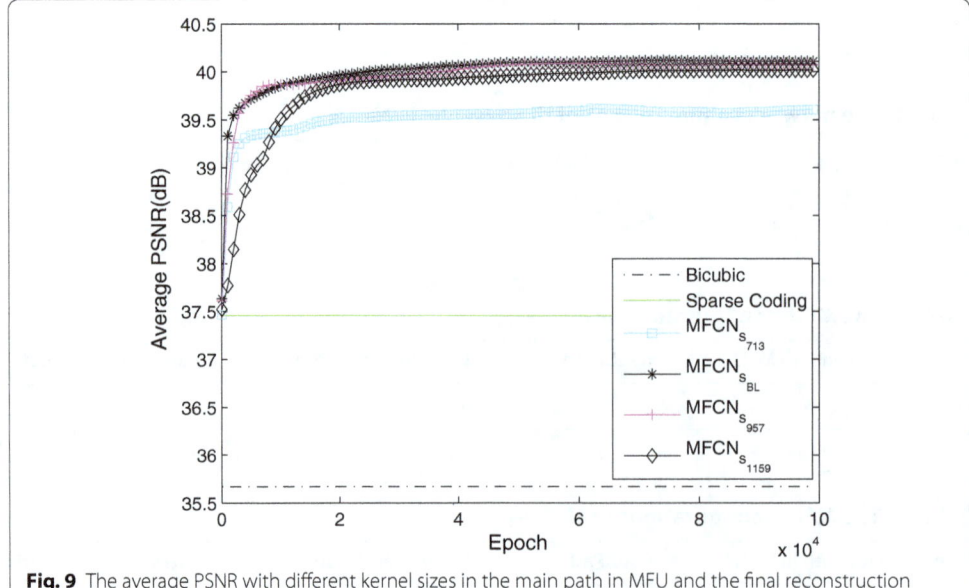

Fig. 9 The average PSNR with different kernel sizes in the main path in MFU and the final reconstruction layer

Table 5 The average SNR, PSNR, SSIM and reconstruction time of each slice with a different kernel size in the main path in MFU and the final reconstruction layer at the 10^5 iteration

Network	SNR	PSNR	SSIM	Time (s)
$MFCN_{s_{713}}$	26.574	39.6064	0.9875	0.009123
$MFCN_{BL}$	27.0741	40.1064	0.9891	0.019142
$MFCN_{s_{957}}$	27.0415	40.0738	0.989	0.020472
$MFCN_{s_{1159}}$	26.9858	40.0181	0.9885	0.027577

networks with bigger kernel size need more training time to converge to achieve a better reconstruction performance, as shown in Table 5. Consequently, increasing the kernel size properly was helpful for achieving superior performance, but considering the balance between the reconstruction performance and the computational efficiency, bigger kernel size is not always good.

Kernel numbers Generally, increasing the kernel number will improve performance. Based on the baseline network with 32 and 64 kernel numbers in the main path in MFU, we increased the kernel number to 64 and 96 and maintained the kernel number in the last reconstruction layer 1, called $MFCN_{n_{64961}}$. We also investigated fewer kernel numbers in the main path in MFCN with 16 and 32, referred to as $MFCN_{n_{16321}}$. The detailed configuration is shown in Table 6.

Table 6 The different kernel number configurations

Network	s_k/n_k in nMFU	s_k/n_k in reconstruction layer
$MFCN_{BL}$	9/32, 3/64	5/1
$MFCN_{n_{16321}}$	9/16, 3/32	5/1
$MFCN_{n_{64961}}$	9/64, 3/96	5/1

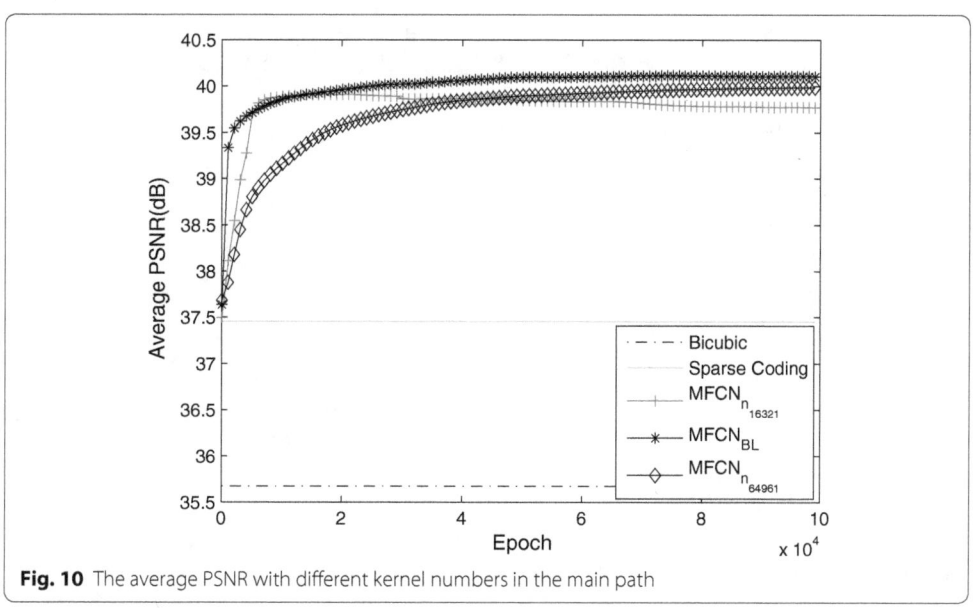

Fig. 10 The average PSNR with different kernel numbers in the main path

Table 7 The different kernel number configurations

Network	s_k/n_k of sub-path within MFU
$MFCN_{BL}$	1/32, 1/64
$MFCN_{S_3}$	3/32, 3/64

These results are shown in Fig. 10. The $MFCN_{n_{16321}}$ network had the worst performance. In the initial iteration, the $MFCN_{n_{64961}}$ network had a worse performance than the baseline $MFCN_{BL}$ network. The performance of the $MFCN_{n_{64961}}$ network improved as the iteration number increased. It is possible to surpass the baseline network with additional training time likely because the $MFCN_{n_{64961}}$ network requires more learning of the network parameters. This network fails to converge during the 10^5 epochs; therefore, it is not superior to $MFCN_{BL}$ with 32 and 64 kernel numbers.

Sub-path parameter discussion

In this section, we discuss the influence of the sub-paths (e.g., the kernel size and the number of convolution layers in each sub-path) and the effects of preserving an ReLU layer before its addition to the MFU.

Kernel size of convolutional layers in the sub-paths

First, we discuss the kernel size of the convolution layers. As shown in Table 7, we attempted to enlarge the kernel size from 1×1 in the baseline network to 3×3 in the convolution layers in the sub-path ($MFCN_{S_3}$). The result is shown in Fig. 11. A superior performance was achieved in $MFCN_{S_3}$. As discussed in "Sub-path parameter discussion" section, the same conclusion was reached, i.e., a wider kernel size is helpful for improving performance.

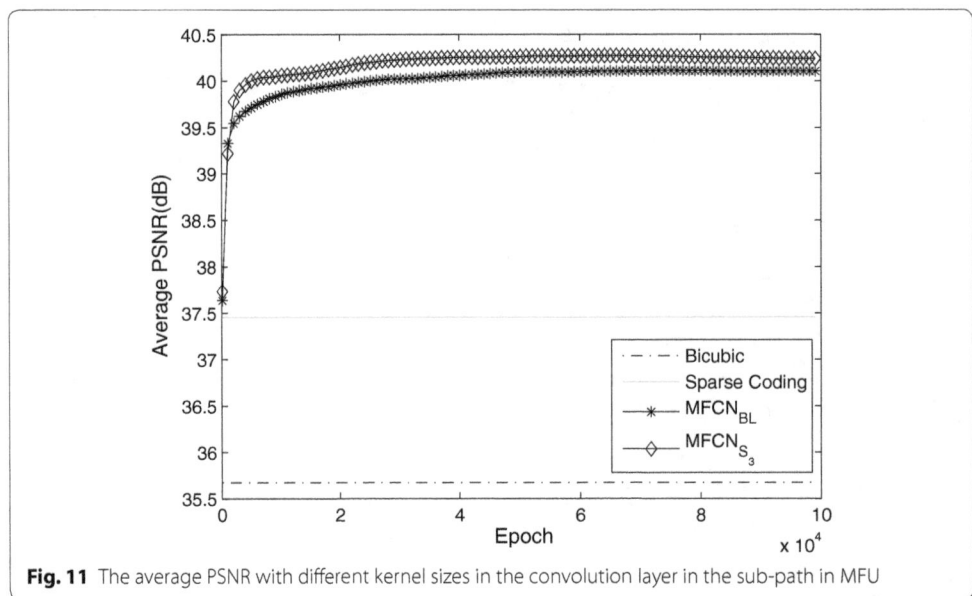

Fig. 11 The average PSNR with different kernel sizes in the convolution layer in the sub-path in MFU

Table 8 **The different kernel number configurations**

Network	nLayer in each sub-path	s_k/n_k of sub-path within MFU_1	s_k/n_k of sub-path within MFU_2
$MFCN_{BL}$	1	1/32	1/64
$MFCN_{L_2}$	2	1/32, 1/32	1/64, 1/64

Number of convolution layers in the sub-paths

In the $MFCN_{BL}$, the sub-path in each MFU contains a convolution layer. Increasing the number of convolution layers in the sub-path is helpful for adding depth to the network. Therefore, we further examined networks with more convolution layers and set two convolution layers in the sub-path of each unit. The detailed configuration is shown in Table 8. Although the average PSNR shown in Fig. 12 demonstrated that the baseline network with one convolution layer ($MFCN_{BL}$) was superior to the network with two convolution layers ($MFCN_{L_2}$), the performance of $MFCN_{L_2}$ approached that of $MFCN_{BL}$ near the 10^5 iteration and could potentially surpass the baseline networks with higher iterative numbers. $MFCN_{L_2}$ may need to learn more parameters and converges more slowly than $MFCN_{BL}$. Consequently, a balance between the reconstruction performance and convergence speed is needed.

ReLU before the fusion layer

Previous studies [33] have shown that the "residual" unit should be in the range of $(-\infty, +\infty)$ and suggested removing the ReLU before the addition of the fusion layer to achieve a lower error in the image classification. To confirm the performance of ReLU in MFU for MRI super-resolution reconstruction, we also investigated a network structure by adding ReLU before the addition of each MFU. The other settings remained the same as those in the baseline network ($MFCN_{BL}$). As shown in Fig. 13, removing ReLU before adding the fusion layer exceeded the performance compared to when ReLU was maintained.

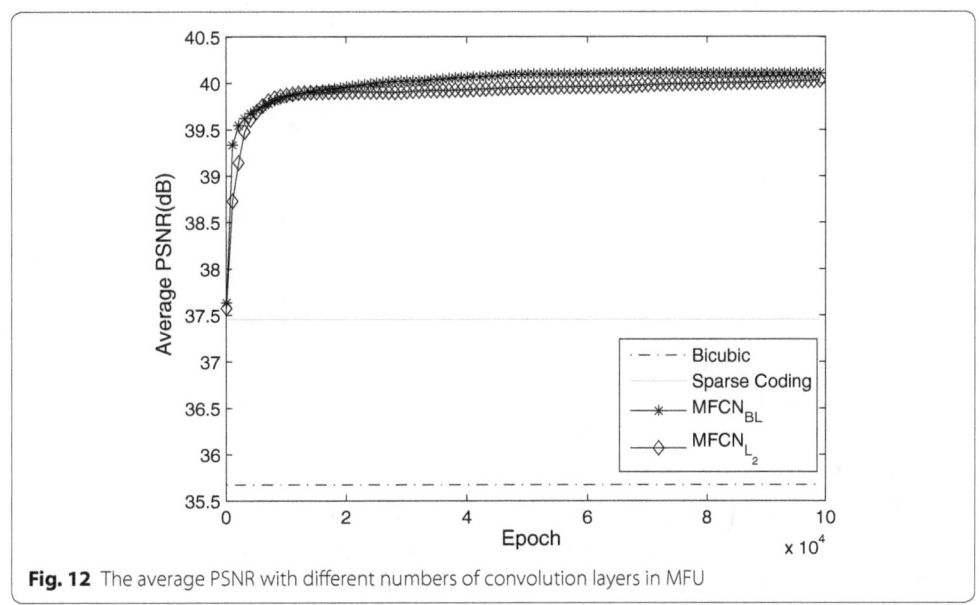

Fig. 12 The average PSNR with different numbers of convolution layers in MFU

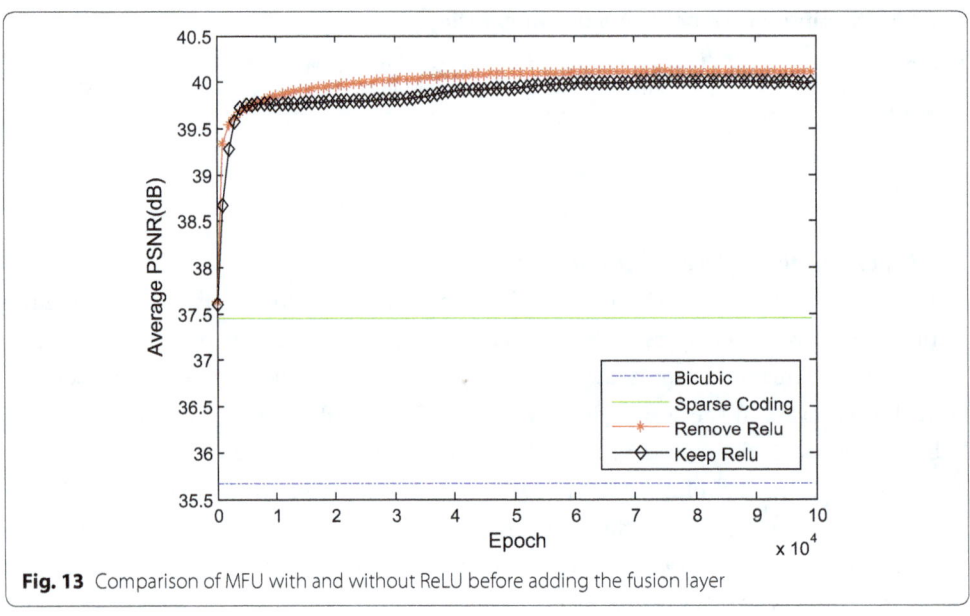

Fig. 13 Comparison of MFU with and without ReLU before adding the fusion layer

Table 9 The different MFUs configurations

Network	s_k/n_k of sub-path within MFU_1	s_k/n_k of sub-path within MFU_2	s_k/n_k of sub-path within MFU_3
$MFCN_{BL}$	1/32	1/64	N/A
$MFCU_{U_1}$	1/32	N/A	N/A
$MFCU_{U_3}$	1/32	1/32	1/64

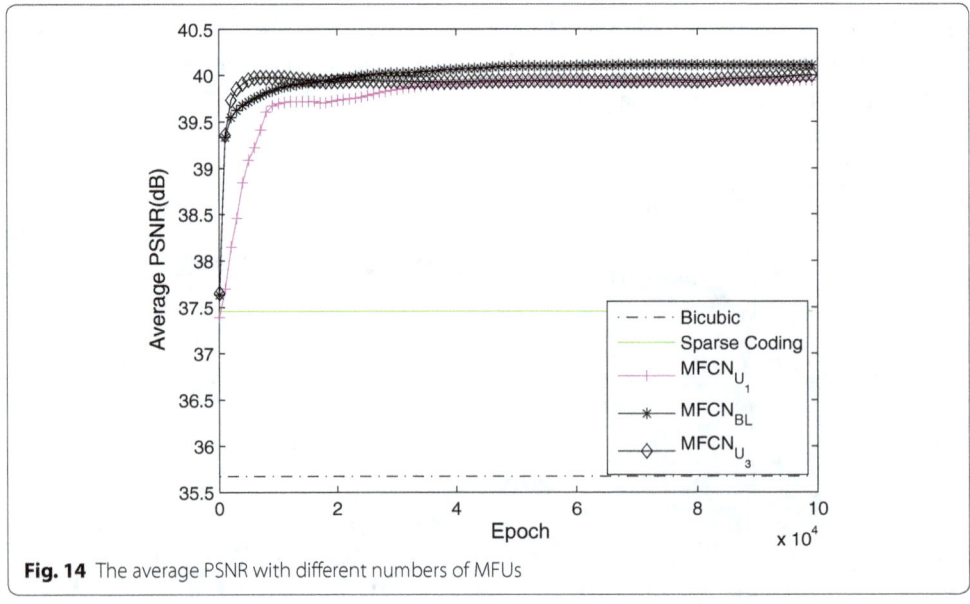

Fig. 14 The average PSNR with different numbers of MFUs

Number of MFUs

Several deep learning image recognition and classification experiments have demonstrated that performance can benefit from increasing the network depth. However, previous studies [26] have claimed that deeper networks do not always achieve an improved performance. In addition to increasing the number of convolution layers in the sub-path in "Sub-path parameter discussion", we attempted to deepen the network by adding several MFUs. The detailed configuration is shown in Table 9. As shown in Fig. 14, a network with one MFU ($MFCU_{U_1}$) had a worse performance than the baseline network with two MFUs ($MFCN_{BL}$). Initially, the networks with three MFUs ($MFCU_{U_3}$) were superior to the baseline network, but their performance worsened after approximately 20K iterations, and the curve increased by nearly 10^5 iterative numbers. Therefore, it is difficult to achieve the same conclusion as [26]. We believe that this trend does not oppose the advantage of the network depth. Deeper

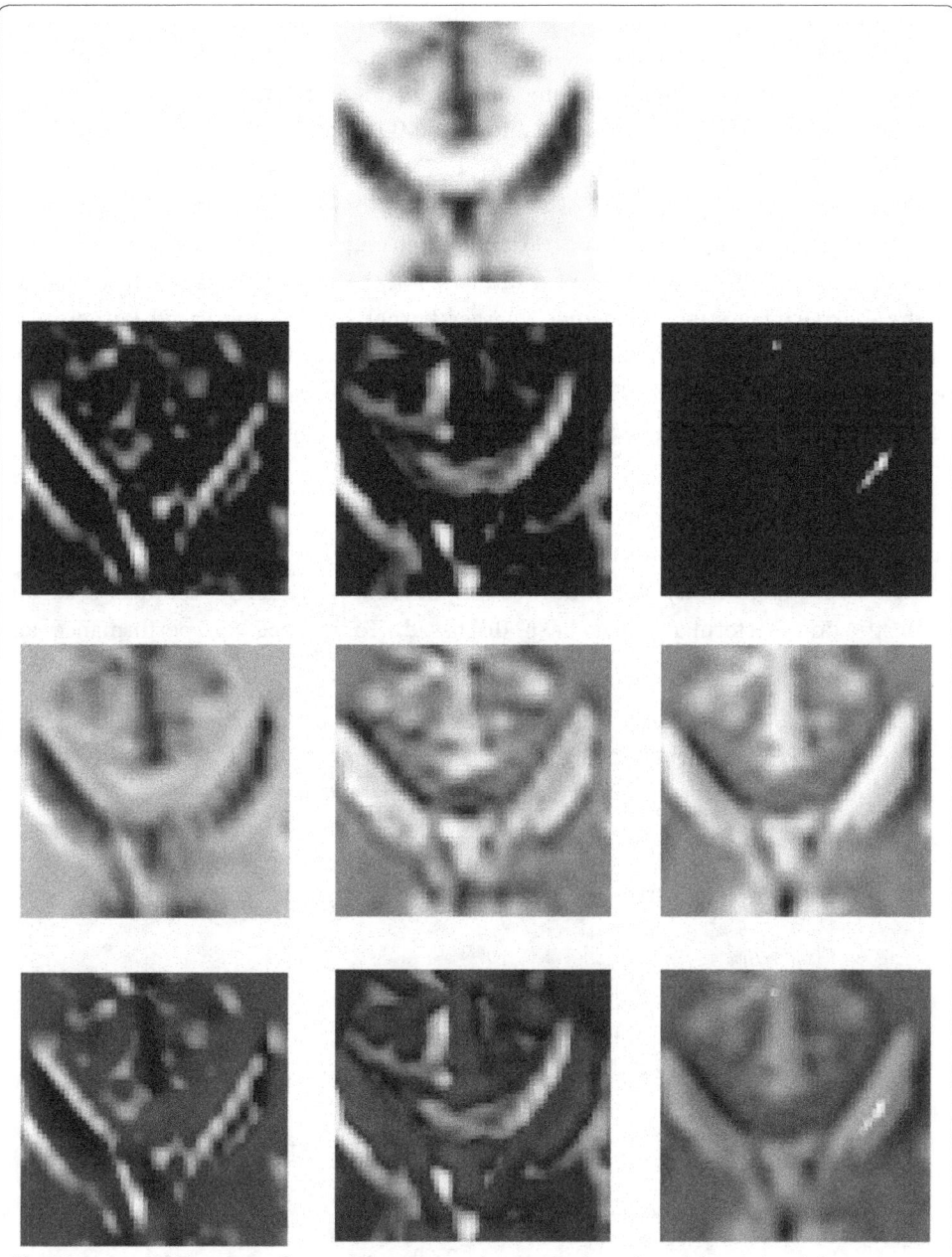

Fig. 15 Low-resolution image (first row), feature maps learned by the main path (second row), feature maps learned by a sub-path in MFU (third row) and feature maps after the addition to an MFU (fourth row)

networks cannot converge within 10^5 iterations due to the requirement of more learned parameters, which leads to a worse performance than that of the baseline network from 2×10^4 to 10^5 iterations.

Learned feature maps

To investigate why the proposed network is capable of super-resolution reconstructions, some feature maps were studied using different layers and are shown in Fig. 15. As shown in Fig. 15, different kernels in the main path extract distinct information from low-resolution images, such as different directions, as shown in the second row. Convolution layers in the sub-path recover different modalities based on the feature maps in the second row as shown in the third row. The feature maps in the second and third rows are complementary, and the final fusion layer with the addition operation in MFU is helpful for combining the complementary information as shown in the fourth row. For a better understanding of MFU, we further compared $MFCN_{BL}$ and the traditional convolution network (SRCNN) using the same configurations. The results are shown in Fig. 16. The MFCN was always superior to the SRCNN.

In summary, we investigated the parameter settings of the proposed network and decomposed MFU to visualize the feature maps of the main path and sub-path. Each of these experiments indicated that a larger kernel size, an increased kernel number in the convolution layers, and a deeper network are helpful for improving the reconstruction performance. However, many parameters need to be learned, and the convergence is, therefore, slow. Consequently, we must compromise between performance and efficiency.

Comparisons to state-of-the-art approaches

In previous experiments, the influence of different parameters on networks and reconstruction performance has been discussed. To balance the performance and

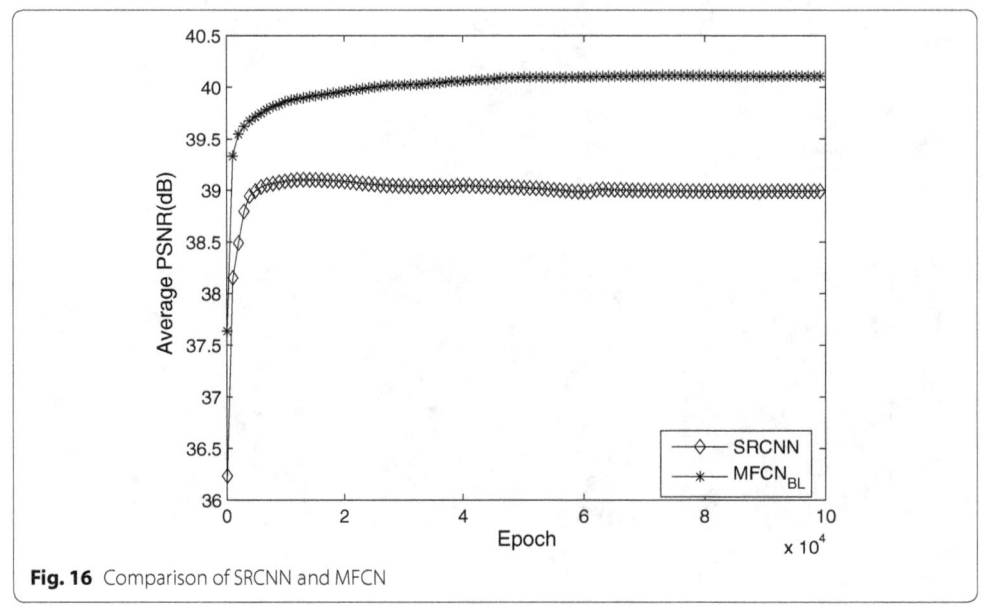

Fig. 16 Comparison of SRCNN and MFCN

Table 10 Quantitative evaluation (RMSE, SNR, PSNR, and SSIM) of different up-sampling factors using BrainWeb MR images

Eval. met	Scale	Bicubic	NLM	Sparse coding	SRCNN	MFCN
RMSE	2	2.5077	2.1112	1.7747	1.4836	1.2026
	3	4.2038	3.7707	3.4262	2.8937	2.5415
	4	5.849	5.5588	5.0577	4.6946	4.5196
SNR	2	27.1376	28.636	30.1495	31.7244	33.5375
	3	22.6394	23.5894	24.422	25.9595	27.0741
	4	19.767	20.2123	21.0326	21.7213	22.0571
PSNR	2	40.1699	41.6684	43.1819	44.7567	46.5698
	3	35.6717	36.6218	37.4262	38.9918	40.1064
	4	32.7993	33.2446	34.056	34.7536	35.0895
SSIM	2	0.9891	0.9923	0.9945	0.9963	0.9975
	3	0.9678	0.9743	0.9788	0.9864	0.9891
	4	0.9375	0.9434	0.9529	0.9639	0.9662

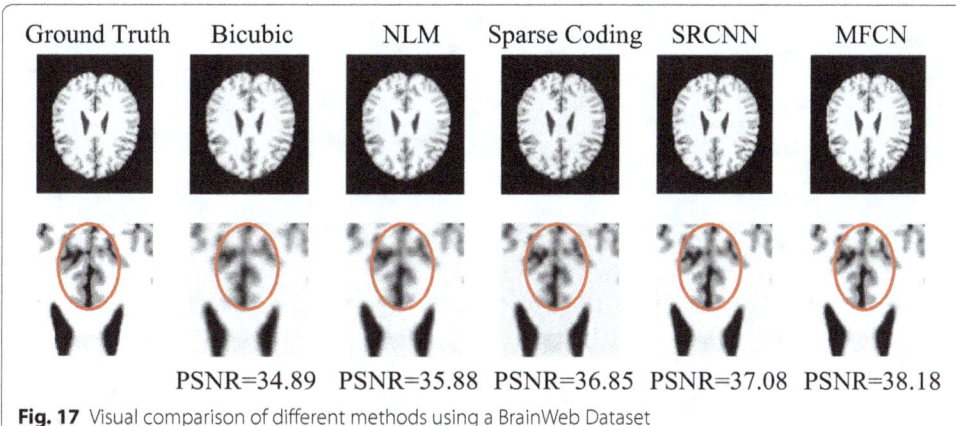

Ground Truth	Bicubic	NLM	Sparse Coding	SRCNN	MFCN
	PSNR=34.89	PSNR=35.88	PSNR=36.85	PSNR=37.08	PSNR=38.18

Fig. 17 Visual comparison of different methods using a BrainWeb Dataset

computational efficiency, we adopted the above-mentioned baseline network due to its good performance-speed trade-off. Once the network architecture was fixed, the super-resolution reconstruction experiments were carried out to validate the performance of the proposed method. In this section, quantitative and qualitative results of the proposed method were compared with results of certain classical methods for different up-sampling factors f, including $f = 2$, 3 and 4. The implementation of existing methods was achieved using publicly available codes provided by the authors. For the MFCN and SRCNN, we trained the network using 10^5 iterations.

Different up-sampling factors

As shown in Table 10, the proposed method always yielded the best scores with different evaluation metrics. Figure 17 also illustrates the reconstructed MR images using different methods in a single slice. Notably, within the red circle of Fig. 17, it can be found that the reconstructions based on MFCN were able to restore more detailed information for the MR images than those based on the other classical methods.

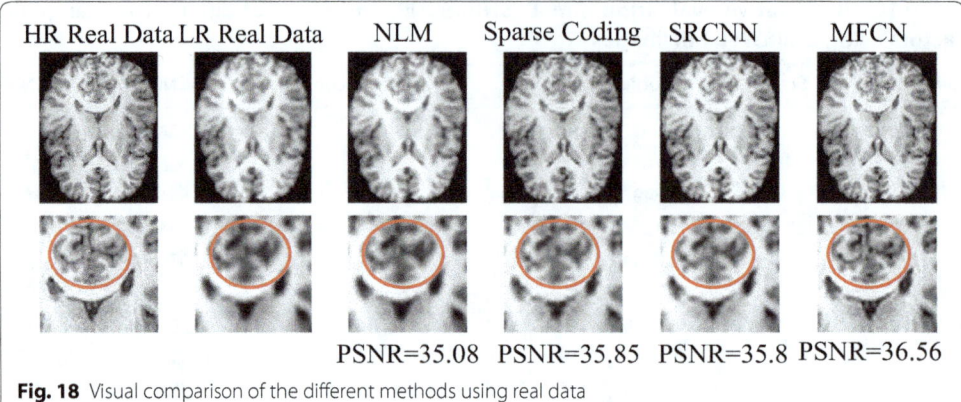

Fig. 18 Visual comparison of the different methods using real data

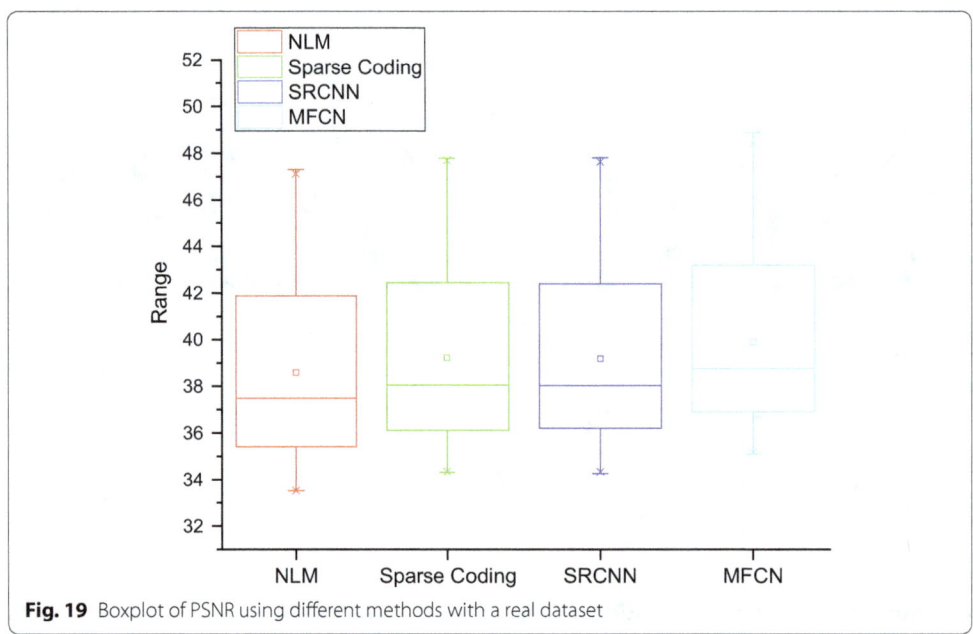

Fig. 19 Boxplot of PSNR using different methods with a real dataset

Evaluation of real data

We further examined the performance of the MFCN using a real dataset. We selected fifteen subjects as the training data and the remaining subjects as the test data. Figure 18 shows representative image reconstruction results using various methods. From left to right, the first row shows the high-resolution image, the corresponding low-resolution image, and the results of NLM, sparse coding, SRCNN, and MFCN. The close-up views of the selected regions are also shown for better visualization. The results of NLM show severe blurring artifacts, and the results of sparse coding are better than those of NLM. The contrast is enhanced in the SRCNN results, while the proposed MFCN is the best for preserving edges and achieving the highest PSNR value as shown in Fig. 18. The quantitative results using the real datasets are illustrated in Fig. 19. As shown in Fig. 19, the total distribution of PSNRs for MFCN are better than others; The mean (small square in the box) and the median (the horizontal line in the box) of PSNR for MFCN are also greater than other one. Therefore, the proposed method significantly outperformed all compared methods.

Discussion

It is well known that the convolution neural network has a large number of network parameters and is needed for training with a large dataset to avoid over-fitting. However, due to limited MRI training data, it is difficult to achieve superior reconstruction using a standard convolution network. In this work, we developed an MFCN for MRI super-resolution reconstruction and achieving end-to-end (one-to-one) mapping between low and high-resolution images. Instead of a traditional convolution network, the network is stacked by MFUs. Each MFU consists of a main path and several sub-paths, and all paths are finally added to the fusion layer to fuse multi-scale information. We conducted several experiments and demonstrated that when the training data are limited, the proposed network always achieves superior reconstruction results using both simulated and real data compared with traditional SR methods, such as bi-cubic, NLM, sparse coding, and SRCNN.

An additional concern is the slow convergence speed caused by the traditional convolution network structure. Regarding fusing multi-scale information from the main path and sub-paths in MFU, we found that the proposed network achieves faster convergence speed than the traditional convolution network SRCNN. As shown in Fig. 16, with the same parameter settings, the proposed network converges after 3000 epochs while SRCNN converges after 5000 epochs. Furthermore, the proposed network achieves a higher PSNR value in the same epoch. Moreover, the proposed network can recover more detailed information and has better visual effects as shown in Figs. 17 and 18.

Finally, currently used convolution networks for image super-resolution usually extract detailed information on a single scale, and the back propagation process fails to utilize prior knowledge of the high-resolution images.

According to previous research on SR [38], the extraction of multi-scale information improves the reconstruction results. Using the proposed convolution network, we also experimentally validated that differently sized convolution kernels can acquire multi-scale information as shown in Fig. 3. We found that multi-scale information can be merged and transmitted from one MFU to the next as shown in Fig. 15. Thus, the proposed network can recover detailed information and achieve better reconstruction performance.

Our results are inconsistent with the conclusion reached using SRCNN [26] that "deeper is not better", and many experiments investigating parameter settings have illustrated that incremental network depths and kernel sizes are helpful for improving the reconstruction results. Generally, we should seek a balance between computational efficiency and reconstruction performance. Using both simulated and real data, the proposed network has demonstrated visually and quantitatively prominent performance for MRI super-resolution reconstruction.

Conclusion

In this paper, we demonstrated an MFCN for MRI super-resolution. The network is able to learn end-to-end mapping from low/high-resolution images. Simultaneously, due to the fusion of different paths in MFU, the network can extract multi-scale information to recover detailed information and accelerate the convergence speed. The extensive experiments using simulated and real data have also demonstrated that this approach is superior to other traditional methods. In addition, the proposed network architecture and experimental framework can be applied to other medical super-resolution reconstructions, such as in CT and diffusion-weighted MR imaging.

Authors' contributions

CL conceived and designed this study and is responsible for the manuscript, XW participated in the analysis of the MRI dataset, XY, YT, JZ and JZ participated in the conception and design of this work and helped to draft the manuscript. All authors read and approved the final manuscript.

Author details

[1] Department of Information Technology and Engineering, Chengdu University, Chengdu 610106, China. [2] The Clinical Hospital of Chengdu Brain Science Institute, MOE Key Lab for Neuroinformation, Center for Information in Medicine, University of Electronic Science and Technology of China, Chengdu 610054, China. [3] School of Life Science and Technology, University of Electronic Science and Technology of China, Chengdu 610054, China. [4] Key Laboratory of Pattern Recognition and Intelligent Information Processing in Sichuan, Chengdu 610106, China. [5] Department of Computer Science, Chengdu University of Information Technology, Chengdu 610225, China. [6] Faculty of Science and Technology, University of Macau, Macau, China. [7] School of Physics and Electronic Engineering, Sichuan Normal University, Chengdu, China.

Acknowledgements

The authors would like to thank all participants for the valuable discussions regarding the content of this article.

Competing interests

The authors declare that they have no competing interests.

Consent for publication

Each participant agreed that the acquired data can be further scientifically used and evaluated. For publication, we made

Funding

This work is supported by the National Natural Science Funds of China (Grant No. 61502059), the China Postdoctoral Science Foundation (Grant No. 2016M592656), Sichuan Science and Technology Program (Grant No. 2018JY0272), the Educational Commission of Sichuan Province of China (Grant No. 15ZA360).

References

1. Thévenaz P, Blu T, Unser M. Interpolation revisited medical images application. IEEE Trans Med Imag. 2000;19(7):739–58.
2. Lehmann TM, Gönner C, Spitzer K. Survey: interpolation methods in medical image processing. IEEE Trans Med Imag. 1999;18(11):1049–75.
3. Shilling RZ, Robbie TQ, Bailloeul T, Mewes K, Mersereau RM, Brummer ME. A super-resolution framework for 3-d high-resolution and high-contrast imaging using 2-d multislice MRI. IEEE Trans Med Imag. 2009;28(5):633–44.
4. Herment A, Roullot E, Bloch I, Jolivet O, De Cesare A, Frouin F, Bittoun J, Mousseaux E. Local reconstruction of stenosed sections of artery using multiple MRA acquisitions. Magn Reson Med. 2003;49(4):731–42.
5. Peled S, Yeshurun Y. Superresolution in MRI: application to human white matter fiber tract visualization by diffusion tensor imaging. Magn Reson Med. 2001;45(1):29–35.
6. Greenspan H, Oz G, Kiryati N, Peled S. MRI inter-slice reconstruction using super-resolution. Magn Reson Imag. 2002;20(5):437–46.

7. Carmi E, Liu S, Alon N, Fiat A, Fiat D. Resolution enhancement in MRI. Magn Reson Imag. 2006;24(2):133–54.
8. Manjón JV, Coupé P, Buades A, Fonov V, Collins DL, Robles M. Non-local mri upsampling. Med Image Anal. 2010;14(6):784–92.
9. Rousseau F, Initiative ADN. A non-local approach for image super-resolution using intermodality priors. Med Image Anal. 2010;14(4):594–605.
10. Elad M, Aharon M. Image denoising via sparse and redundant representations over learned dictionaries. IEEE Trans Med Imag. 2006;15(12):3736–45.
11. Mairal J, Elad M, Sapiro G. Sparse representation for color image restoration. IEEE Trans Image Process. 2008;17(1):53–69.
12. Donoho DL. Compressed sensing. IEEE Trans Inform Theory. 2006;52(4):1289–306.
13. Yang J, Wright J, Huang TS, Ma Y. Image super-resolution via sparse representation. IEEE Trans Image Process. 2010;19(11):2861–73.
14. Zeyde R, Elad M, Protter M. On single image scale-up using sparse-representations. In: International conference on curves and surfaces. Berlin: Springer; 2010. p. 711–30.
15. Rueda A, Malpica N, Romero E. Single-image super-resolution of brain mr images using overcomplete dictionaries. Med Image Anal. 2013;17(1):113–32.
16. Wohlberg B. Efficient convolutional sparse coding. In: IEEE international conference on acoustics, speech and signal processing (ICASSP). New Jersey: IEEE; 2014 , p. 7173–77.
17. Bristow H, Eriksson A, Lucey S. Fast convolutional sparse coding. In: IEEE conference on computer vision and pattern recognition (CVPR). New Jersey: IEEE; 2013. p. 391–8.
18. Osendorfer C, Soyer H, van der Smagt P. Image super-resolution with fast approximate convolutional sparse coding. In: Neural information processing. Berlin: Springer; 2014. p. 250–7.
19. Krizhevsky A, Sutskever I, Hinton GE. Imagenet classification with deep convolutional neural networks. In: Advances in neural information processing systems; 2012. p. 1097–1105.
20. LeCun Y, Bengio Y, Hinton G. Deep learning. Nature. 2015;521(7553):436–44.
21. Cui Z, Chang H, Shan S, Zhong B, Chen X. Deep network cascade for image super-resolution. In: Computer vision–ECCV 2014. Berlin: Springer; 2014. p. 49–64.
22. Wang Z, Liu D, Yang J, Han W, Huang T. Deeply improved sparse coding for image super-resolution. arXiv preprint arXiv:1507.08905. 2015.
23. Ji S, Xu W, Yang M, Yu K. 3d convolutional neural networks for human action recognition. IEEE Trans Patt Anal Mach Intell. 2013;35(1):221–31.
24. Bengio Y. Learning deep architectures for ai. Foundations and trends® in Mach Learn. 2009;2(1):1–127.
25. Hinton GE, Osindero S, Teh Y-W. A fast learning algorithm for deep belief nets. Neural comput. 2006;18(7):1527–54.
26. Dong C, Loy CC, He K, Tang X. Image super-resolution using deep convolutional networks. In: IEEE transactions on pattern analysis and machine intelligence; 2015.
27. Kim J, Lee JK, Lee KM. Accurate image super-resolution using very deep convolutional networks. arXiv preprint arXiv :1511.04587. 2015.
28. LeCun Y, Jackel L, Bottou L, Brunot A, Cortes C, Denker J, Drucker H, Guyon I, Muller U, Sackinger E. Comparison of learning algorithms for handwritten digit recognition. In: International conference on artificial neural networks; 1995. p. 53–60.
29. LeCun Y, Jackel L, Bottou L, Cortes C, Denker JS, Drucker H, Guyon I, Muller U, Sackinger E, Simard P. Learning algorithms for classification: a comparison on handwritten digit recognition. Neural Netw. 1995;261:276.
30. Simonyan K, Zisserman A. Very deep convolutional networks for large-scale image recognition. arXiv preprint arXiv :1409.1556. 2014.
31. Girshick R, Donahue J, Darrell T, Malik J. Rich feature hierarchies for accurate object detection and semantic segmentation. In: Proceedings of the IEEE conference on computer vision and pattern recognition; 2014. p. 580–7.
32. Taigman Y, Yang M, Ranzato M, Wolf L. Deepface: closing the gap to human-level performance in face verification. In: Proceedings of the IEEE conference on computer vision and pattern recognition; 2014. p. 1701–8.
33. He K, Zhang X, Ren S, Sun J. Identity mappings in deep residual networks. arXiv preprint arXiv:1603.05027. 2016.
34. Jia Y, Shelhamer E, Donahue J, Karayev S, Long J, Girshick R, Guadarrama S, Darrell T. Caffe: convolutional architecture for fast feature embedding. In: Proceedings of the ACM international conference on multimedia. New York: ACM; 2014. p. 675–8.
35. Cocosco CA, Kollokian V, Kwan RKS, Pike GB, Evans AC. Brainweb: online interface to a 3d MRI simulated brain database. In: NeuroImage. Kyoto: Citeseer; 1997.
36. Smith SM. Fast robust automated brain extraction. Hum Brain Map. 2002;17(3):143–55.
37. Wang Z, Bovik AC, Sheikh HR, Simoncelli EP. Image quality assessment: from error visibility to structural similarity. IEEE Trans Image Process. 2004;13(4):600–12.
38. Sun J, Zheng NN, Tao H, Shum HY. Image hallucination with primal sketch priors. In: Proceedings of IEEE computer society conference on computer vision and pattern recognition, 2003. New York: IEEE; 2003. p. 729.

Effects on the pulmonary hemodynamics and gas exchange with a speed modulated right ventricular assist rotary blood pump: a numerical study

Feng Huang[1,2]* , Zhe Gou[2], Yang Fu[3] and Xiaodong Ruan[2]

*Correspondence:
hf@cjlu.edu.cn
[1] College of Metrology &
Measurement Engineering,
China Jiliang University,
Xueyuan Road 258,
Hangzhou, China
Full list of author information
is available at the end of the
article

Abstract

Rotary blood pumps (RBPs) are the newest generation of ventricular assist devices. Although their continuous flow characteristics have been accepted widely, more and more research has focused on the pulsatile modulation of RBPs in an attempt to provide better perfusion. In this study, we investigated the effects of an axial RBP serving as the right ventricular assist device on pulmonary hemodynamics and gas exchange using a numerical method with a complete cardiovascular model along with airway mechanics and a gas exchange model. The RBP runs in both constant speed and synchronized pulsatile modes using speed modulation. Hemodynamics and airway O_2 and CO_2 partial pressures were obtained under normal physiological conditions, and right ventricle failure conditions with or without RBP. Our results showed that the pulsatile mode of the RBP could support right ventricular assist to restore most hemodynamics. Using speed modulation, both pulmonary arterial pressure and flow pulsatility were increased, while there was only very little effect on alveolar O_2 and CO_2 partial pressures. This study could provide basic insight into the influence of pulmonary hemodynamics and gas exchange with speed modulated right ventricular assist RBPs, which is concerned when designing their pulsatile control methods.

Keywords: Speed modulation, Hemodynamics, Pulmonary gas exchange, Rotary blood pump, Right ventricular assist

Background

Rotary blood pumps (RBPs) have become the most popular ventricular assist device (VAD) due to their numerous advantages [1, 2]. Among the existing RBPs, most are designed to assist the left ventricle because it is subjected to a heavy systemic circulatory load and is more likely to fail than the right side. However, there are still some scenarios in which a right ventricular assist device (RVAD) is required, such as severe right ventricular failure after implantation of a left ventricular assist device (LVAD) [3]. Some studies have focused on the development of RVADs, in particular, in which two up-to-date prototypes are RBPs [2].

Throughout the extensive usage of RBPs, other than the constant speed operation mode, the idea to develop pulsing control methods to make RBPs "beat" more like the natural heart has been put forward [4–9]. The physiological effects of pulsatile versus continuous blood flow have been studied extensively [10–12]. It has been demonstrated that pulsatile flow would not only unload the heart better, but would also not induce the typical "stiffening" of peripheral arteries typically found in a continuous flow state [12]. Besides, pulsatile perfusion has the advantages of causing less vital organ injury and systemic inflammation [10].

Up to now, almost all of the pulsatile operation methods have focused on RBPs used as LVADs. Along with the development of specific RBPs used as RVADs, the same issue also arises if to make them pulsatile. Taking the benefits of the pulsatile perfusion mentioned before into account, and considering the fact that even in the pulmonary vein the flow pulsatility is still much obvious [13], it is desirable to design pulsatile control methods for right ventricular assist RBPs.

Before the control method design, it is necessary to investigate the physiological effect with pulsatile operation of right ventricular assist RBPs. With different purposes, there are several methods that could be used. To investigate a specific part of the cardiovascular system or three-dimensional structures of blood vessels, CFD is thought to be a good tool and adopted by many researchers [14–16]. Apart from CFD, some researchers newly introduced state-space approaches when studying the artery wall [17–19]. These are all effective methods to investigate the effect of pulsatile flow in cardiovascular system. However, when regarding the whole circulation system and two-dimensional hemodynamics, a simple system model could be effective and computational saving, and will be adopted in this study [20].

Generally, the pulsatile operation is realized by rotary speed modulation synchronized with the heartbeat. For LVADs, researchers usually only consider the resultant hemodynamic effects. However, for pulmonary circulatory assist another factor could also be taken into account. There is a unique important function of the pulmonary circulation that it is where oxygen (O_2) and carbon dioxide (CO_2) gas exchange take place between the blood in the lungs and the atmosphere. Hence, besides hemodynamics, we are going to also include the effect investigation of pulmonary gas exchange when implementing the pulsatile modulation operation for a right ventricular assist RBP.

In this study, we investigated the effects of a right ventricular assist RBP running in both constant and synchronized pulsatile mode on pulmonary hemodynamics and O_2 and CO_2 gas exchange using a numerical method. In addition to the complete lumped parameter cardiovascular model, which includes both systemic and pulmonary circulation, models for airway mechanics and gas exchange for O_2 and CO_2 between the lung and the atmosphere were also obtained. Using an RBP model in various speed patterns to assist the failing right ventricle, hemodynamics and airway O_2 and CO_2 partial pressures were obtained, revealing the effects of these speed modulations.

Methods

Cardiovascular model

To investigate pulmonary hemodynamics and gas exchange with an implanted right RBP, a mathematical model of the complete cardiovascular system, including the systemic and pulmonary circulation, was adopted from a previous study by Colacino et al. [20] (Fig. 1). The four heart chambers are described using the nonlinear time-varying

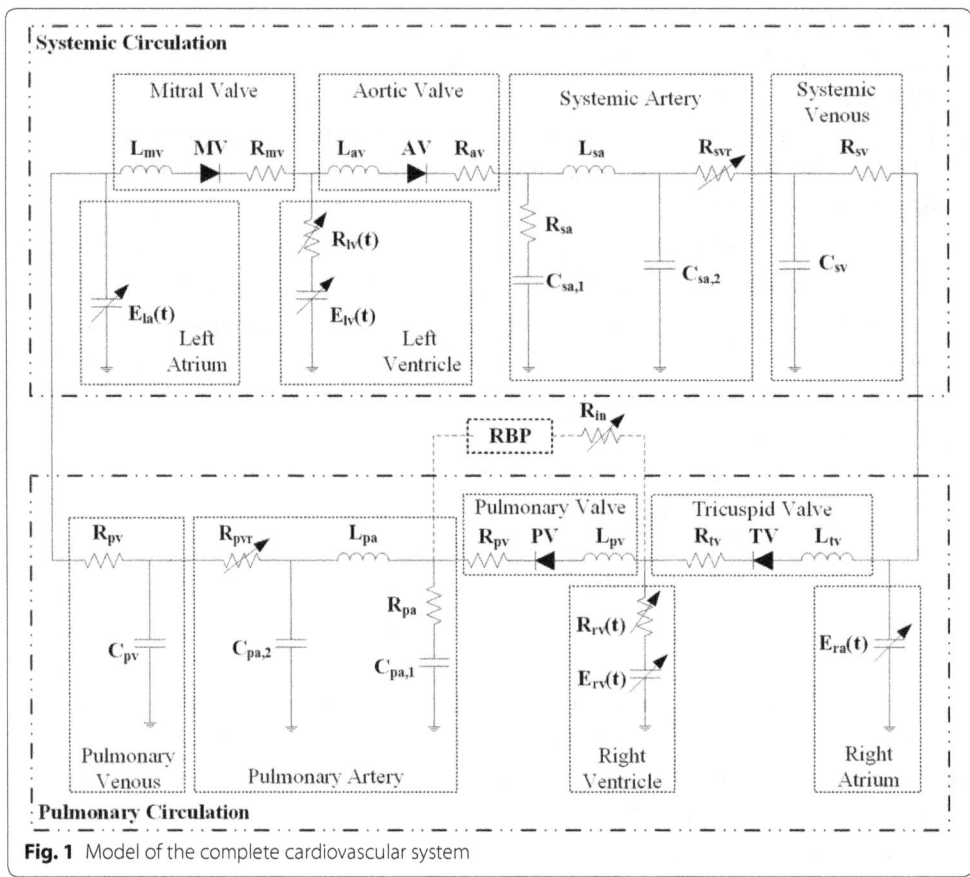

Fig. 1 Model of the complete cardiovascular system

elastance model, with different elastance values between ventricles and atria. In addition, an internal resistance of each ventricle was also included to take into account the energy dissipation during ejection. The heart valve was modeled by an ideal diode representing the one-way function, coupled with resistance and inertance.

The five-elements model was applied to both the systemic and pulmonary arterial systems as parts of the cardiovascular mathematic model. It can reproduce the main arterial characteristics over the entire frequency range of interest [21]. The systemic and pulmonary venous systems are both characterized by resistance and compliance.

Airway and lung mechanics model

Air from the atmosphere enters the alveoli of the lung, where gas exchange takes place. The airway and lung mechanics model included in this study was adopted from a previous study by Lu et al. [22] (Fig. 2, pneumatic circuit representation). The driving pressure of the air flow is the intrathoracic pleural pressure generated by the respiratory muscles (P_{mus}) and the recoil of the chest wall (P_{CW}), which implies that the frequency of breathing is determined by the intrathoracic pleural pressure in the model. The upper, middle, and small airways are characterized by a nonlinear flow-dependent resistor, a nonlinear collapsible-segment volume-dependent resistance and nonlinear P–V relationship (P_{TM}), and a nonlinear alveolar volume-dependent resistance, respectively [22, 23].

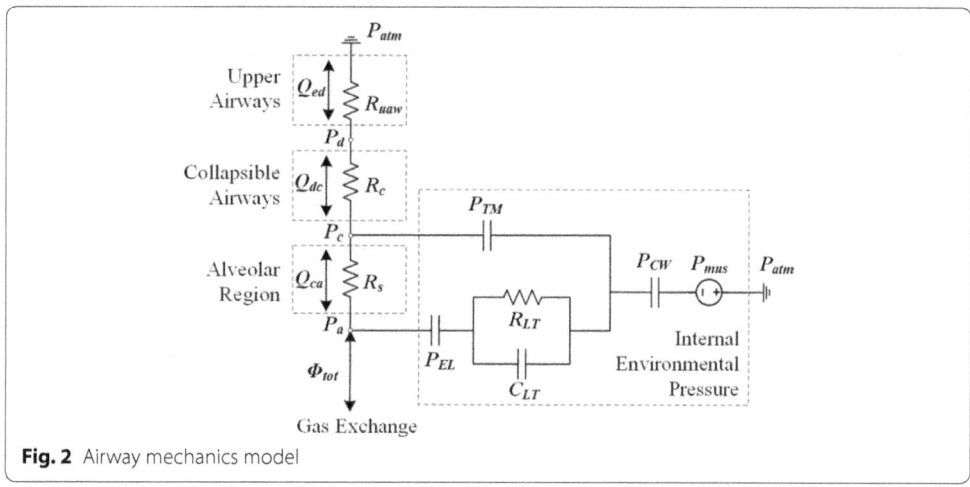

Fig. 2 Airway mechanics model

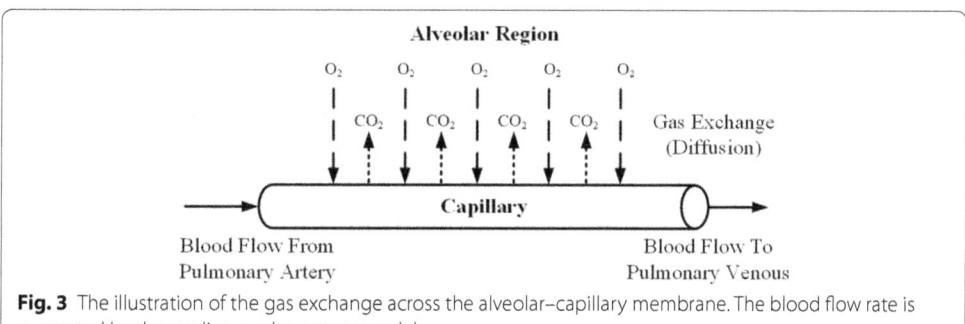

Fig. 3 The illustration of the gas exchange across the alveolar–capillary membrane. The blood flow rate is generated by the cardiovascular system model

Gas exchange model

The gaseous species considered in this study are mainly O_2 and CO_2. Gases come from the airway to the alveolar region (Fig. 2) and diffuse across the alveolar–capillary membrane (Fig. 3), where they are thought to equilibrate instantaneously. Specifically, O_2 is taken up by blood flowing in the capillary, while CO_2 is removed from the blood. The capillary is considered as a single tube. The relationship between species content and their corresponding equilibrium partial pressures is described by the empirical dissociation curves. The diffusion for a specific gas, which is assumed here to be the only mode of gas transport across the membrane, is characterized by a lumped diffusing capacity. The dynamics of the species concentration in the pulmonary capillary is described by a partial differential equation. For modeling simplicity, some of the same assumptions are also made as detailed by Lu [22] and Liu [23], such that the gaseous content obeys the ideal gas law and blood is characterized as a uniform homogeneous medium. Using the species conservation law, the dynamic partial pressures of O_2 and CO_2 in the airways and blood can be described by formulas for both inspiration and expiration processes [22, 23].

Model of the right ventricular assist RBP

The RBP serving as the right ventricular assistance is connected to the blood circulatory system via cannulation from the right ventricle to the pulmonary artery (Fig. 1). There is a worldwide lack of existing RBPs with hydraulic characteristics designed especially for right ventricle assisting; therefore, there is no ready-to-use model of a right RBP. Here we used the method proposed by Krabatsch et al. [24], which enables the use of an already available LVAD for right ventricular assistance. The main problem of using an LVAD as an RVAD is the excessive flowrate against the much lower pulmonary resistance compared with the systemic resistance, even at the lowest rotary speed. The key point of the proposed method was to add an additional resistance in series with the left ventricular assist RBP in order to decrease the flow delivered into the pulmonary circulation.

In our simulation, the left ventricular assist RBP chosen to act as an RVAD was an axial flow pump reported previously by Choi et al. [25]. The dynamic hydraulic characteristic of the original pump can be described using the following equation [25], in which the pressure head (H) depends on the flow rate (Q), its derivative $(\dot{Q})$, and the rotary speed (ω).

$$H = a_0 Q + a_1 \dot{Q} + a_2 \omega^2 \tag{1}$$

where $a_0 = -0.296$, $a_1 = -0.027$, and $a_2 = 0.0000933$ are the coefficients identified using dynamic experimental data [25]. In the above equation, the units for the flow rate, pressure head, and rotary speed are mL/s, mmHg, and rad/s, respectively.

To adopt the axial pump as the RVAD, an additional resistance (0.6 mmHg s/mL) term was added in the pump model. The corresponding working areas of the axial pump before and after the added resistance on the static H–Q plane can be seen in Fig. 4.

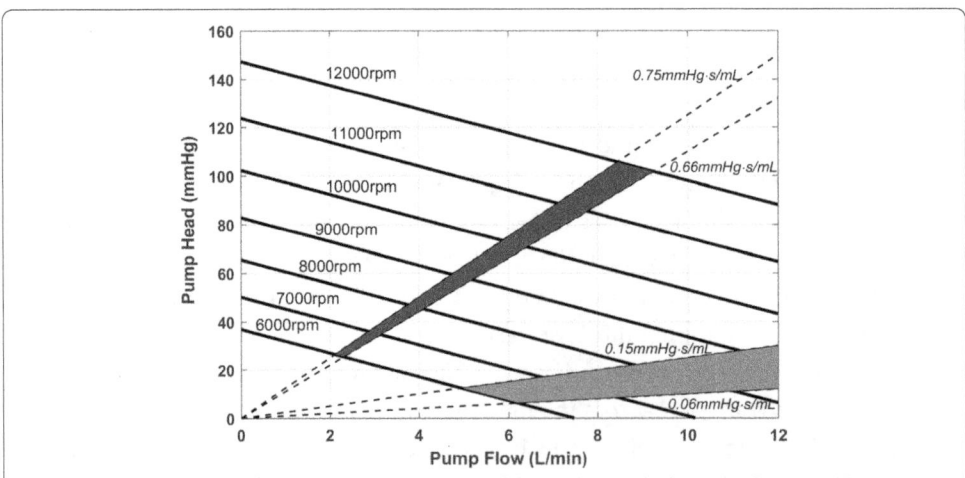

Fig. 4 The static H–Q curves and the operational areas of the axial pump before and after the added resistance 0.6 mmHg s/mL. The normal range of the pulmonary resistance is found between 0.06 and 0.15 mmHg s/mL

Computational settings

All of the models were programmed using Simulink/Matlab software (The MathWorks Inc., Natick, MA, USA). The first-order spatial derivative in the gas exchange model was approximated by using a four-point upwind biased formula [26], after which all the equations of the models were ODEs that can be implemented in Simulink. The Runge–Kutta solver and a 10^{-3} s fixed step were chosen in the simulation.

First, simulations without the RVAD were carried out for modeling validation in both normal and right heart failure conditions. Basic hemodynamics and airway gaseous partial pressures were obtained. The right heart failure condition was achieved by setting the right ventricular contractility to 10% of the normal value and the heart rate to 90 bpm. Then the RBP was connected in parallel to the failing right ventricle. Constant rotary speed and sinusoidal modulated speeds were compared. The mean value of the sinusoidal speed profile was the same as the constant speed, which was set to 9000 rpm. The amplitudes of the sine speed profiles were set to 1000, 2000, and 3000 rpm, respectively. The sinusoidal modulated speed was synchronized with the heartbeat, meaning that the pump was running in copulsation mode. During all the simulations, the breathing and gas exchange were always considered normal.

Results

Hemodynamics

Basic hemodynamics in both normal and right heart failure conditions without the RBP were obtained and are shown in Figs. 5 and 6, respectively. Pressures in normal physiological conditions at the main positions of the blood circulation are depicted in Fig. 5a, while those during right ventricular failure can be seen in Fig. 6a. Normal left ventricular pressure (LVP) ranges from 3 to 129 mmHg, compared with a range of 2–82 mmHg in right heart failure conditions. Systemic arterial pressures (SAP) in the two conditions are

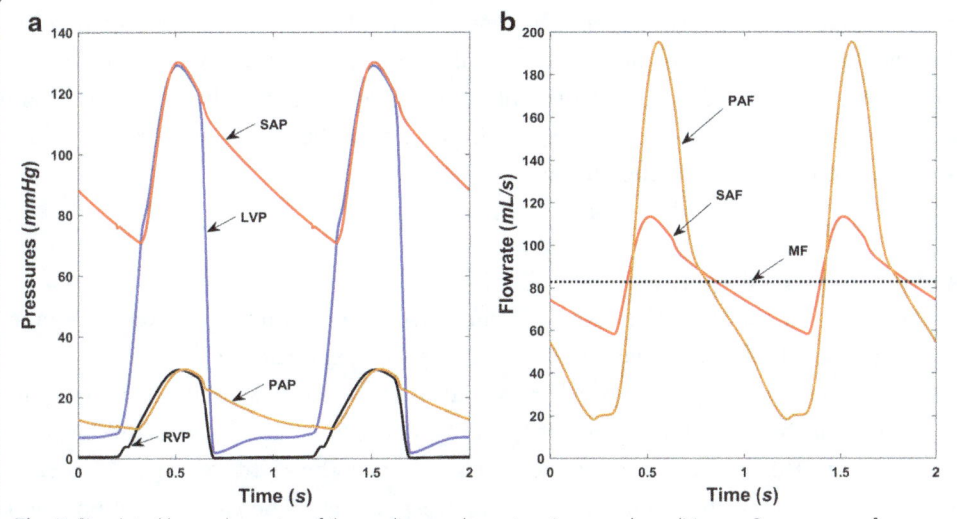

Fig. 5 Simulated hemodynamics of the cardiovascular system in normal conditions. **a** Pressure waveforms and **b** flow waveforms. *LVP* left ventricular pressure, *SAP* systemic arterial pressure, *RVP* right ventricular pressure, *PAP* pulmonary arterial pressure, *SAF* systemic arterial flowrate, *PAF* pulmonary arterial flowrate, *MF* mean flowrate of the systemic circulation

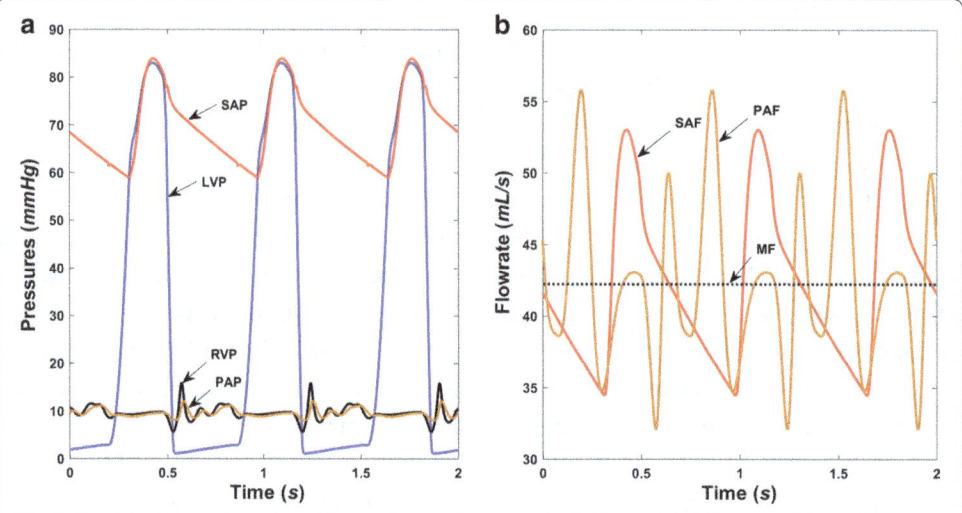

Fig. 6 Simulated hemodynamics of the cardiovascular system in right ventricle failure condition. **a** Pressure waveforms and **b** flow waveforms. *LVP* left ventricular pressure, *SAP* systemic arterial pressure, *RVP* right ventricular pressure, *PAP* pulmonary arterial pressure, *SAF* systemic arterial flowrate, *PAF* pulmonary arterial flowrate, *MF* mean flowrate of the systemic circulation

72–129 mmHg and 59–82 mmHg, respectively. The LVP waveform is followed well by the SAP waveform in the systolic period. The corresponding systemic arterial flowrate (SAF) measured at the node R_{svr} in Fig. 1 is also pulsatile, with a range of 58–113 mL/s and a mean value of 83 mL/s for normal conditions, and 34–53 mL/s with a mean value of 42 mL/s for right heart failure conditions (Figs. 5b and 6b). These hemodynamic results are in accordance with those reported in previous studies [20].

Normally, the right ventricular pressure (RVP), which ranges from 0 to 29 mmHg, and the pulmonary arterial pressure (PAP), which ranges from 10 to 29 mmHg, are consistent with physiological values. During right ventricle failure, RVP and PAP fluctuate within a narrow range, and the waveforms becomes disordered (Fig. 6a), as does the pulmonary arterial flowrate (PAF) waveform measured at the node R_{pvr}. It is worth noting that the flow pulsatile in the pulmonary circulation is much larger than in the systemic circulation, and both decrease significantly during right heart failure. Another point worth noting is the change in heart period from 60 to 90 bpm between the two figures, which is in accordance with the settings.

Airway gaseous partial pressure

O_2-enriched air fills into the alveoli during inspiration, where gas exchange takes place, taking away O_2 and leaving CO_2, followed by expiration where the air is discharged into the atmosphere again. The variation in airway gas composition in terms of changes in the partial pressures of O_2 (PO_2) and CO_2 (PCO_2) generated by simulation during this process is depicted in Fig. 7 when the right ventricle is normal, and Fig. 8, when the right ventricle is failing, respectively. The variation of both PO_2 and PCO_2 in the up airways (dead space) is much more dramatic than that in the alveolar region. As shown in Fig. 7, alveolar PO_2 varies from 107 to 113 mmHg and PCO_2 varies from 37 to 40 mmHg, both of which vary within a narrow range. Not all

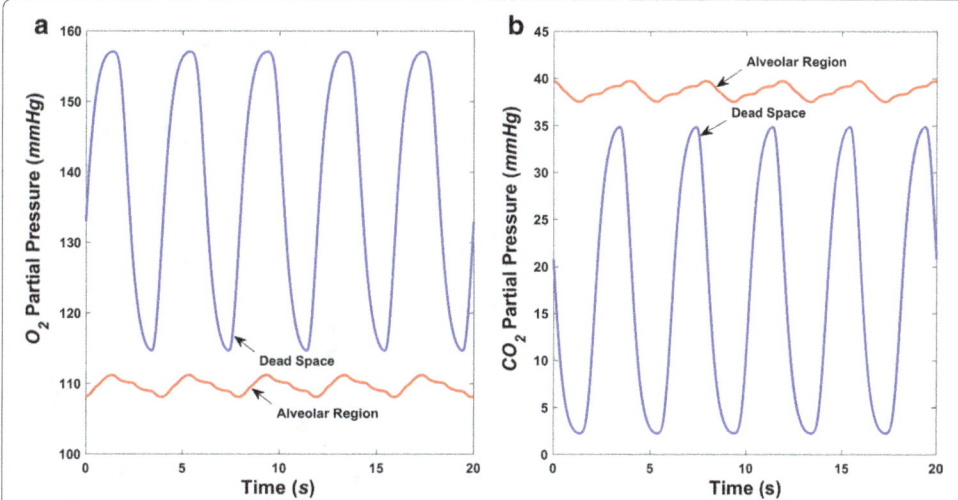

Fig. 7 Simulated airway gaseous partial pressure during normal breathing in normal hemodynamic condition. **a** O_2 partial pressure (PO_2) in the dead space and alveolar region and **b** CO_2 partial pressure (PCO_2) in the dead space and alveolar region

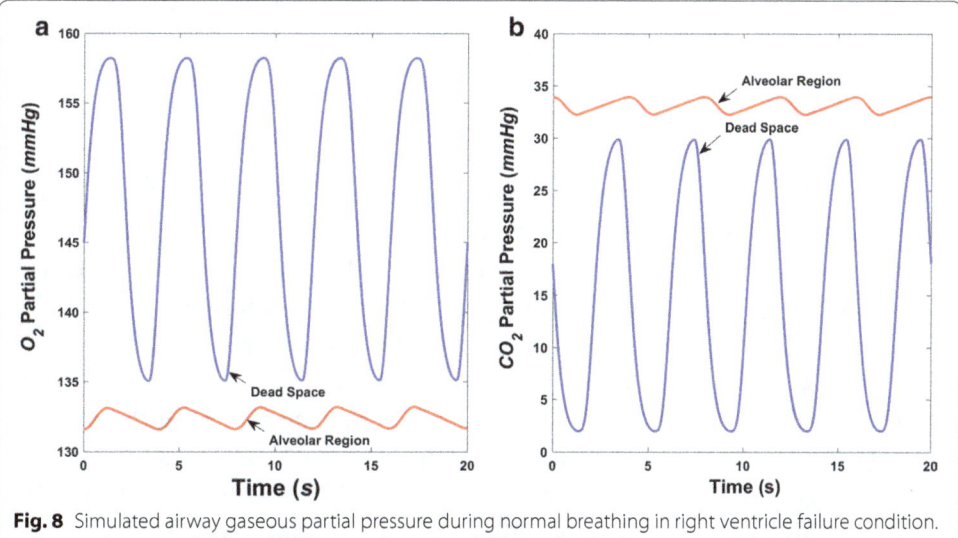

Fig. 8 Simulated airway gaseous partial pressure during normal breathing in right ventricle failure condition. **a** O_2 partial pressure (PO_2) and **b** CO_2 partial pressure (PCO_2) in the dead space and alveolar region

inhaled air enters the alveoli, and inhaled and residual air mix to create variations in alveolar PO_2 and PCO_2 much smaller than those in the dead space. During right ventricle failure, alveolar PO_2 rises to a range of 131–134 mmHg, which is caused by the decreasing of the volume of oxygen that diffuses into the blood. The total flux of O_2 across the alveolar–capillary membrane is around 1.7 mL/s, compared with 3.5 mL/s in normal conditions. Meanwhile, alveolar PCO_2 declines to between 32 and 34 mmHg. It is important to note that breathing function was assumed to be normal during the simulation.

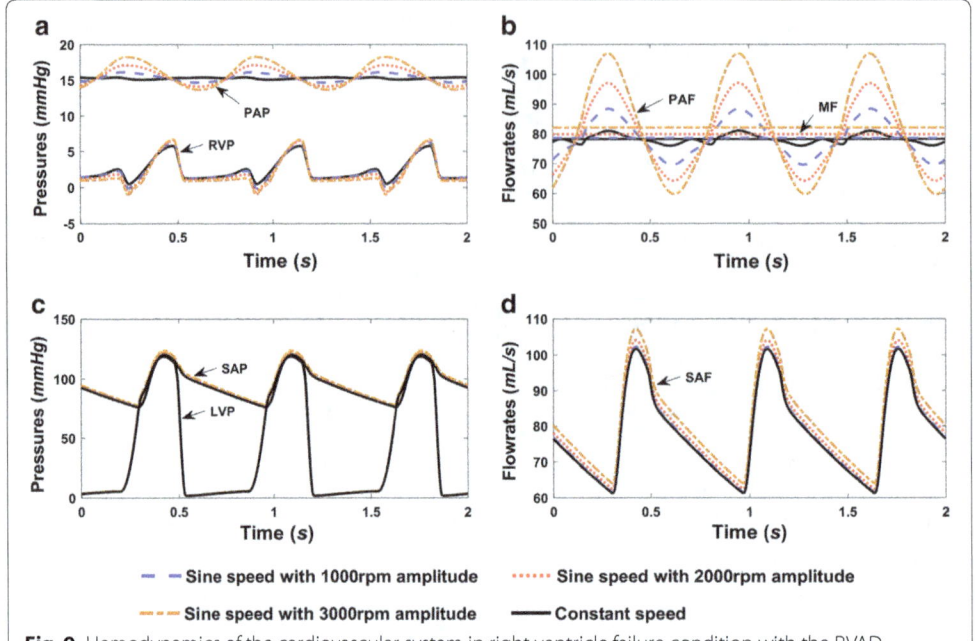

Fig. 9 Hemodynamics of the cardiovascular system in right ventricle failure condition with the RVAD assisting in constant and sinusoidal speed modes. **a** Pressure waveforms of the pulmonary circulation; **b** flow waveforms of the pulmonary circulation; **c** pressure waveforms of the systemic circulation; and **d** flow waveforms of the systemic circulation. *LVP* left ventricular pressure, *SAP* systemic arterial pressure, *RVP* right ventricular pressure, *PAP* pulmonary arterial pressure, *SAF* systemic arterial flowrate, *PAF* pulmonary arterial flowrate, *MF* mean flowrate of the systemic circulation

Effects of speed modulation

In this study, the axial blood pump serving as a RVAD during right ventricular failure was run in constant speed mode and sinusoidal modulated speed mode with three different amplitudes. As mentioned previously, the offset of the sinusoidal speed is the same as the constant speed value, which was 9000 rpm in our simulation.

The hemodynamic results generated by the numerical model are shown in Fig. 9. Compared with Fig. 6, it is clear that the PAP increased from 10 mmHg to nearly 15 mmHg with the contribution of the assist pump in all of the running modes. Furthermore, the corresponding PAF fluctuates around a mean value of about 80 mL/s (4.8 L/min), which is about twice as great as without the assist pump in Fig. 6b. With the help of the RVAD, regardless of the mode, the systemic circulatory hemodynamics (LVP, SAP, and SAF) returned to normal values, demonstrating the assisting capacity of the pump. Unlike normal physiological conditions, RVP was much lower than PAP during pump assisting, implying most pump function was contributed by the axial blood pump.

In systemic hemodynamics, the speed modulation of the right RBP has very little effect. As can be seen from Fig. 9c, d, the systemic pressure and flow waveforms with different RBP speed profiles (including constant speed) almost overlap. Although all of the running modes could undertake the right ventricular assisting task, a different influence was found in the pulmonary circulation between them. Along with increasing the sinusoidal modulation amplitude (the amplitude of the constant speed could be regarded as zero), PAP and PAF become more pulsatile (Fig. 9a, b); however, these were still much weaker than the pulsatility in the normal physiological conditions (Fig. 5).

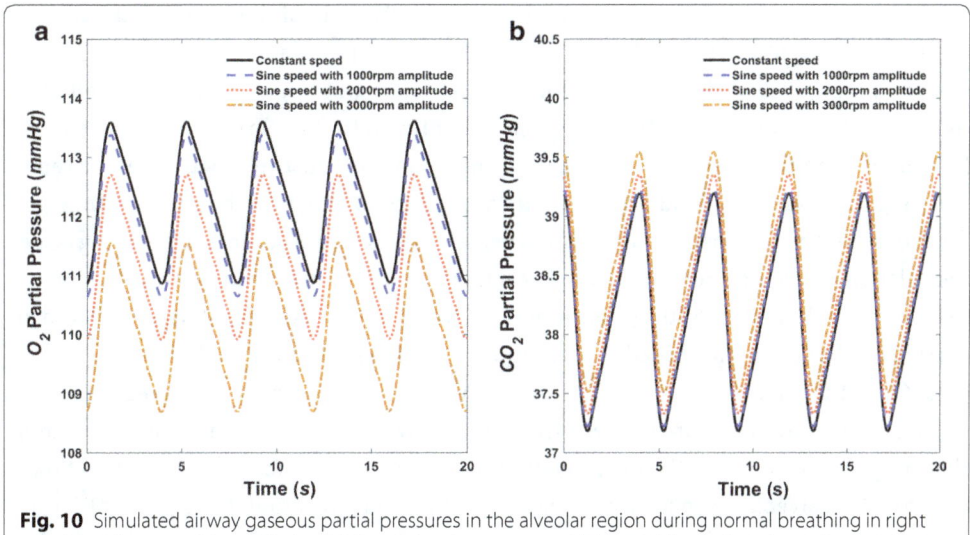

Fig. 10 Simulated airway gaseous partial pressures in the alveolar region during normal breathing in right ventricle failure condition with the RVAD assisting in constant and sinusoidal speed modes. **a** O_2 partial pressure (PO_2) and **b** CO_2 partial pressure (PCO_2)

Figures 7 and 8 show that, when only right ventricle failure occurs but breathing is normal, the alveolar PO_2 will rise to a relatively high value of more than 131 mmHg, while alveolar PCO_2 decreases to less than 34 mmHg. After using the assist pump, the alveolar PO_2 and PCO_2 are restored to normal ranges, which are around 110 and 38 mmHg, respectively (Fig. 10). With different speed modulations, the alveolar PO_2 and PCO_2 have slight differences. The larger the modulation amplitude, the lower the mean value of PO_2 and the greater the mean value of PCO_2, whereas the value of change is very small. Along with the increase of amplitude, a trend for the separation between the adjacent PO_2 waveforms became more obvious (Fig. 10).

Discussion

Many studies have focused on the pulsatile speed modulation of RBPs as LVADs, whereas there have been few reports on RVADs. Flow pulsation in the pulmonary circulation is remarkable (Fig. 5b). Previous reports have shown the peak flow velocity in pulmonary vein is 52.1 ± 16.4 cm/s [13], and the average pulmonary vein ostium diameter can reach 20 mm [27], which corresponds to a peak volume flow of about 164 ± 51 mL/s. In the present study, the peak flow results agreed with previous studies, side demonstrating the effectiveness of the numerical model. Besides blood transportation, another important function of the pulmonary circulation is gas exchange. Therefore, this study aimed to investigate the effects on both hemodynamics and gas exchange when a right ventricular assist RBP runs in pulsatile mode.

The basic hemodynamics under normal and right ventricle failure conditions generated by the numerical model fit the physiological and pathological results well, demonstrating the effectiveness of the used model. Then sinusoidal speed modulations with different amplitudes under right ventricle failure condition are implemented, revealing that with a sufficient mean rotary speed the hemodynamics could recover back to normal and meanwhile enhance the pulsatility of the pulmonary circulation. The airway gaseous partial pressures also return to normal with the help of the RVAD. Enlarging the modulation amplitude will influence alveolar PO_2 and PCO_2 inversely, decreasing PO_2 and increasing PCO_2. However, this influence is slight and not as obvious as the hemodynamic effect. Considering the change in mean flowrate (Fig. 9b), this influence might be caused from this mean flow difference. This finding is consistent with a recent report claiming that pulsatile blood flow has only a small to no effect on the gas exchange performance in an oxygenator [28]. Therefore, it implies pulsatile operation using speed modulation would not influence the gas exchange much while could improve the hemodynamics (including pulsatility) enormously. More experimental validations could be carried out further with a breath sensor and a wearable sensor system [29, 30].

As O_2 diffusion across the alveolar–capillary membrane decreases, alveolar PO_2 rises during right ventricle failure (Fig. 8). This is due to the low flowrate in the lung, meaning that oxygen diffusing into the blood cannot be taken away immediately. The total flux of O_2 across the alveolar–capillary membrane decreasing from about 3.5 to nearly 1.7 mL/s could also support this. Lower O_2 diffusion would mean lower O_2 partial pressure in the arterial blood after gas exchange, in line with the actual situation in which a patient with right heart failure would have low blood O_2 partial pressure. Breathing function was regarded as normal during the simulations, whereas generally, heart failure is accompanied by abnormal breathing, such as faster breath frequency. Further models describing the relationship between heart failure and respiratory function are required to improve the numerical simulation.

Another limitation of this study is that it does not include tissue O_2 consumption and CO_2 generation model. The initial O_2 and CO_2 partial pressures from the venous blood are set as constant values according to the study by Lu et al. [22]. Introducing a tissue O_2 and CO_2 exchange model to construct a whole closed-loop cardiovascular–pulmonary tissue model would extend the capability of the numerical study. More physiological conditions, such exercise, can be simulated and it would be beneficial to designing control algorithms of RBPs in use.

The speed modulation method of the RBP adopted in this study was limited to sinusoidal modulation. Only amplitude varied, while the modulation frequency was set at a constant heart rate. Besides, the RBP was run in copulsation mode to enhance the pulsation, and no other modes, such as counterpulsation mode was implemented in this study. To give a more comprehensive understanding of speed modulation of continuous flow RVAD, investigations using more rotary speed modulated waveforms, such as square wave, and more modulation modes, such as counterpulsation mode, and even asynchronous modulation with different heartbeats, could be carried out in the future.

Conclusions

In this study, we investigated the effects of a right ventricular assist RBP running in both constant and synchronized pulsatile modes on the pulmonary hemodynamics and gas exchange using a numerical method. Basic hemodynamics and airway O_2 and CO_2 partial pressures in both normal physiological conditions and during right ventricle failure were obtained. Results showed that the pulsatile run mode of the assist pump was able to recover most hemodynamics to normal levels during right ventricular failure, and that speed modulation could obviously increase the flow and pressure pulsatility in the pulmonary circulation while with only little effect on the pulmonary gas exchange. This study could provide basic insight of the influence of speed modulation of a right ventricular assist RBP when designing pulsatile control algorithms for them.

Authors' contributions

FH proposed the method and established the model in this research, and was a major contributor in writing the manuscript. ZG analyzed the results and was a minor contributor in writing the manuscript. YF provided the numerical implementation of the model. XR overviewed the research and provided some discussions. All authors read and approved the final manuscript.

Author details

[1] College of Metrology & Measurement Engineering, China Jiliang University, Xueyuan Road 258, Hangzhou, China. [2] State Key Laboratory of Fluid Power and Mechatronic Systems, Zhejiang University, Hangzhou, China. [3] School of Mechanical and Automotive Engineering, Zhejiang University of Science and Technology, Hangzhou, China.

Acknowledgements

This work is supported by National Natural Science Foundation of China (Grant No. 51505455) and funded by Open Foundation of the State Key Laboratory of Fluid Power and Mechatronic Systems (No. GZKF-201713).

Competing interests

The authors declare that they have no competing interests.

Consent for publication

Not applicable.

Disclosures

This work is supported by National Natural Science Foundation of China (Grant No. 51505455) and funded by Open Foundation of the State Key Laboratory of Fluid Power and Mechatronic Systems (No. GZKF-201713).

Funding

This work is supported by National Natural Science Foundation of China (Grant No. 51505455) and Open Foundation of the State Key Laboratory of Fluid Power and Mechatronic Systems (No. GZKF-201713).

References

1. Moscato F, Steinseifer U. From rotary blood pumps to mechanical circulatory support systems. Artif Organs. 2016;40:821–2.
2. Hsu P, Parker J, Egger C, Autschbach R, Schmitz-Rode T, Steinseifer U. Mechanical circulatory support for right heart failure: current technology and future outlook. Artif Organs. 2012;36:332–47. http://www.ncbi.nlm.nih.gov/pubmed/22150419.
3. Lampert BC, Teuteberg JJ. Right ventricular failure after left ventricular assist devices. J Heart Lung Transplant. 2015;34:1123–30. https://doi.org/10.1016/j.healun.2015.06.015.
4. Moazami N, Dembitsky WP, Adamson R, Steffen RJ, Soltesz EG, Starling RC, et al. Does pulsatility matter in the era of continuous-flow blood pumps? J Heart Lung Transplant. 2015;34:999–1004. https://doi.org/10.1016/j.healun.2014.09.012.
5. Kleinheyer M, Timms DL, Greatrex NA, Masuzawa T, Frazier OH, Cohn WE. Pulsatile operation of the BiVACOR TAH—Motor design, control and hemodynamics. In: The 38th annual international conference of the IEEE engineering in medicine and biology society. 2014;5659–62.
6. Huang F, Ruan X, Fu X. Pulse-pressure-enhancing controller for better physiologic perfusion of rotary blood pumps based on speed modulation. ASAIO J. 2014;60:269–79. http://www.ncbi.nlm.nih.gov/pubmed/24614360.
7. Ising M, Warren S, Sobieski MA, Slaughter MS, Koenig SC, Giridharan GA. Flow modulation algorithms for continuous flow left ventricular assist devices to increase vascular pulsatility: a computer simulation study. Cardiovasc Eng Technol. 2011;2:90–100.
8. Amacher R, Ochsner G, Schmid Daners M. Synchronized pulsatile speed control of turbodynamic left ventricular assist devices: review and prospects. Artif Organs. 2014;38:867–75.
9. Bozkurt S. Physiologic outcome of varying speed rotary blood pump support algorithms: a review study. Australas Phys Eng Sci Med. 2016;39:13–28.

10. Alkan T, Akçevin A, Ündar A, Türkoğlu H, Paker T, Aytaç A. Benefits of pulsatile perfusion on vital organ recovery during and after pediatric open heart surgery. ASAIO J. 2007;53:651–4. http://content.wkhealth.com/linkback/openurl?sid=WKPTLP:landingpage&an=00002480-200711000-00001.

11. Ündar A, Masai T, Yang SQ, Eichstaedt HC, McGarry MC, Vaughn WK, et al. Pulsatile perfusion improves regional myocardial blood flow during and after hypothermic cardiopulmonary bypass in a neonatal piglet model. ASAIO J. 2002;48:90–5. http://www.embase.com/search/results?subaction=viewrecord&from=export&id=L3457 5775%5Cnhttp://sfx.library.uu.nl/utrecht?sid=EMBASE&issn=10582916&id=doi:&atitle=Pulsatile+perfu sion+improves+regional+myocardial+blood+flow+during+and+after+hypothermic+cardiop.

12. Hornick P, Taylor K. Pulsatile and nonpulsatile perfusion: the continuing controversy. J Cardiothorac Vasc Anesth. 1997;11:310–5.

13. Takaya T, Arakawa MTT. Pulmonary vein blood flow velocity waveform—with special reference to pulmonary "systolic runoff" in patients with atrial septal defect. Jpn Circ J. 1986;5:405–15.

14. Wong KKL, Wang D, Ko JKL, Mazumdar J, Le TT, Ghista D. Computational medical imaging and hemodynamics framework for functional analysis and assessment of cardiovascular structures. Biomed Eng Online. 2017;16:1–23.

15. Liu X, Gao Z, Xiong H, Ghista D, Ren L, Zhang H, et al. Three-dimensional hemodynamics analysis of the circle of Willis in the patient-specific nonintegral arterial structures. Biomech Model Mechanobiol. 2016;15:1439–56.

16. Wong KKL, Cheung SCP, Yang W, Tu J. Numerical simulation and experimental validation of swirling flow in spiral vortex ventricular assist device. Int J Artif Organs. 2010;33:856–67. http://www.ncbi.nlm.nih.gov/pubmed/21186467.

17. Gao Z, Li Y, Sun Y, Yang J, Xiong H, Zhang H, et al. Motion tracking of the carotid artery wall from ultrasound image sequences: a nonlinear state-space approach. IEEE Trans Med Imaging. 2017;0062:1–11.

18. Zhao S, Gao Z, Zhang H, Xie Y, Luo J, Ghista D, et al. Robust segmentation of intima-media borders with different morphologies and dynamics during the cardiac cycle. IEEE J Biomed Health Inform. 2017;2194:1–11.

19. Gao Z, Xiong H, Liu X, Zhang H, Ghista D, Wu W, et al. Robust estimation of carotid artery wall motion using the elasticity-based state-space approach. Med Image Anal. 2017;37:1–21. https://doi.org/10.1016/j.media.2017.01.004.

20. Colacino FM, Moscato F, Piedimonte F, Arabia M, Danieli GA. Left ventricle load impedance control by apical VAD can help heart recovery and patient perfusion: a numerical study. ASAIO J. 2007;53:263–77. http://content.wkhealth.com/linkback/openurl?sid=WKPTLP:landingpage&an=00002480-200705000-00002.

21. Toy SM, Melbin J, Noordergraaf A. Reduced models of arterial systems. IEEE Trans Biomed Eng. 1985;32:174–6. http://www.ncbi.nlm.nih.gov/pubmed/3997173.

22. Lu K, Clark JW, Ghorbel FH, Ware DL, Bidani A. A human cardiopulmonary system model applied to the analysis of the valsalva maneuver. Am J Physiol Heart Circ Physiol. 2001;281:H2661–79.

23. Liu CH, Clark JW, Niranjan SC, San KY, Zwischenberger JB, Bidani A. Airway mechanics, gas exchange, and blood flow in a nonlinear model of the normal human lung. J Appl Physiol. 1998;84:1447–69.

24. Krabatsch T, Hennig E, Stepanenko A, Schweiger M, Kukucka M, Huebler M, et al. Evaluation of the HeartWare HVAD centrifugal pump for right ventricular assistance in an in vitro model. ASAIO J. 2011;57:183–7. http://www.ncbi.nlm.nih.gov/pubmed/21336105.

25. Choi S, Boston JR, Thomas D, Antaki JF. Modeling and identification of an axial flow blood pump. Proc Am Control Conf. 1997;6:3714–5.

26. Schiesser WE. Computational mathematics in engineering and applied science, ODEs, DAEs, and PDEs. Boca Raton: CRC Press; 1994.

27. Wittkampf FHM, Vonken EJ, Derksen R, Loh P, Velthuis B, Wever EFD, et al. Pulmonary vein ostium geometry: analysis by magnetic resonance angiography. Circulation. 2003;107:21–3. http://circ.ahajournals.org/cgi/doi/10.1161/01.CIR.0000047065.49852.8F.

28. Schraven L, Kaesler A, Flege C, Kopp R, Schmitz-Rode T, Steinseifer U, et al. Effects of pulsatile blood flow on oxygenator performance. Artif Organs. 2018. https://doi.org/10.1111/aor.13088.

29. Xiong Y, Ye Z, Xu J, Zhu Y, Chen C, Guan Y. An integrated micro-volume fiber-optic sensor for oxygen determination in exhaled breath based on iridium(III) complexes immobilized in fluorinated xerogels. Analyst. 2013;138:1819–27.

30. Wu W, Pirbhulal S, Zhang H, Mukhopadhyay SC. Quantitative assessment for self-tracking of acute stress based on triangulation principle in a wearable sensor system. IEEE J. Biomed Health Inform. 2018;1:1. https://doi.org/10.1109/JBHI.2018.2832069.

3

Dynamic stability and spatiotemporal parameters during turning in healthy young adults

Chuan He[1,2†] ⓘ, Rui Xu[1,2†], Meidan Zhao[3], Yongming Guo[3], Shenglong Jiang[1,2], Feng He[1,2] and Dong Ming[1,2*]

*Correspondence:
richardming@tju.edu.cn
†Chuan He and Rui Xu
equally contributed to this
work
[1] Lab of Neural Engineering
& Rehabilitation,
Department of Biomedical
Engineering, College
of Precision Instruments
and Optoelectronics
Engineering, Tianjin
University, Tianjin, China
Full list of author information
is available at the end of the
article

Abstract

Background and purpose: Turning while walking has a frequent occurrence in daily life. Evaluation of its dynamic stability will facilitate fall prevention and rehabilitation scheme. This knowledge is so limited that we set it as the first aim of this study. Another aim was to investigate spatiotemporal parameters during turning.

Methods: Fifteen healthy young adults were instructed to perform straight walking, 45° step turn to the left and 45° spin turn to the right at natural speed. Dynamic stability was measured by margin of stability (MoS) in anterior, posterior, left and right direction at each data point where significant differences were detected using 95% bootstrap confidence band. Common spatiotemporal parameters were computed in each condition subdivided into approach, turn and depart phases.

Results: Results showed that minimum anterior MoS appeared at middle of swing while minimum lateral MoS at contralateral heel strike in all conditions. Posterior MoS decreased before middle of turn phase in spin whereas after middle of turn phase in step. Lateral MoS and stride width declined in turn phase of spin while in depart of step. Spin had a long step and stride length. Long swing phases were observed in turns.

Conclusions: These data help explain that people are most likely to fall forward at middle of swing and to fall toward the back and the support side at heel strike. Our findings demonstrate that instability mainly exist in turn phase of spin and depart phase of step turn.

Keywords: Gait analysis, Turning, Dynamic stability, Step turn, Spin turn, Margin of stability, Gait stability, Spatiotemporal parameters

Background

Turning during ambulation is crucial to daily life. Turning steps constitute a considerable percentage (about 20–50%) of steps taken during activities of daily living [1, 2]. Turning increases the risk of falling caused by slipping compared with that caused by slipping while straight walking [3]. Falling while turning causes a 7.9 times increase in hip fractures than falling during straight walking [4]. Giving insight into turning gait may contribute to the evaluation and the development of rehabilitation program for movement disorders, design of assistive devices, gait planning of biped robot and animation design in computer animation industry.

According to whether there are obvious transition steps or not, turns can be divided into "steady-state" turns, such as circular walking, and "transient" turns, such as 90° turns [5, 6]. The latter can be performed using two turning strategies: (1) step turn, turning toward the contralateral side of the stance limb (outside leg strategy); (2) spin turn, turning to the ipsilateral direction of the stance limb (inside leg strategy) [7].

Researches show that there exists turning preference. Dixon et al. [8, 9] suggested that both typically developing children (n = 54) and cerebral palsy children (n = 22) preferred spin turns, while step turns were singly performed by only 1/54 and none respectively. Patla et al. [10] reported that healthy young adults performed step turns during more than 80% of trials when executing 60° turns while walking. Akram et al. [11] found that healthy older adults (n = 19, age = 66 ± 4.2 years) preferred spin turns, whereas step turns were preferred only in 90° turns while walking fast. But Conradsson et al. [12] revealed that step and spin turns followed a nearly 50:50 distribution during 90° turns both for 19 individuals with Parkinson's disease (age = 72 ± 4 years) and for 17 age and gender matched healthy controls. Now, the reason for the turning preference is still unknown, although some studies [8, 12–14] have investigated the kinematics and kinetics during turning in people of different age groups. It is all-important to understand the intrinsic difference between step and spin turns.

Because turning increases risk of falling [3] and injury [4], we focused on the gait stability of turns in this study. We found that the stability evaluation of turning was limited, even though the assessment method of gait stability had been greatly developed [15, 16]. Stability is the capacity to maintain balance during perturbations [17]. Static stability, which is generally defined as 'center of mass (CoM) should be vertically projected within base of support (BoS)' [18], is widely accepted to assess stability of posture. Gait stability is usually evaluated by dynamic stability because gait is a dynamic movement. Dynamic stability can be divided into local dynamic stability and global dynamic stability depending on small and large perturbations respectively from internal and external sources [15, 16, 19]. Bruijn et al. [15] assessed four levels of validity of common gait stability measures and found that measures with high validity included largest Lyapunov exponent and variability measures in local dynamic stability, and foot placement estimator and margin of stability in global dynamic stability. Local dynamic stability, derived from dynamical systems theory, requires collecting kinematic data in a large number of continuous strides (usually around 110 strides [20]), which essentially determines that the measures derived from dynamical systems theory might not be practical for assessing the dynamic stability of a transient turn because it is only a small part of the walking trajectory. Global dynamic stability describes the probability of falling [21]. Foot placement estimator is high validity but in need of a full body marker set [15]. Margin of stability (MoS), proposed by Hof et al. [18], is extensively used due to high validity [15] and simple operation (requiring a few continuous gait cycle samples and no force plate). Based on the inverted pendulum model of balance, MoS can be used to assess gait stability for straight walking, for starting and stopping walking and for turning [22], and is much more convincing than the static method by considering both the relative position and velocity between CoM and BoS [18]. However, most studies investigating turning gait stability either assessed static stability, such as distance between the ankles [13, 23], stride width [12,

24] and distance between CoM and BoS [9, 13, 23, 24], or did not compare dynamic stability between two turning strategies and straight walking [9, 23]. For example, spin turns are generally considered less stable and demanding a higher biomechanical cost than step turns, mainly because Taylor et al. [13] indicated that the spin turn had a narrow BoS leading to CoM outside displacement and required increased joint moments and power, compared to the step turn and straight gait.

Furthermore, both turning strategies demand modification of spatiotemporal parameters from those of straight walking, which can somewhat reflect biomechanical adaptations simply and effectively [8]. Previous study [25] revealed that young adults decreased velocity and stride length, increased stance time and stride width during step turns compared to those of straight walking. However, the current evidence is limited.

There had been no previous study to compare MoS of healthy young adults during straight line walking, step turn and spin turn and this was the first aim of this study. In addition, spatiotemporal parameters were also comprehensively compared in three conditions. We hypothesized that both MoS and stride width of step turn would be greater while those of spin turn smaller than straight walking [8, 13]. We also hypothesized that for both turn conditions, walking speed, step length and stride length would be smaller whereas temporal parameters larger than for straight walking.

Methods

Participants

Fifteen healthy young adults [six males, nine females; age, 23.6 (3.1) years; mass, 55.6 (12.2) kg; height, 1.686 (0.105) m; body mass index, 19.4 (2.3) kg·m^{-2}; leg length (measured as the distance between the greater trochanter and the floor), 0.850 (0.064) m; mean (standard deviation)] from Tianjin University were included in this analysis. All participants are right-footed determined by kicking a ball [26]. All participants had no history of any neuromusculoskeletal disorders and did not undergo any strenuous physical exercise during 3 days before the experiment. The study was approved by the local ethical committee and all participants gave written informed consent prior to their participation.

Experimental procedures and data collection

Participants were instructed to perform practice sessions and then complete three barefoot walking tasks at the self-selected preferred walking speed: straight walking, 45° step turn to the left and 45° spin turn to the right (Right lower limb, the dominant leg, was the turning limb in both turn conditions), see Fig. 1 and the additional videos (Additional file 1). A 45° turning angle was chosen due to moderate intensity [27] and easy implementation. Each task consisted of four trials. Marker position data were collected using a 10-camera VICON motion capture system (Vicon Motion Systems Ltd, Oxford, UK) at 100 Hz. Kinetic data were obtained from four AMTI force plates (Advanced Mechanical Technology, Inc., Watertown, MA, USA) sampling at 1000 Hz. Finally, C3D files were imported into MATLAB (R2012b, The MathWorks Inc., Natick, MA, USA) by using the powerful open-source Biomechanical ToolKit [28]. Data processing was performed by a custom program written in MATLAB.

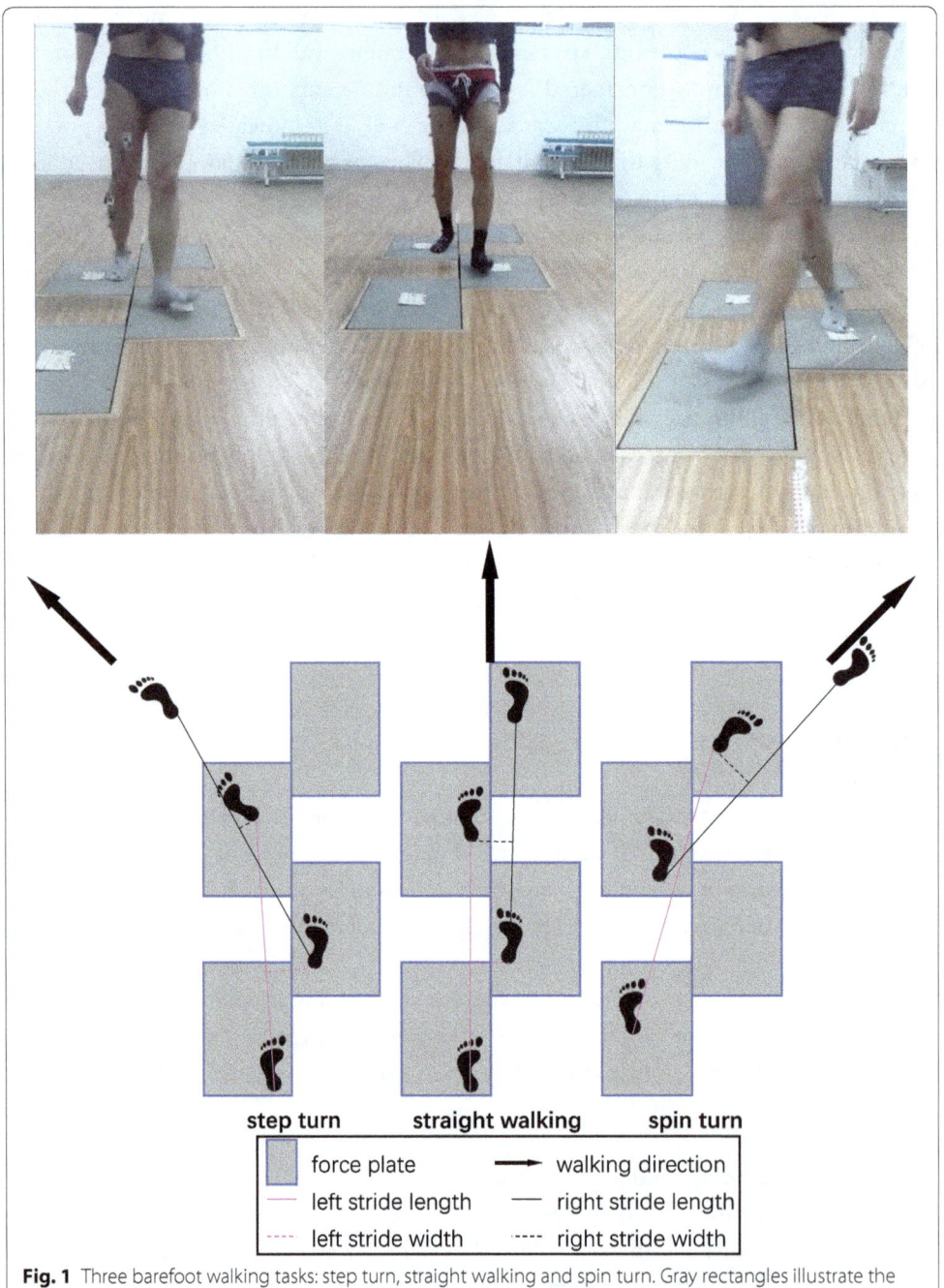

Fig. 1 Three barefoot walking tasks: step turn, straight walking and spin turn. Gray rectangles illustrate the relative position of four force plates (40 × 60 cm) embedded in the middle of the walkway (2.5 × 10 m). Arrows in the floor surface pointed to the walking direction

Data processing

The present study only analyzed the trajectory data of 10 reflective markers placed bilaterally over anatomical landmarks on anterior and posterior superior iliac spine (ASIS and PSIS), lateral malleolus, heel and tip of the toe. Raw marker position data were low-pass filtered using a fourth-order zero-lag Butterworth filter with a cut-off frequency of 4 Hz. All gait data were analyzed in the duration from the first heel strike (HS1) to the third toe off (TO3) (Fig. 1). Heel strike (HS) and toe off (TO) events were identified using force plate data, while the HS4 events of turning conditions were detected as the instant where the heel marker reached local minimum in the vertical direction.

Based on assumptions of inverted pendulum model, extrapolated center of mass (XCoM) and MoS were computed as follows [18]:

$$\mathbf{XCoM} = \mathbf{CoM} + \frac{\mathbf{v}_{\mathrm{CoM}}}{\omega_0}$$
$$\omega_0 = \sqrt{g/l}$$
$$\mathbf{MoS} = \mathbf{BoS} - \mathbf{XCoM}$$

(1)

where **CoM** and $\mathbf{v}_{\mathrm{CoM}}$ are the instantaneous position and velocity of whole-body CoM, ω_0 is the (angular) eigenfrequency of the pendulum, g is the earth gravitational acceleration ($g = 9.801$ m/s^2 in Tianjin, China), l is the equivalent length of pendulum defined as the distance between CoM and the lateral malleolus, and **BoS** is the anterior, posterior, left or right boundary of BoS (during double support phase) or virtual BoS (during swing phase) [10]. Anterior MoS (MoS$_\mathrm{A}$) was computed as the distance from XCoM to the line through left and right toe markers, while posterior MoS (MoS$_\mathrm{P}$) the distance from XCoM to the line through bilateral heel markers. Left and right MoS (MoS$_\mathrm{L}$ and MoS$_\mathrm{R}$) were defined as the distance from left and right lateral malleolus to the line between CoM and XCoM. Positive and negative MoS values suggested stable and unstable states (XCoM inside and outside BoS), respectively. We computed MoS at every time point, mean MoS over whole stance phase (including double support period) and MoS at key events (heel strike and mid stance [29]). Considering the convenience (a few markers required) and high validity, we estimated whole-body CoM position using center of pelvis model, as the centroid of the triangle formed by two ASIS markers and the midpoint between two PSIS markers [30]. Instantaneous velocity of CoM ($\mathbf{v}_{\mathrm{CoM}}$) was computed using a first order central difference method. Walking speed was computed as average speed of CoM, viz. the distance travelled divided by the duration of the interval.

Turning gait can be commonly separated into approach, turn, and depart phases [8, 25, 31]. In our work, we defined each phase as stance phase (i.e. approach: HS1 ~ TO1; turn: HS2 ~ TO2; depart: HS3 ~ TO3), excluding swing phase, to compare between matching steps of different walking tasks and between different steps of the same task.

Walking direction was defined as the angle between the direction of $\mathbf{v}_{\mathrm{CoM}}$ and the x-axis of the laboratory coordinate system [32] then averaged over every stance phase. The positive values indicated left turns whereas negative values right turns. Spatiotemporal parameters were computed according to standard methods in [33] and non-dimensionally normalized based on [34]. Periods from the HS1 to the HS3 then to the TO3 were normalized to 101 points (0–100) and 63 points (100–162) [8].

Statistical analysis

A one-way repeated measures analysis of variance was conducted to test for significant differences between matching steps of different walking tasks and between different steps of the same task. Post hoc analyses using the Bonferroni adjustment were carried out to detect the significance of pairwise comparisons. When to determine the difference between two steps of the same task, a paired-samples t-test was used. Variability of spatiotemporal parameters was estimated using 95% confidence interval (CI). These aforementioned statistical analyses were performed using SPSS (version 20.0, Chicago, IL, USA). Variability of gait variables varying throughout gait cycle was estimated using 95% bootstrap confidence band (CB) [8, 35] in MATLAB. All tests were applied at the $\alpha = 0.05$ level.

Results

Walking direction and speed

The walking direction in straight walking appeared a cosine curve with an amplitude of 10° (Fig. 2). The mean walking direction over every stance phase was closed to zero in three steps of straight walking and approach phase of turns (Additional file 2). The signed walking direction was compared between turns and straight walking (Fig. 2A, B; Additional file 2). The absolute walking direction, i.e. the turning angle magnitude,

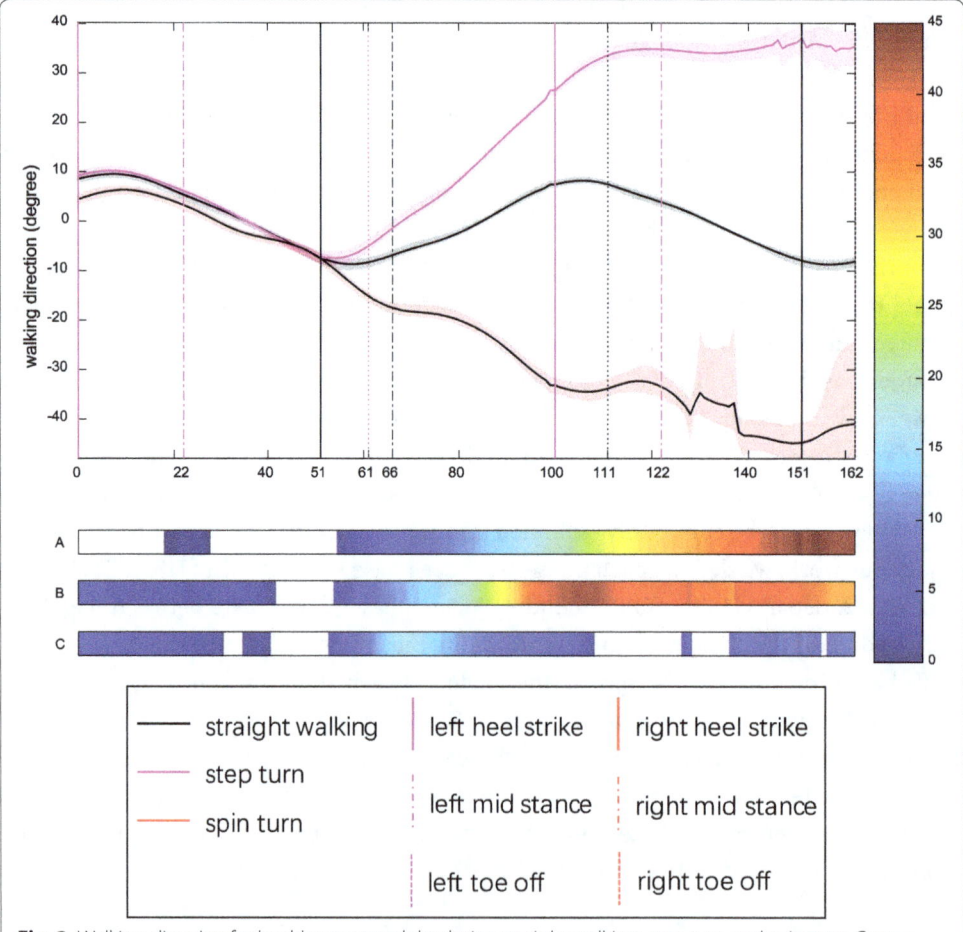

Fig. 2 Walking direction for healthy young adults during straight walking, step turn and spin turn. Curves and shadows represent mean and 95% bootstrap CBs. The positive values indicate left turns while negative values right turns. Horizontal colorbars display areas where significant differences ($p < 0.05$) exist (A: step vs straight; B: spin vs straight; C: absolute value of walking direction in step vs spin). Meanwhile, intensity of colorbar indicates difference size from small (blue) to large (red). Color vertical lines show key events: heel strike (HS), toe off (TO), mid stance (MS). Results of time normalization are: HS1—0, MS1—22%, HS2—51%, TO1—61%, MS2—66%, HS3—100%, TO2—111%, MS3—122%, HS4—151%, TO3—162%

was compared between step and spin turn, and showed smaller in turn ($p < 0.0005$) and depart ($p = 0.045$) phases of step turn than matching phases of spin turn (Additional file 2†; Fig. 2C). Walking speed was significantly lower in approach and turn phases of step turn compared to counterparts of straight (both $p \leq 0.026$) and spin (both $p < 0.0005$). There was no significant difference among the third steps of straight walking and turns.

MoS_A and MoS_P

The curves MoS_A and MoS_P had opposite shape: an inverted peak vs a peak in each step (Fig. 3). In most conditions, MoS_A was negative while MoS_P positive. The minimum MoS_A occurred at the mid-point of each step whereas the minimum MoS_P about heel strike. MoS_P of spin turn was smaller than straight walking and step turn in the first and second steps while MoS_P of step turn was the smallest of three conditions in the third step. For step turn, mean MoS_P over depart was less than turn and approach (both $p < 0.0005$) (Additional file 2). For spin turn, mean MoS_P over depart was more than turn and approach (both $p < 0.0005$).

MoS_L and MoS_R

For MoS_L, spin turn was smaller than other conditions in the first step whereas step turn was the smallest in the second and third step. MoS_R of spin turn was the smallest in all three steps. For step turn, mean MoS_L over depart was less than turn phase ($p < 0.0005$) which in turn was less than approach ($p < 0.0005$) (Additional file 2). Minimum MoS_R appeared at right pre-swing (left loading response) (Fig. 4). For spin turn, mean MoS_R over turn phase was less than approach ($p < 0.0005$) which in turn was less than depart ($p = 0.003$). Minimum MoS_L existed at left pre-swing (right loading response). Example trajectories of CoM, XCoM and lateral malleoli are shown in Fig. 5.

Spatiotemporal parameters

Compared to straight gait, stride width was wider in turn phase ($p < 0.0005$) and much narrower ($p < 0.0005$) in depart of step turn (Additional file 2). For spin turn, on the other hand, stride width was far narrower in turn ($p < 0.0005$) and little wider in depart ($p = 1.000$). Negative values indicated that the stance foot landed laterally to the swing

(See figure on next page.)
Fig. 3 MoS_A and MoS_P for healthy young adults during straight walking, step turn and spin turn. Curves and shadows represent mean and 95% bootstrap CBs. Horizontal colorbars display areas where significant differences ($p < 0.05$) exist (A: step vs straight; B: spin vs straight; C: step vs spin). Meanwhile, intensity of colorbar indicates difference size from small (blue) to large (red). Color vertical lines show key events: heel strike (HS), toe off (TO), mid stance (MS). Results of time normalization are: HS1—0, MS1—22%, HS2—51%, TO1—61%, MS2—66%, HS3—100%, TO2—111%, MS3—122%, HS4—151%, TO3—162%. For step turn, both MoS_A and MoS_P were almost consistent with straight walking in the first step. Then MoS_A decreased and MoS_P increased slightly in the first half of the second step. Finally, MoS_A increased and MoS_P decreased noticeably. Minimum MoS_P appeared in initial and terminal double support phases of the third step. For spin turn, MoS_A was greater and MoS_P was smaller than straight walking in all three steps. Before the middle of turn phase, MoS_A was greater and MoS_P was smaller than step turn. Then to the middle of depart, MoS_A was smaller and MoS_P was greater. Finally, both MoS_A and MoS_P were almost consistent with step turn. Minimum MoS_P occurred at MS2 in turn phase

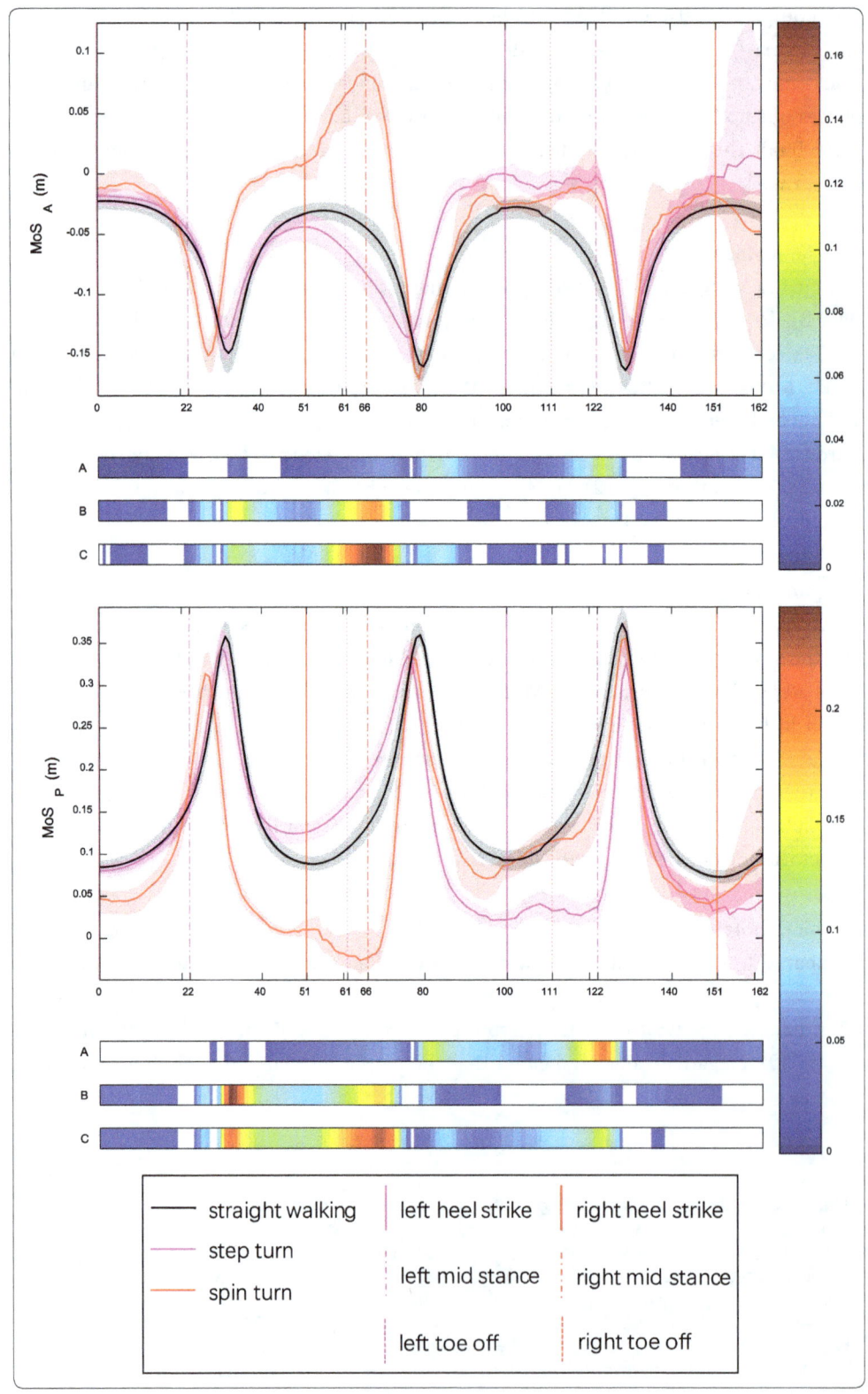

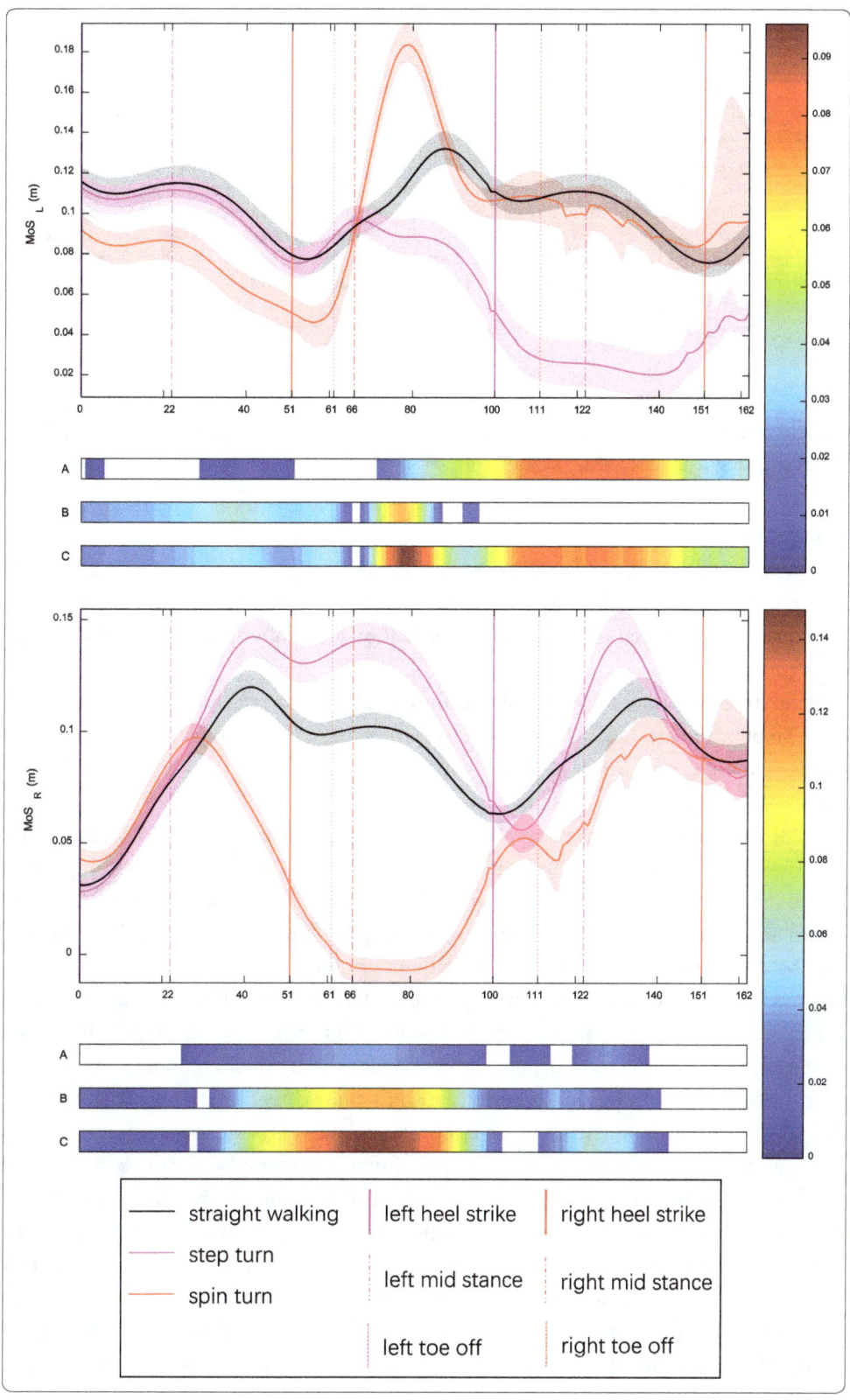

(See figure on previous page.)

Fig. 4 MoS_L and MoS_R for healthy young adults during straight walking, step turn and spin turn. Curves and shadows represent mean and 95% bootstrap CBs. Horizontal colorbars display areas where significant differences ($p < 0.05$) exist (A: step vs straight; B: spin vs straight; C: step vs spin). Meanwhile, intensity of colorbar indicates difference size from small (blue) to large (red). Color vertical lines show key events: heel strike (HS), toe off (TO), mid stance (MS). Results of time normalization are: HS1—0, MS1—22%, HS2—51%, TO1—61%, MS2—66%, HS3—100%, TO2—111%, MS3—122%, HS4—151%, TO3—162%. For step turn, MoS_L was in line with straight walking before MS2 then fell to bottom in depart. MoS_R tracked straight walking before MS1 then increased much in turn phase and finally fluctuated around straight walking. For spin turn, MoS_L was smaller than straight walking and step turn before MS2 then jumped to a peak in late turn phase and finally varied in concert with straight walking during depart. MoS_R was slightly larger than straight walking and step turn before MS1, then plunged to negative bottom in turn phase and finally showed an upturn less than straight walking and step turn in depart

foot stride vector. Step and stride length were longer in spin condition than in step and straight (all $p \le 0.002$). The stance time, single support time, and cycle time were all significantly longer in turns than straight (all $p \le 0.016$).

Discussion

The present study investigated the dynamic stability, as measured by MoS, and spatiotemporal parameters of healthy young adults during straight line walking, step turn and spin turn. This is crucial in rehabilitation evaluation and training for people with risk of falling.

Accurate estimation of the CoM position and velocity is a major factor enabling the computation of MoS. There exist several types of methods for estimation of CoM movement that differ in underlying assumptions and limitations [36]. Generally speaking, the segment-based approach and ground reaction force-based approach [37, 38] have similar performance [36, 39] and are considered as gold standards [38, 40]. Havens et al. [40] computed anteroposterior and mediolateral MoS through four simplified CoM models during straight walking and turning tasks, and assessed their biases and sources of bias via comparing to the gold standard. Their results indicated that bias was larger during turning tasks than straight walking and also showed that bias was smallest when using lower limb plus trunk segment model and center of pelvis. Considering the convenience (a few markers required) and high validity, we estimated the CoM position using center of pelvis model then computed the MoS of every data point in anterior, posterior, left and right directions. The variability was described using 95% bootstrap CB, a continuous data analysis procedure, rather than point-by-point Gaussian-based CI resulting in increased type I error rates [8, 35].

Fifteen subjects were instructed to perform straight walking and turns at comfortable speed. Results showed that the real turn angle in spin turn was slightly greater than in step turn. Broadly speaking, step turn was a bit slower whereas spin turn faster than straight walking. These are inconsistent with what we hypothesized and previous studies [8, 13] and reveal that spin turn might be a high efficiency strategy to change the direction of progression largely and quickly in healthy young adults [41].

MoS_A was an inverted peak in each step and the minimum value occurred at the mid-point of each stance phase, which is also the middle of swing phase on contralateral side. It suggests that people are most likely to fall forward at the middle of swing

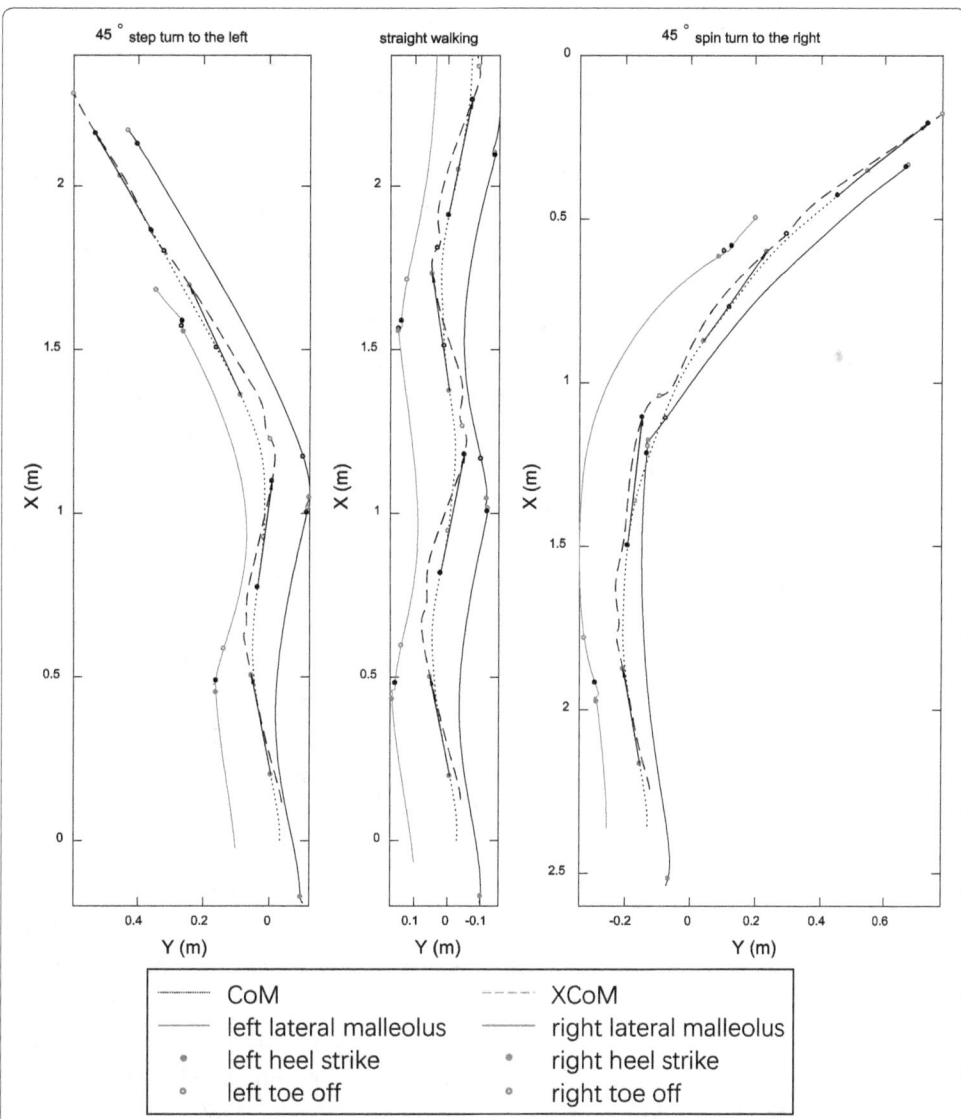

Fig. 5 Representative examples of CoM, XCoM and lateral malleolus trajectories for a subject. Black arrows point from CoM to XCoM at HS. Color marker specifiers show key events. The CoM for spin turn was found to be outside the right lateral malleolus during single support (TO1 ~ HS3) of turn phase but XCoM not

phase in straight walking and turns. This result is in agreement with a recent study [42] which indicated that the greatest risk of trip-related falls occurred at the middle of swing phase. In contrast, MoS_P presented a peak in each step and the minimum value appeared at heel strike. It indicates that people are most likely to fall backward at heel strike. This result is in line with [43] which reported that the greatest risk of slip occurred at the time shortly after heel strike and resulted in backward fall in normal level walking. The negative MoS_A leads to step forward because the maximum MoS_A will occur at heel strike. In other words, walking cannot be stopped during the step period, especially at the middle of swing phase where the minimum MoS_A occurred, which indicates that the negative MoS_A maybe characterize the difficulty of stopping within one step [15]. The negative value also reveals that people prefer falling forward instead of backward when balance is threatened [44]. This may be due to body inertia, or probably because people could better reduce the risk of head and spine impacting on lower surfaces to protect central nervous system with the help of vision and upper limbs when falling forward.

For straight walking, lateral MoS grew down during stance phase (the first and third step on left side and the second step on right side, Fig. 4) in which both CoM and XCoM were close to stance limb (Fig. 5). Conversely, lateral MoS went up during swing phase (the first and third step on the right and the second step on the left). Minimum lateral MoS was located at contralateral heel strike during straight walking, turn stance phase of spin turn and depart stance phase of step turn. This suggests that people are most likely to fall toward the support side at contralateral heel strike during straight walking, turn phase of spin turn and depart phase of step turn.

In general, there were almost the same size relationships of MoS, represented by four common characterizations, mean and minimum MoS and MoS at heel strike and mid stance, between three conditions (Additional file 2). We also found that the midpoint of stance phase or the middle of swing phase was vital for MoS_A and MoS_P.

For stride width, step turn phase was a good bit wider while depart a lot narrower. In contrast, spin turn phase was a great deal narrower whereas depart somewhat wider. This is in keeping with [8, 12]. We believe that this is the two natural cooperating strategies for the bilateral lower limbs to complete the turn within only three steps. For step length and stride length, spin turn was the longest. This implies that spin turn takes a long path to succeed in altering direction. The stance time, single support time, and cycle time increased in turns than straight. Precisely speaking, it is mainly because that the single support time or swing time gets longer in turns than straight walking.

Two limitations may have affected this current study. First, the embedded force plates may affect the subjects' natural walking pattern [45]. Future studies should consider choosing portable force plates and blinding their locations to the participants, or using cutting-edge technology such as [45, 46]. Second, the estimation of CoM position using center of pelvis model rather than gold standard might be not precise enough. This simplified CoM model would result in similar overestimation of MoS_A whereas similar underestimation of MoS_P between straight walking, step turn and spin turn, but underestimation of the lateral MoS during stance phase only in spin turn [40]. Therefore, future research should include consideration of more accurate estimation methods.

Conclusions

In conclusion, MoS_A reaches a minimum at the middle of swing phase in straight walking and turns. The backward instability threatens the first half of spin turn and the last half of step turn. Both backward and contralateral MoS appear a minimum at heel strike. Great lateral instability and negative stride width occur in the depart phase of step and the turn phase of spin. Spin turn takes a long step and stride length. Both turns undergo long swing phases.

For people with poor balance function, turning gait might expose potential problems which are undetectable in straight walking. This work could provide a valuable reference for rehabilitation evaluation and training for patients with cerebral palsy [9], post-stroke hemiparesis [46, 47], Parkinson's disease [12, 23], or multiple sclerosis [29]. For example, they could be trained from straight walking to step turn and spin turn, for gradual adaptation of gait stability and spatiotemporal parameters. Additionally, they should be instructed to pay attention to posterior and contralateral stability at heel strike, such as avoid slipping, and anterior stability at the middle of swing phase, such as avoid tripping.

Future studies overcoming the limitations are required to further understand preference strategy, kinematics, kinetics and bilateral asymmetry during planned and unplanned turning at different angles and velocities in older adults and people with cerebral palsy, post-stroke hemiparesis, Parkinson's disease, or multiple sclerosis.

Additional files

Additional file 1. Example videos of three barefoot walking tasks: step turn, straight walking and spin turn. Participants were instructed to perform practice sessions and then complete three barefoot walking tasks (wearing socks if cold) at the self-selected preferred walking speed: straight walking, 45° step turn to the left and 45° spin turn to the right (Right lower limb is the turning limb in both turn conditions). Four force plates were used in straight walking while three in turns.

Additional file 2. Spatiotemporal parameters, MoS_A, MoS_P, MoS_L and MoS_R for healthy young adults ($n = 15$). Mean (95% confidence interval). Statistically significant difference ($p < 0.05$) from the first phase of same condition (a), the second phase of same condition (t), straight walking (*), between turn conditions (†). ‡: Statistically significant difference ($p < 0.05$) between the absolute mean value of walking direction in spin turn and that in step turn.

Abbreviations
BoS: base of support; CoM: center of mass; v_{CoM}: velocity of center of mass; XCoM: extrapolated center of mass; MoS: margin of stability; MoS_A: anterior margin of stability; MoS_P: posterior margin of stability; MoS_L: left margin of stability; MoS_R: right margin of stability; ASIS: anterior superior iliac spine; PSIS: posterior superior iliac spine; HS: heel strike; HS1: the first heel strike; HS2: the second heel strike; HS3: the third heel strike; HS4: the fourth heel strike; TO: toe off; TO1: the first toe off; TO2: the second toe off; TO3: the third toe off; MS: mid stance; MS1: the first mid stance; MS2: the second mid stance; MS3: the third mid stance; CI: confidence interval; CB: confidence band.

Authors' contributions
Study concept: CH, RX, DM; literature review: CH, RX, MZ, SJ; experimental design: CH, RX, DM, MZ, YG, SJ, FH; data collection: RX, CH, MZ, SJ; data processing: CH and RX; statistical analysis: CH and RX; interpretation of data: CH and RX; study review: DM, FH, YG, MZ, RX, SJ, CH; manuscript preparation: CH, RX, MZ, MD. All authors read and approved the final manuscript.

Author details
[1] Lab of Neural Engineering & Rehabilitation, Department of Biomedical Engineering, College of Precision Instruments and Optoelectronics Engineering, Tianjin University, Tianjin, China. [2] Tianjin International Joint Research Center for Neural Engineering, Academy of Medical Engineering and Translational Medicine, Tianjin University, Tianjin, China. [3] College of Acupuncture and Massage, Tianjin University of Traditional Chinese Medicine, Tianjin, China.

Acknowledgements
The authors would like to thank all participants in this study.

Competing interests
The authors declare that they have no competing interests.

Consent for publication
Not applicable.

Funding
This work was supported by the National Key Research and Development Program of China (Grant Number 2017YFB1300300), National Natural Science Foundation of China (Grant Numbers 81630051, 91520205, 61603269, 81601565, 81571762 and 31500865), and Tianjin Key Technology R&D Program (Grant Number 15ZCZDSY00930).

References

1. Sedgman R, Goldie P, Iansek R. Development of a measure of turning during walking. In: Advancing rehabilitation: Proceedings of the inaugural conference of the faculty of health sciences La Trobe University. 1994.
2. Glaister BC, Bernatz GC, Klute GK, Orendurff MS. Video task analysis of turning during activities of daily living. Gait Posture. 2007;25(2):289–94.
3. Yamaguchi T, Yano M, Onodera H, Hokkirigawa K. Effect of turning angle on falls caused by induced slips during turning. J Biomech. 2012;45(15):2624–9.
4. Cumming RG, Klineberg RJ. Fall frequency and characteristics and the risk of hip fractures. J Am Geriatr Soc. 1994;42(7):774–8.
5. Dixon PC. The biomechanics of turning gait in children with cerebral palsy. Doctor of Philosophy Doctoral thesis. University of Oxford. 2015.
6. Peyer KE, Brassey CA, Rose KA, Sellers WI. Locomotion pattern and foot pressure adjustments during gentle turns in healthy subjects. J Biomech. 2017;60:65–71.
7. Hase K, Stein R. Turning strategies during human walking. J Neurophysiol. 1999;81(6):2914–22.
8. Dixon PC, Stebbins J, Theologis T, Zavatsky AB. Spatio-temporal parameters and lower-limb kinematics of turning gait in typically developing children. Gait Posture. 2013;38(4):870–5.
9. Dixon PC, Stebbins J, Theologis T, Zavatsky AB. The use of turning tasks in clinical gait analysis for children with cerebral palsy. Clin Biomech. 2016;32:286–94.
10. Patla AE, Prentice SD, Robinson C, Neufeld J. Visual control of locomotion: strategies for changing direction and for going over obstacles. J Exp Psychol Hum Percept Perform. 1991;17(3):603–34.
11. Akram SB, Frank JS, Chenouri S. Turning behavior in healthy older adults: is there a preference for step versus spin turns? Gait Posture. 2010;31(1):23–6.
12. Conradsson D, Paquette C, Lökk J, Franzén E. Pre- and unplanned walking turns in Parkinson's disease—effects of dopaminergic medication. Neuroscience. 2017;341(Supplement C):18–26.
13. Taylor MJD, Dabnichki P, Strike SC. A three-dimensional biomechanical comparison between turning strategies during the stance phase of walking. Hum Mov Sci. 2005;24(4):558–73.
14. Dixon PC, Stebbins J, Theologis T, Zavatsky AB. Ground reaction force and center of mass velocity during turning. Gait Posture. 2014;39:S11–2.
15. Bruijn S, Meijer O, Beek P, Van Dieën J. Assessing the stability of human locomotion: a review of current measures. J R Soc Interface. 2013;10(83):20120999.
16. van Emmerik REA, Ducharme SW, Amado AC, Hamill J. Comparing dynamical systems concepts and techniques for biomechanical analysis. J Sport Health Sci. 2016;5(1):3–13.
17. Hamacher D, Singh NB, Van Dieen JH, Heller MO, Taylor WR. Kinematic measures for assessing gait stability in elderly individuals: a systematic review. J R Soc Interface. 2011;8(65):1682–98.
18. Hof A, Gazendam M, Sinke W. The condition for dynamic stability. J Biomech. 2005;38(1):1–8.
19. Piórek M, Josiński H, Michalczuk A, Świtoński A, Szczęsna A. Quaternions and joint angles in an analysis of local stability of gait for different variants of walking speed and treadmill slope. Inf Sci. 2017;384:263–80.
20. Mehdizadeh S. The largest Lyapunov exponent of gait in young and elderly individuals: a systematic review. Gait Posture. 2018;60:241–50.
21. Bruijn SM, Bregman DJJ, Meijer OG, Beek PJ, van Dieën JH. Maximum Lyapunov exponents as predictors of global gait stability: a modelling approach. Med Eng Phys. 2012;34(4):428–36.
22. Hof AL. The 'extrapolated center of mass' concept suggests a simple control of balance in walking. Hum Mov Sci. 2008;27(1):112–25.
23. Mellone S, Mancini M, King LA, Horak FB, Chiari L. The quality of turning in Parkinson's disease: a compensatory strategy to prevent postural instability? J Neuroeng Rehabil. 2016;13(1):39.
24. Conradsson D, Paquette C, Franzén E. Medio-lateral stability during walking turns in older adults. PLoS ONE. 2018;13(6):e0198455.
25. Strike SC, Taylor MJD. The temporal–spatial and ground reaction impulses of turning gait: is turning symmetrical? Gait Posture. 2009;29(4):597–602.
26. Sadeghi H, Allard P, Prince F, Labelle H. Symmetry and limb dominance in able-bodied gait: a review. Gait Posture. 2000;12(1):34–45.
27. Tateuchi H, Tsukagoshi R, Fukumoto Y, Akiyama H, So K, Kuroda Y, Ichihashi N. Compensatory turning strategies while walking in patients with hip osteoarthritis. Gait Posture. 2014;39(4):1133–7.
28. Barre A, Armand S. Biomechanical ToolKit: open-source framework to visualize and process biomechanical data. Comput Methods Programs Biomed. 2014;114(1):80–7.
29. Peebles AT, Reinholdt A, Bruetsch AP, Lynch SG, Huisinga JM. Dynamic margin of stability during gait is altered in persons with multiple sclerosis. J Biomech. 2016;49(16):3949–55.
30. Whittle MW. Three-dimensional motion of the center of gravity of the body during walking. Hum Mov Sci. 1997;16(2):347–55.
31. Glaister BC, Orendurff MS, Schoen JA, Bernatz GC, Klute GK. Ground reaction forces and impulses during a transient turning maneuver. J Biomech. 2008;41(14):3090–3.
32. van Meulen FB, Weenk D, van Asseldonk EH, Schepers HM, Veltink PH, Buurke JH. Analysis of balance during functional walking in stroke survivors. PLoS ONE. 2016;11(11):e0166789.
33. Huxham F, Gong J, Baker R, Morris M, Iansek R. Defining spatial parameters for non-linear walking. Gait Posture. 2006;23(2):159–63.
34. Dixon PC, Bowtell MV, Stebbins J. The use of regression and normalisation for the comparison of spatio-temporal gait data in children. Gait Posture. 2014;40(4):521–5.
35. Duhamel A, Bourriez JL, Devos P, Krystkowiak P, Destée A, Derambure P, Defebvre L. Statistical tools for clinical gait analysis. Gait Posture. 2004;20(2):204–12.

36. Schepers HM, Van Asseldonk EH, Buurke JH, Veltink PH. Ambulatory estimation of center of mass displacement during walking. IEEE Trans Biomed Eng. 2009;56(4):1189–95.

37. Cavagna GA. Force platforms as ergometers. J Appl Physiol. 1975;39(1):174–9.

38. Pavei G, Seminati E, Cazzola D, Minetti AE. On the estimation accuracy of the 3D body center of mass trajectory during human locomotion: inverse vs. forward dynamics. Front Physiol. 2017;8:129.

39. Lafond D, Duarte M, Prince F. Comparison of three methods to estimate the center of mass during balance assessment. J Biomech. 2004;37(9):1421–6.

40. Havens KL, Mukherjee T, Finley JM. Analysis of biases in dynamic margins of stability introduced by the use of simplified center of mass estimates during walking and turning. Gait Posture. 2018;59(Supplement C):162–7.

41. Torre D. Effects of direction time constraints and walking speed on turn strategies and gait adaptations in healthy older and young adults. Ph.D. Health Sciences Dissertation. Seton Hall University, Health and Medical Sciences. 2017.

42. Schulz BW. A new measure of trip risk integrating minimum foot clearance and dynamic stability across the swing phase of gait. J Biomech. 2017;55:107–12.

43. Grönqvist R, Roine J, Järvinen E, Korhonen E. An apparatus and a method for determining the slip resistance of shoes and floors by simulation of human foot motions. Ergonomics. 1989;32(8):979–95.

44. Hak L, Houdijk H, van der Wurff P, Prins MR, Mert A, Beek PJ, van Dieën JH. Stepping strategies used by post-stroke individuals to maintain margins of stability during walking. Clin Biomech. 2013;28(9):1041–8.

45. Karatsidis A, Bellusci G, Schepers H, de Zee M, Andersen M, Veltink P. Estimation of ground reaction forces and moments during gait using only inertial motion capture. Sensors. 2017;17(1):75.

46. van Meulen FB, Weenk D, Buurke JH, van Beijnum B-JF, Veltink PH. Ambulatory assessment of walking balance after stroke using instrumented shoes. J Neuroeng Rehabil. 2016;13(1):48.

47. Vistamehr A, Kautz SA, Bowden MG, Neptune RR. Correlations between measures of dynamic balance in individuals with post-stroke hemiparesis. J Biomech. 2016;49(3):396–400.

4

The climatic factors affecting dengue fever outbreaks in southern Taiwan: an application of symbolic data analysis

Yi-Horng Lai*

From International Conference on Biomedical Engineering Innovation (ICBEI) 2016 Taichung, Taiwan.

*Correspondence:
FL006@mail.oit.edu.tw
Department of Health Care
Administration, Oriental
Institute of Technology,
No. 58, Sec. 2, Sichuan Rd.,
Banqiao Dist., New Taipei
City 22061, Taiwan

Abstract

Background: Dengue fever is a leading cause of severe illness and hospitalization in Taiwan. This study sought to elucidate the linkage between dengue fever incidence and climate factors.

Results: The result indicated that temperature, accumulated rainfall, and sunshine play an important role in the transmission cycles of dengue fever. A predictive model equation plots dengue fever incidence versus temperature, rainfall, and sunshine, and it suggests that temperature, rainfall, and sunshine are significantly correlated with dengue fever incidence.

Conclusions: The data suggests that climate factors are important determinants of dengue fever in southern Taiwan. Dengue fever viruses and the mosquito vectors are sensitive to their environment. Temperature, rainfall and sunshine have well-defined roles in the transmission cycle. This finding suggests that control of mosquito by climatic factor during high temperature seasons may be an important strategy for containing the burden of dengue fever.

Keywords: Dengue fever, Climate factors, Interval-valued data, Symbolic data analysis

Background

Dengue Fever is a common epidemiological mosquito-borne disease in subtropical and tropical regions and has become one of the public health's biggest challenges. Dengue is a febrile illness caused by one of the antigenically different serotypes of dengue viruses and mainly transmitted to human through the bite of vectors, including *Aedes aegypti* and *Aedes albopictus* [1]. One area that has received particular attention is the association between climatic factors and vector-borne diseases [2].

Among the 193 WHO member countries, more than 70% of the populations are at risk of dengue. WHO reported that 390 million dengue infections occurred every year before 2013 [3]. The dengue infected regions include South-East Asia and Western Pacific, with the most vulnerable area in developing countries. The outbreaks do not only occur in rural areas but also in urban areas.

Taiwan is located in the Pacific Ocean region and is a hotbed of dengue vectors because of its high temperature and humidity [4]. The risk of dengue fever has increased gradually in southern Taiwan and has become a major public health issue that affects the quality of life and the health of Taiwan's residents. During the first half of the twentieth century, there were three dengue fever outbreaks in Taiwan (1915, 1931, and 1942). After almost 40 years of dormancy, a dengue fever outbreak reoccurred in 2002 in southern Taiwan. The total number of indigenous cases in this outbreak was 5336, including 241 cases of dengue hemorrhagic fever (DHF) that caused 19 deaths. After that, the indigenous dengue cases were less than 400 in 2003–2005. Since 2006, Taiwan has faced dengue fever outbreaks of different scales every year; the cases were concentrated mainly in southern Taiwan, including Kaohsiung City, Tainan City, and Pingtung County. In 2015, Taiwan battled one of the most severe dengue outbreaks in history with over 42,000 dengue cases—22,741 cases in Tainan City, 18,933 cases in Kaohsiung City, and 373 cases in Pingtung City—and 228 deaths found to be associated with dengue infection [5].

Previous studies have been carried out on the correlation between climate factors and dengue fever using a wide spectrum of mathematical and statistical modeling methods [6–10]. Findings from most previous studies in other parts of the world also showed that climatic variables have an effect on dengue fever transmission. Studies in Taiwan [6, 11–13], Singapore [14], Vietnam [8, 15], Thailand [7, 16], China [9, 17], Trinidad [18, 19], Malaysia [20], Puerto Rico [21], Cambodia [10], and Saudi Arabia [22] showed a significant correlation between dengue fever incidence and temperatures, precipitation, and sunshine.

As temperature increases, the *Aedes aegypti* mosquito displays shorter periods of development in all stages of their life cycle, which leads to increased population growth. The mosquito feeding rate also increases; and dengue fever viruses in adult *Aedes aegypti* mosquitoes require shorter incubation periods to migrate to salivary glands [6–10, 14, 23]. Specifically, increasing temperatures increases the available habitat for the dengue fever vector, the *Aedes aegypti* mosquito, while concurrently increasing both the longevity of the virus and the mosquito [14]. Higher temperatures can also shorten the duration of virus replication, and increase mosquito reproduction and contacts with humans [9]. If temperature increases by approximately 3 °C, mean incidence rates during epidemics can double [24]. Warmer temperatures can increase the transmission rates of dengue fever in various ways. It may allow vectors to survive and reach maturity much faster than at lower temperatures [25]. Moreover, it may also reduce the size of mosquito larvae resulting in smaller adults that have high metabolism rates, requiring more frequent blood meal and need to lay eggs more often [25].

Some studies reported that rainfall can lead to increases in dengue fever transmission. They suggested that rainfall creates abundant outdoor breeding sources for *Aedes aegypti*, and the water storage containers also can serve as breeding habitats. Bhatt, Gething, Brady, Messina, Farlow, Moyes, Drake, Brownstein, Hoen, Sankoh, Myers, George, Jaenisch, Wint, Simmons, Scott, Farrar, and Hay paired the resulting risk map with detailed longitudinal information from dengue fever cohort studies, and they predicted dengue fever to be ubiquitous throughout the tropics, with local spatial variations influenced strongly by rainfall [23]. Choi, Tang, McIver, Hashizume, Chan, Abeyasinghe, Iddings, and Huy developed negative binomial models using monthly average maximum, minimum, mean temperatures and monthly cumulative rainfall, and they also claimed that rainfall significantly increased the dengue fever incidence [10]. When more consecutive wet days occurred in a period, dengue fever incidence increased. Rainfall leads to an increase in breeding sites of the mosquito vector, which would contribute to the increase in dengue fever occurrence [21]. On the contrary, however, some other studies showed that heavy rainfall can possibly lower dengue fever transmission by reducing the survival rate of the *Aedes aegypti* mosquito. Wegbreit analyzed weekly dengue fever morbidity data from the twin-island country of Trinidad and Tobago, and he suggested that there is a slightly negative correlation with the precipitation [26]. Thammapalo, Chongsuwiwatwong, McNeil, and Geater determine the independent effects of rainfall [16] in Thailand, and they also found that increased rainfall is associated with a decreased incidence of dengue fever cases in some provinces. Alshehri [22] aimed to address the effects of heavy rainfall on *Aedes aegypti* mosquito density in Saudi Arabia, and he argued that dengue fever has negative correlation with rainfall and humidity.

Sunshine is also closely linked to other ecological factors such as temperature and humidity and thereby might affect the dengue fever incidence [15]. Correlation studies carried out on monthly dengue fever cases have found the risk of dengue fever to be inversely associated with duration of sunshine [8]. With the monthly data in Vietnam, Vu, Okumura, Hashizume, Tran, and Yamamoto indicated that there is a significant negative association between dengue fever cases and the hours of sunshine [15]. Wongkoon, Jaroensutasinee, Jaroensutasinee investigated the effect of seasonal variation on the abundance of *Aedes aegypti* mosquito larvae and explored the impact of weather variability on dengue fever transmission in Thailand, and they concluded that maximum temperature, sunshine and evaporation are negatively correlated with dengue fever incidence [19].

However, while most studies claimed that climate is a determinant of dengue fever, some other studies argued that climate factor has no obvious correlation with this disease. They suggested that temperature [18, 27–29] and rainfall [13, 27–29] did not affect dengue fever incidence. The weekly average maximum temperature, total rainfall and the total number of dengue fever cases from 2005 to 2011 were used as time series data in Goto, Kumarendran, Mettananda, Gunasekara, Fujii, and Kaneko's study [27]. They found that weekly average maximum temperatures and the weekly total rainfall did not significantly affect dengue fever incidence in three geographically different areas of Sri Lanka. Pandey, Nagar, Gupta, Khan, Singh, Mishra, Prakash, Singh, Singh, and Jain reported the annual trend of dengue fever virus infection in north India [28], and they indicated that there is no statistical significant correlation between weather data and increasing dengue fever positive cases. In a population-based study on the effects of climate and mosquito indices on dengue fever in Trinidad, Chadee, Shivnauth, Rawlins, and Chen declared that no significant correlations are observed between temperature and dengue fever [18]. Chang, Lee, Ko, Tsai, Lin, Chen, Lu, and Chen pointed out that climatic factors correlated significantly with case numbers of many diseases, such as murine typhus and Q fever, but neither temperature nor rainfall correlated with the case number of dengue fever [13]. According to the epidemiological investigation, the incidence of dengue fever had no relationship with temperature, or precipitation, and some studies [29] showed a clear relationship only with the sociological factors.

Most of the climatic data are range-type data. Because of the limitation of traditional statistics (e.g., regression analysis), range-type data is difficult to be analyzed. Most of them are analyzed by minimum value (e.g., minimum temperature), maximum value (e.g., maximum temperature), mean value (e.g., mean temperature, mean rainfall), and cumulative value (e.g., cumulative rainfall, cumulative sunshine). Most studies developed linear regression model [6–10] or negative binomial regression model [10] using monthly average temperatures [14, 17], maximum temperatures [7, 27], minimum temperatures [10], mean rainfall [14, 17], cumulative rainfall [27], and cumulative sunshine [8, 15, 19] over the period for the relationship between dengue fever incidence and climatic data. However, the major drawback of the traditional statistical methods is that when the correlation between dengue fever and each of the above-mentioned value is not consistent, it will be difficult to draw a conclusion [10, 19]. For example, in Choi, Tang, McIver, Hashizume, Chan, Abeyasinghe, Iddings, and Huy's study [10], mean temperature is significantly associated with dengue fever incidence, but dengue fever incidence did not correlate well with the maximum temperature and minimum temperature. Wongkoon, Jaroensutasinee, Jaroensutasinee's study also had the same problem. They investigated the effect of seasonal variation on the abundance of the *Aedes aegypti* mosquito larvae and explored the impact of weather variability on dengue fever transmission in Thailand, and they found that mean temperature and minimum temperatures are positively associated with dengue fever incidence, but maximum temperature is negatively correlated with dengue fever incidence [19].

With the advent of information technology, very large datasets have become routine. Traditional statistical methods do not have the power or flexibility to analyze these efficiently and extract the required knowledge. Symbolic data analysis is to summarize a large dataset in such a way that the resulting summary dataset is of a manageable size and yet retains as much of the knowledge in the original dataset as possible [30, 31]. One consequence of this is that the data may no longer be formatted as single values, but be represented by lists, intervals, distributions, etc. The summarized data have their own internal structure, which must be taken into account in any analysis.

High peaks for dengue outbreak is reported on summer in Taiwan. This suggests that climatic factors are likely to exert potential impact on dengue fever outbreak in tropical or subtropical regions [32, 33]. This study is aimed for investigate the relationship between climatic factors and the outbreaks of dengue fever in southern Taiwan with symbolic data analysis and to compare the differential effects of climatic factors on the incidence of dengue fever in southern Taiwan.

Materials and research method

Climatic factors, such as temperature, rainfall, and sunshine play an important role in the spread of dengue fever viruses. The dengue fever data and climatic data of Kaohsiung city from January 2005 to March 2014 were analyzed with symbolic data analysis for the interval-valued data in this study.

Study area

Kaohsiung City is the largest metropolitan city of southern Taiwan with an estimated population of 2,777,784 in 2016. According to the computerized database from the surveillance system of Taiwan's Center for Disease Control, Kaohsiung City had 5543 confirmed dengue fever cases from January 2005 to March 2014, accounting for most of the total cases in Taiwan. Dengue fever transmission has been active in this area, and the latest large-scale outbreak occurred in the end of 2014.

Data collection

Since 1988, dengue fever has been announced as a Class III Notified Disease in Taiwan, and the data are collected continuously and systematically by Taiwan's Ministry of Health and Welfare with the Taiwan National Infectious Disease Statistics System [34]. The data collection mechanism has been stable over time, and this routinely-collected data can be used for analyzing factors affecting the occurrence of dengue fever. Because the most recent data is subject to update, this study focuses only on the data from January 2005 to March 2014.

Meteorological data on the monthly maximum temperature, minimum temperature, amount of rainfall, and amount of sunshine were obtained from the Climate Statistics Database provided by Taiwan's Central Weather Bureau [35].

The variables that correlate with dengue fever are then submitted to symbolic linear regression analysis. Symbolic data analysis is employed to explore and identify statistically significant risk indicators [33].

Symbolic data analysis

Symbolic data analysis is a relatively new field that provides a range of methods for analyzing complex datasets. Traditional statistical methods do not have the power or flexibility to make sense of very large datasets, and symbolic data analysis techniques can be developed in order to extract knowledge from such data. The analysis of symbolic data differs from that of the traditional. Rather than identifying points of interest in the data, symbolic data methods allow the user to build models of the data and make predictions about future events [32].

Dengue fever data of Kaohsiung city from January 2005 to August 2014 were analyzed with symbolic linear regression analysis for interval-valued data with R 3.3.2 (with Package RSDA) with SparkR 2.1 in symbolic data analysis with center-method [36, 37]. The regression model equation plots dengue fever incidence (cases) versus temperature (°C), accumulated rainfall (mm), and accumulated sunshine (hours) is as follows:

$$
\begin{aligned}
\text{Dengue fever incidence} = \beta_0 &+ \beta_1 \times \text{temperature} \\
&+ \beta_2 \times \text{accumulated rainfall} \\
&+ \beta_3 \times \text{accumulated sunshine}
\end{aligned} \tag{1}
$$

Billard and Diday proposed an approach for a constrained linear regression model on the midpoints and range of the interval values [30]. The prediction of the lower and upper boundaries of the interval value of the dependent variable is accomplished from its midpoint and range, which are estimated from the fitted linear regression models applied to the midpoint and range of each interval value of the independent variables.

Based on Billard and Diday's study [30], the estimate of the parameters β is based only on the midpoint of the intervals according to the criterion considered. Let $E = \{e_1,...,e_n\}$ be a set of examples that are described by $p+1$ interval-valued variables y, $x_1,...,x_p$. Each example is represented as an interval quantitative feature vector $z_i = (x_i, y_i)$, $xi = (x_{i1},...,x_{ip})$, $x_{ij} = [a_{ij}, b_{ij}] \in \hat{s} = \{[a,] : a, b \in R, a \leq b\}$ where $(j=1,...,p)$ and $y_i = [y_{Li}, y_{Ui}] \in \hat{s}$ are, respectively, the observed values of x_j and y.

It can be considered that $X_1,...,X_p$ related to Y according to the linear regression relationship:

$$y_{Li} = \beta_0 + \beta_1 a_{i1} + \cdots + \beta_p a_{ip} + \varepsilon_{Li}$$

$$y_{Ui} = \beta_0 + \beta_1 b_{i1} + \cdots + \beta_p b_{ip} + \varepsilon_{Ui} \tag{2}$$

From Eq. (2), the sum of the squares of deviations in this first approach is as follows:

$$S_{cm} = \sum_{i=1}^{n} (\epsilon_{Li} + \epsilon_{Ui})^2 = \sum_{i=1}^{n} (y_{Li} - \beta_0 - \beta_1 \alpha_{i1} - \cdots \beta_p \alpha_{ip} + y_{Ui} - \beta_0 - \beta_1 b_{i1} - \cdots \beta_p b_{ip})^2 \tag{3}$$

which represents the sum of the square of the sum of the lower and upper boundary errors.

Lima Neto and de Carvalho present the estimates of the vector of parameters β in matrix notation for the center method [37], which can be rewritten in the simplest form as

$$y^C = X^C \beta + \varepsilon^C \tag{4}$$

where $y^C = (y_1^C, \ldots y_n^C)^T$, $X^C = (x_1^C)^T, \ldots (x_n^C)^T)^T$, $(x_i^C)^T = (1, x_{i1}^C, \ldots x_{ip}^C)$ $i = 1, \ldots, n$ $\beta = (\beta_0, \ldots, \beta_p)^T$, and $\varepsilon^C = (\varepsilon_1^C, \ldots, \varepsilon_n^C)^T$.

If X_c has full rank $(p+1) \leq n$, the least square estimate of β in Eq. (3) is given by

$$\hat{\beta} = ((X^C)^T X^C)^{-1} (X^C)^T y^C \tag{5}$$

Given a new example e, described by $z = (x, y)$, where $x = (x_1,...,x_p)$ with $x_j = [a_j, b_j]$ $(j=1,...,p)$, the value $y = [y_L, y_U]$ of Y will be predicted by $\hat{y} = [\hat{y}_L, \hat{y}_U]$ as follows:

$$\hat{y}_L = (x_L)^T \hat{\beta} \text{ and } \hat{y}_U = (x_U)^T \hat{\beta} \tag{6}$$

where $(x_L)^T = (1, a_1, \ldots, a_p)$,

$$(x_U)^T = (1, b_1, \ldots, b_p).$$

The determination coefficient (R^2) represents a goodness-of-fit measure commonly used in regression analysis to capture the adjustment quality of a model. The determination coefficient (R^2) for the CM method is easily established as

$$R_{cm}^2 = \frac{\sum_{i=1}^{n} (\hat{y}_i^C - \bar{y}^C)^2}{\sum_{i=1}^{n} (y_i^C - \bar{y}^C)^2} \tag{7}$$

However, note that $y^C = (y_L + y_U)$. Thus, the Eq. (7) can be replaced by

Table 1 **The characteristics of the research data in dengue fever data**

	N (months)	Total (years)	Mean (months)	S. E. (months)
2005/Jan.–2005/Dec.	12	144	12.00	15.82
2006/Jan.–2006/Dec.	12	956	79.67	100.36
2007/Jan.–2007/Dec.	12	202	16.83	20.60
2008/Jan.–2008/Dec.	12	443	36.92	48.67
2009/Jan.–2009/Dec.	12	773	64.42	102.70
2010/Jan.–2010/Dec.	12	1106	92.17	122.07
2011/Jan.–2011/Dec.	12	1183	98.58	143.43
2012/Jan.–2012/Dec.	12	532	44.33	57.98
2013/Jan.–2013/Dec.	12	102	8.50	9.74
2014/Jan.–2014/Mar.	3	7	2.33	3.47

Table 2 **The characteristics of the research data in climatic data**

	N (months)	Temperature (°C) U[lower, upper]	Rainfall (mm) U[lower, upper]	Sunshine (hours) U[lower, upper]
2005/Jan.–2005/Dec.	12	U[8.7, 34.9]	U[0, 1030.0]	U[0, 259.4]
2006/Jan.–2006/Dec.	12	U[11.3, 35.7]	U[0, 901.5]	U[0, 224.6]
2007/Jan.–2007/Dec.	12	U[9.2, 35.4]	U[0, 1229.3]	U[0, 290.6]
2008/Jan.–2008/Dec.	12	U[10.2, 35.0]	U[0, 1199.7]	U[0, 240.1]
2009/Jan.–2009/Dec.	12	U[9.3, 35.0]	U[0, 934.5]	U[0, 256.0]
2010/Jan.–2010/Dec.	12	U[10.9, 35.9]	U[0, 853.0]	U[0, 238.6]
2011/Jan.–2011/Dec.	12	U[11.3, 35.8]	U[0, 543.0]	U[0, 240.9]
2012/Jan.–2012/Dec.	12	U[11.0, 36.4]	U[0, 832.5]	U[0, 231.8]
2013/Jan.–2013/Dec.	12	U[12.0, 35.7]	U[0, 765.7]	U[0, 248.1]
2014/Jan.–2014/Mar.	3	U[11.1, 30.6]	U[0, 67.0]	U[0, 254.8]

$$R_{cm}^2 = \frac{\sum_{i=1}^{n} \left((\hat{y}_{Li} + \hat{y}_{Ui}) - (\bar{y}_{Li} + \bar{y}_{Ui}) \right)^2}{\sum_{i=1}^{n} \left((y_{Li} + y_{Ui}) - (\bar{y}_{Li} + \bar{y}_{Ui}) \right)^2} \tag{8}$$

Billard and Diday's method [30] indicate the importance of range-type information in prediction performance as well as the application of inequality constraints to ensure mathematical coherence between the predicted values of the lower and upper boundaries of the interval-value data.

Results

Using monthly aggregated data, dengue fever data of Kaohsiung city from January 2005 to March 2014 were analyzed with symbolic linear regression analysis for the interval-valued data in this study. The climatic data in this study includes temperature, rainfall, and sunshine. The type of these data is interval-valued data, and they all can be presented as U[lower, upper], meaning the range of the data is from minimum (lower) to maximum (upper). Details of the dengue fever incidence in Kaohsiung City and the monthly temperature/rainfall/sunshine measurements are presented in Tables 1 and 2 respectively.

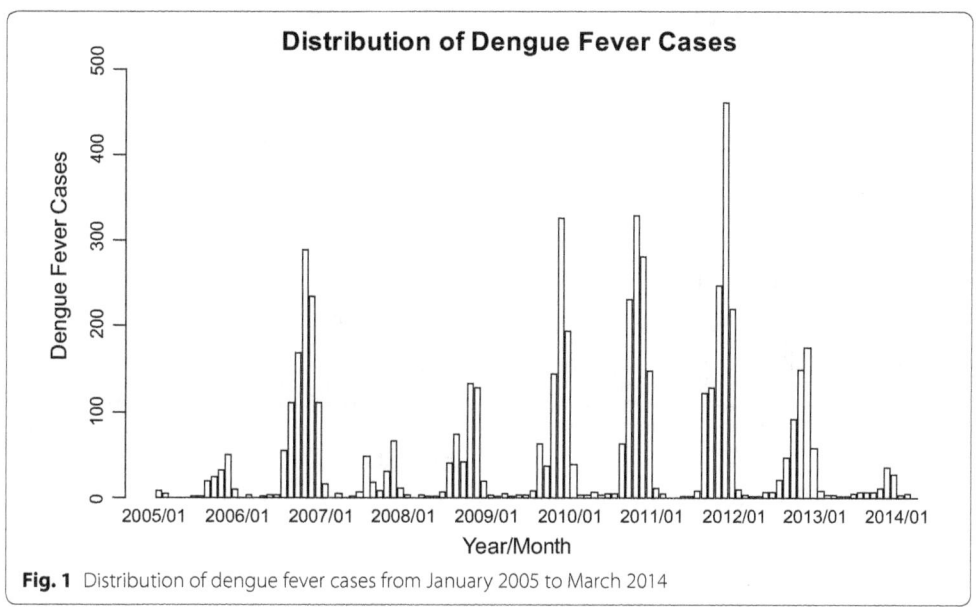

Fig. 1 Distribution of dengue fever cases from January 2005 to March 2014

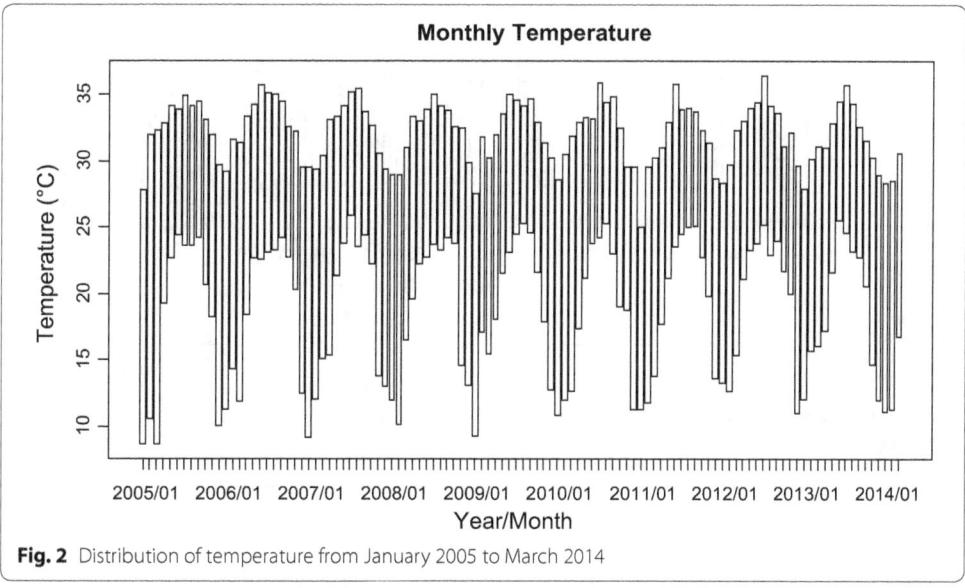

Fig. 2 Distribution of temperature from January 2005 to March 2014

For the 111-month period (January 2005–March 2014), the number of dengue fever cases in Kaohsiung City was from 0 to 462, with the highest one (462 cases) occurred in November 2011 (Fig. 1). The monthly minimum temperature was from 8.7 °C to 12.0 °C, with the lowest recorded minimum temperature in January 2005 and March 2005. The monthly maximum temperature was from 30.6 to 36.4 °C, with the highest recorded minimum temperature in July 2012 (Fig. 2). The monthly accumulated rainfall measurement was from 0 to 1229.3 mm, with the highest recorded measurement in August 2007 (Fig. 3). The monthly accumulated sunshine measurement was from 0 to 290.6 h, with the highest recorded measurement in July 2007 (Fig. 4).

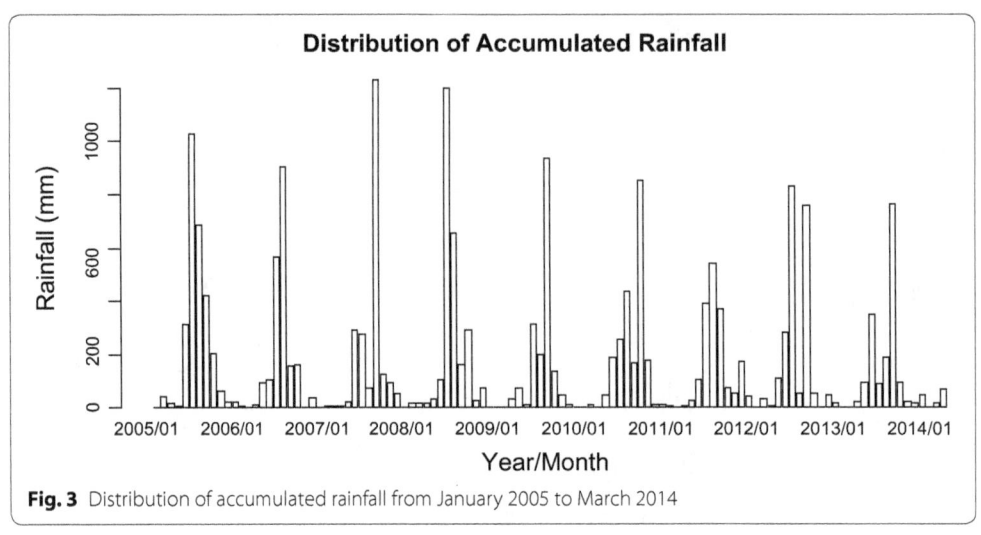

Fig. 3 Distribution of accumulated rainfall from January 2005 to March 2014

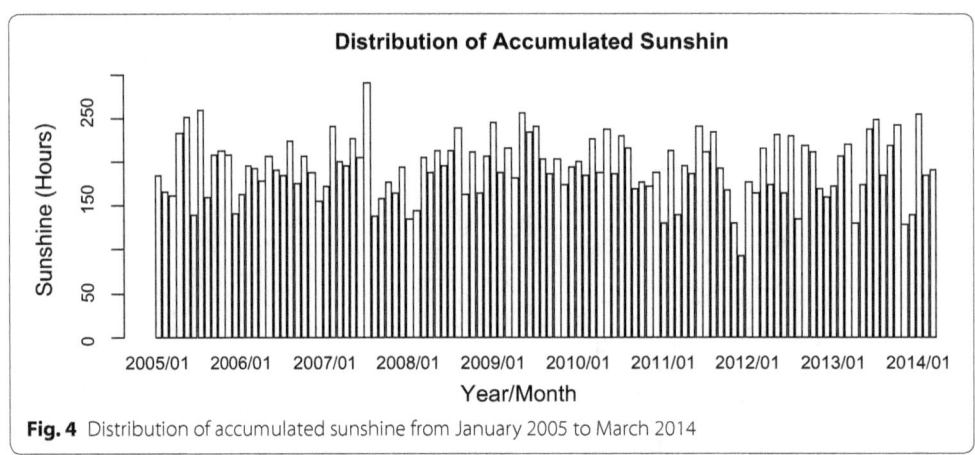

Fig. 4 Distribution of accumulated sunshine from January 2005 to March 2014

Table 3 Regression analysis of climatic factor affecting dengue fever incidence in southern Taiwan

	Estimate	S. E.	t-value	p-value
Intercept	3.118	66.797	0.047	0.963
Temperature	6.289	2.756	2.282	0.024
Rainfall	− 0.144	0.072	− 2.002	0.047
Sunshine	− 1.025	0.368	− 2.790	0.007

The predictive model equation plots dengue fever incidence (cases) versus temperature (°C), accumulated rainfall (mm), and accumulated sunshine (hours) is as follows:

$$\text{Dengue fever incidence} = 3.118 + 6.289 \times \text{temperature} + (-0.144) \times \text{rainfall} + (-1.025) \times \text{sunshine} \tag{9}$$

The results demonstrate that climatic factors are associated with dengue fever cases. Table 3 shows that, first, monthly temperature (t-value = 2.282, p value = 0.024) is positively correlated with the dengue fever cases, and this result is in agreement with

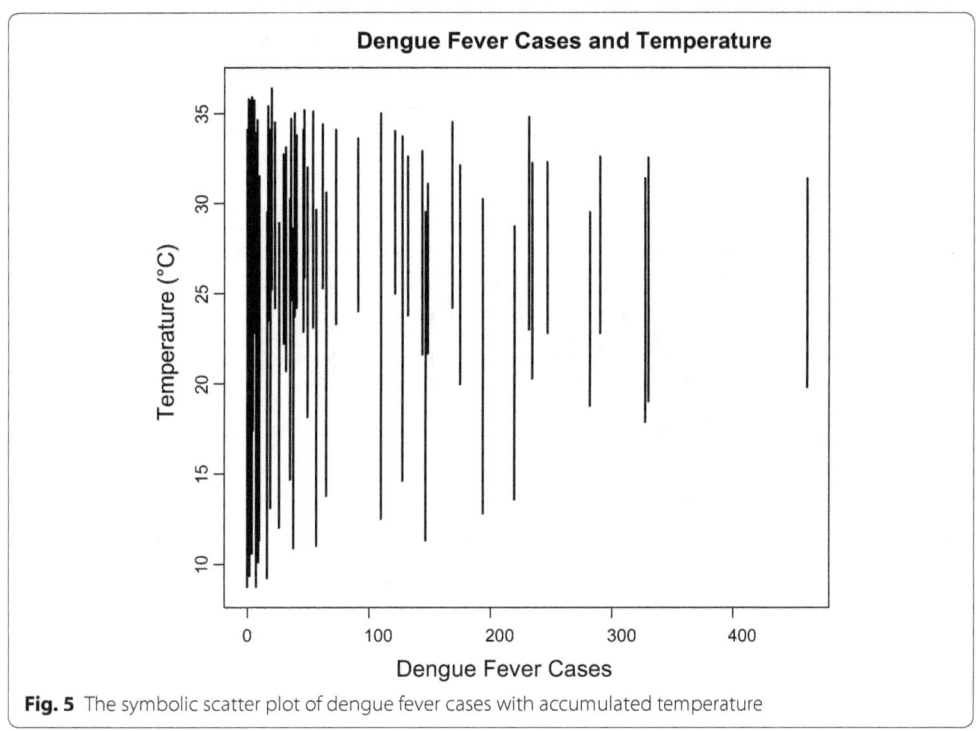

Fig. 5 The symbolic scatter plot of dengue fever cases with accumulated temperature

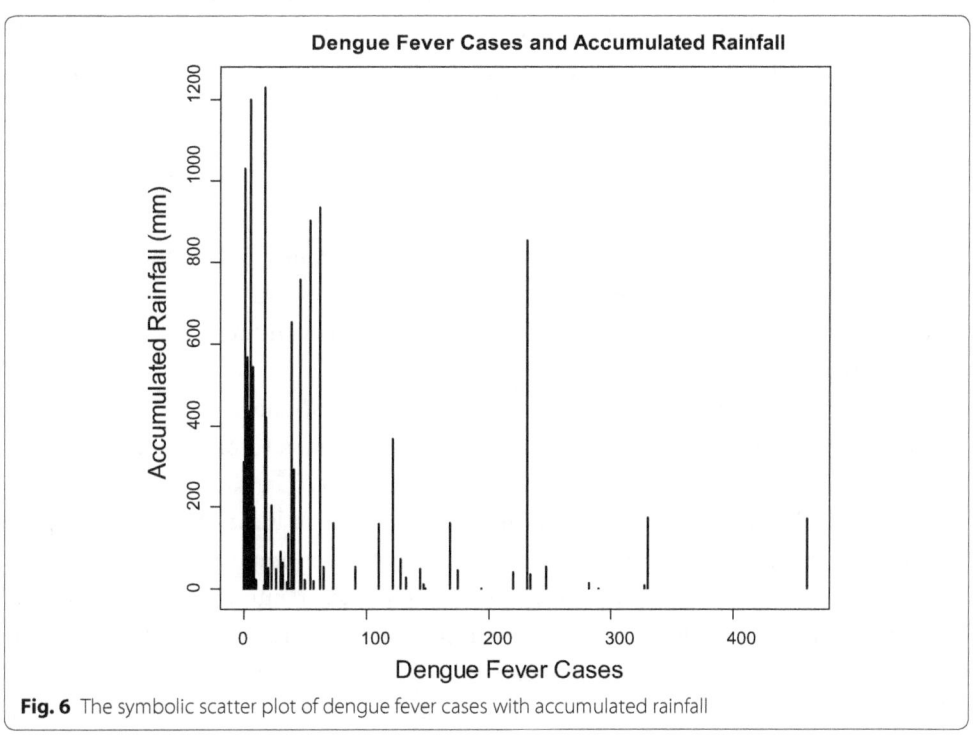

Fig. 6 The symbolic scatter plot of dengue fever cases with accumulated rainfall

previous studies [6–10, 14, 23]. Second, same with Wegbreit's study [26] and Alshehri's study [22], accumulated rainfall (t-value = − 2.002, p-value = 0.047) is negatively correlated with the dengue fever cases. Third, accumulated sunshine (t-value = − 2.790,

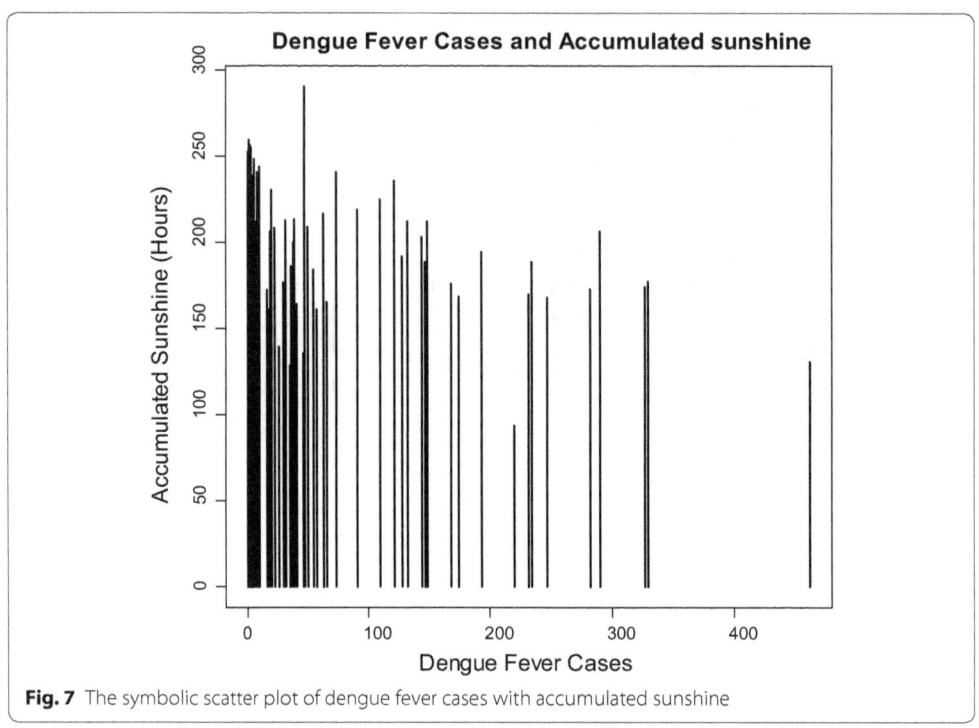

Fig. 7 The symbolic scatter plot of dengue fever cases with accumulated sunshine

p-value $= 0.007$) is also negatively correlated with the dengue fever cases, and this result supports Vu, Okumura, Hashizume, Tran, and Yamamoto's research [15]. The symbolic scatter plot of dengue fever cases with temperature/accumulated rainfall/accumulated sunshine is presented in Figs. 5, 6, and 7. The r-squared value is .138, which indicates that 13.8% of the error is explained by the model.

Discussion

From the study that has been carried out, it is possible to conclude that the risk of dengue fever is positively associated with high temperature, but inversely associated with duration of rainfall and sunshine. This result is consistent with findings of most of the previous studies.

This study shows that temperature is significantly associated with dengue fever incidence. As the temperature increases, the transmission rate of dengue fever also increases. The weight of temperature is larger than others in the predictive model, and it is possible that temperature plays an important role in most of the physiological functions of vectors in southern Taiwan. Temperature is an important climatic factor affecting biological processes of mosquitoes, including their interactions with viruses. Temperature is also positively associated with pre-adult mosquito maturation, oviposition rate, and virus incubation rate in mosquitos [11].

It is found in this study that rainfall and sunshine are both negatively associated with the transmission of dengue fever in southern Taiwan. Figure 2 shows that heavy rainfalls are frequent in southern Taiwan. According to previous studies [16, 22, 26], heavy rainfalls may not favor mosquito density as most of the mosquito eggs and larvae would be washed away from breeding sites [26]. This argument may explain why dengue fever decreases as rainfall increases in southern Taiwan.

Sunshine is also closely linked to other ecological factors such as temperature and thereby might affect the dengue fever incidence. The longer the hours of sunshine, the higher of the temperature is. Although warmer temperatures can increase the transmission rates of dengue fever in various ways, many studies have concluded that high temperature may not favor mosquito density as most of the mosquito larvae would die in heat [19]. This may be the reason why previous studies [10, 19, 20] indicated that maximum temperature and sunshine are negatively correlated with dengue fever incidence.

Conclusions

Dengue fever is ubiquitous throughout the tropics. With spatial variation in different regions, the effects of climate on dengue fever are also different. Dengue viruses and their mosquito vectors are sensitive to their environment. Temperature, rainfall and sunshine have well-defined roles in the transmission cycle. Findings of this paper suggest that control of mosquito by climatic factor during high temperature seasons may be an important strategy for containing the burden of dengue fever.

This study only concentrated on climatic factors. However, some studies [29] showed a clear relationship between sociological factors (e.g., population, urbanization public health policy, and health education) and dengue fever. Further research can continue to test and verify those. In addition, this study focused on Taiwan. Taiwan is a region that is influenced deeply by Chinese culture, so this study can be seen as an example of how climatic factors affect dengue fever in Chinese lifestyle. Future research can further explore how lifestyles in different cultures can influence the occurrence of dengue fever.

Declarations

Author's contributions The author of this manuscript is YHL. The author has made substantial contributions to the conception and design, acquisition, analysis and interpretation of data, and he was involved in drafting the manuscript. The author read and approved the final manuscript.

Acknowledgements

This study is based in part on data from Taiwan National Infectious Disease Statistics System provided by Taiwan's Centers for Disease Control Taiwan. The interpretation and conclusions contained herein do not represent those of Taiwan National Infectious Disease Statistics System or Taiwan's Centers for Disease Control.

Competing interests

The author declare that he has no competing interests.

Funding

Not applicable. This study is not funded by any organization.

References

1. World Health Organization (WHO), Epidemiology. Dengue Control. 2016. http://www.who.int/denguecontrol/en/. Accessed 30 Sept 2017.
2. Hales S, Weinstein P, Souares Y, Woodward A. El Niño and the dynamics of vectorborne disease transmission. Environ Health Perspect. 1999;107(2):99–102.
3. Hales S, de Wet N, Maindonaid J, Woodward A. Potential effect of population and climatic changes on global distribution of dengue fever: an empirical model. Lancet. 2002;360(9336):830–4.
4. Yu HL, Angulo JM, Cheng MH, Wu J, Christakos G. An online spatiotemporal prediction model for dengue fever epidemic in Kaohsiung (Taiwan). Biometrical J. 2014;56(3):428–40.
5. Centers for Disease Control (Taiwan), Dengue Fever. Communicable Diseases & Prevention. 2015. http://www.cdc. gov.tw/english/info.aspx?treeid=e79c7a9e1e9b1cdf&nowtreeid=e02c24f0dacdd729&tid=D76AD76D26365478. Accessed 30 Sept 2017.

6. Tseng YT, Chang FS, Chao DY, Lian IB. Re-model the relation of vector indices, meteorological factors and dengue fever. J Trop Dis. 2016. https://doi.org/10.4172/2329-891X.1000200.

7. Nakhapakorn K, Tripathi NK. An information value based analysis of physical and climatic factors affecting dengue fever and dengue haemorrhagic fever incidence. Int J Health Geogr. 2005; 4(13). http://ij-healthgeographics.biomedcentral.com/articles/10.1186/1476-072X-4-13. Accessed 30 June 2016.

8. Hau VP, Huong TMD, Thao TTP, Nguyen NTM. Ecological factors associated with dengue fever in a central highlands Province, Vietnam. BMC Infect Dis. 2011; 11(172). https://bmcinfectdis.biomedcentral.com/articles/10.1186/1471-2334-11-172. Accessed 30 June 2016.

9. Tong MX, Hansen A, Hanson-Easey S, Xiang J, Cameron S, Liu Q, Liu X, Sun Y, Weinstein P, Han GS, Williams C, Bi P. Perceptions of capacity for infectious disease control and prevention to meet the challenges of dengue fever in the face of climate change: a survey among CDC staff in Guangdong Province, China. Environ Res. 2016;148:295–302.

10. Choi Y, Tang CS, McIver L, Hashizume M, Chan V, Abeyasinghe RR, Iddings S, Huy R. Effects of weather factors on dengue fever incidence and implications for interventions in Cambodia. BMC Public Health. 2016; 16(241). https://bmcpublichealth.biomedcentral.com/articles/10.1186/s12889-016-2923-2. Accessed 30 June 2016.

11. Chen SC, Hsieh MH. Modeling the transmission dynamics of dengue fever: implications of temperature effects. Sci Total Environ. 2012;431:385–91.

12. Wu PC, Guo HR, Lung SC, Lin CY, Su HJ. Weather as an effective predictor for occurrence of dengue fever in Taiwan. Acta Trop. 2007;103(1):50–7.

13. Chang K, Lee NY, Ko WC, Tsai JJ, Lin WR, Chen TC, Lu PL, Chen YH. Identification of factors for physicians to facilitate early differential diagnosis of scrub typhus, murine typhus, and Q fever from dengue fever in Taiwan. J Microbiol Immunol Infect. 2015. https://doi.org/10.1016/j.jmii.2014.12.001.

14. Hii YL, Zhu H, Ng N, Ng LC, Rocklöv J. Forecast of dengue incidence using temperature and rainfall. PLOS Negl Trop Dis. 2012. https://doi.org/10.1371/journal.pntd.0001908.

15. Vu HH, Okumura J, Hashizume M, Tran M, Yamamoto T. Regional differences in the growing incidence of dengue fever in Vietnam explained by weather variability. Trop Med Health. 2014;42(1):25–33.

16. Thammapalo S, Chongsuwiwatwong V, McNeil D, Geater A. The climatic factors influencing the occurrence of dengue hemorrhagic fever in Thailand. Southeast Asian J Trop Med Public Health. 2005;36(1):191–6.

17. Jin X, Lee M, Shu J. Dengue fever in China: an emerging problem demands attention. Emerg Microb Infect. 2015; 4(e3). http://www.nature.com/emi/journal/v4/n1/full/emi20153a.html. Accessed 30 June 2016.

18. Chadee DD, Shivnauth B, Rawlins SC, Chen AA. Climate, mosquito indices and the epidemiology of dengue fever in Trinidad (2002–2004). Ann Trop Med Parasitol. 2007;101(1):69–77.

19. Wongkoon S, Jaroensutasinee M, Jaroensutasinee K. Distribution, seasonal variation and dengue transmission prediction in Sisaket, Thailand. Indian J Med Res. 2013;138(3):347–53.

20. Cheong YL, Burkart K, Leitão PJ, Lakes T. Assessing weather effects on dengue disease in Malaysia. Int J Environ Res Public Health. 2013;10(12):6319–34.

21. Méndez-Lázaro P, Muller-Karger FE, Otis D, McCarthy MJ, Peña-Orellana M. Assessing climate variability effects on dengue incidence in San Juan, Puerto Rico. Int J Environ Res Public Health. 2014;11(9):9409–28.

22. Alshehri MSA. Dengue fever outburst and its relationship with climatic factors. World Appl Sci J. 2013;22(4):506–15.

23. Bhatt S, Gething PW, Brady OJ, Messina JP, Farlow AW, Moyes CL, Drake JM, Brownstein JS, Hoen AG, Sankoh O, Myers MF, George DB, Jaenisch T, Wint GRW, Simmons GP, Scott TW, Farrar JJ, Hay SI. The global distribution and burden of dengue. Nature. 2013;496(7446):504–7.

24. Teurlai M, Menkès CE, Cavarero V, Degallier N, Descloux E, Grangeon JP, Guillaumot L, Libourel T, Lucio PS, Mathieu-Daudé F, Mangeas M. Socio-economic and climate factors associated with dengue fever spatial heterogeneity: a worked example in New Caledonia. PLOS Negl Trop Dis. 2015. https://doi.org/10.1371/journal.pntd.0004211.

25. Promprou S, Jaroensutasinee M, Jaroensutasinee K. Climatic factors affecting dengue haemorrhagic fever incidence in southern Thailand. Dengue Bull. 2005;29:41–8.

26. Wegbreit J. The possible effects of temperature and precipitation on dengue morbidity in Trinidad and Tobago: a retrospective longitudinal study. Population-Environment Dynamics: Issues and Policy (University of Michigan, School of Natural Resources and Environment). 1997. http://www.umich.edu/~csfound/545/1997/weg/. Accessed 30 June 2016.

27. Goto K, Kumarendran B, Mettananda S, Gunasekara D, Fujii Y, Kaneko S. Analysis of effects of meteorological factors on dengue incidence in Sri Lanka using time series data. PLoS ONE. 2013; 8(5). http://journals.plos.org/plosone/article/asset?id=10.1371%2Fjournal.pone.0063717.PDF. Accessed 30 June 2016.

28. Pandey N, Nagar R, Gupta S, Khan OD, Singh DD, Mishra G, Prakash S, Singh KP, Singh M, Jain A. Trend of dengue virus infection at Lucknow, north India (2008–2010): a hospital based study. Indian J Med Res. 2012;136(5):862–7.

29. Yang T, Lu L, Fu G, Zhong S, Ding G, Xu R, Zhu G, Shi N, Fan F, Liu Q. Epidemiology and vector efficiency during a dengue fever outbreak in Cixi, Zhejiang Province, China. J Vector Ecol. 2009;34(1):148–54.

30. Billard L, Diday E. Symbolic data analysis: conceptual statistics and data mining. New York: Wiley-Interscience; 2007.

31. Diday E, Monique NF. Symbolic data analysis and the SODAS software. New York: Wiley-Interscience; 2008.

32. Patz JA, Epstein PR, Burke TA, Balbus JM. Global climate change and emerging infectious diseases. JAMA. 1996;275(3):217–23.

33. Lai YH. Temperature factor affecting dengue fever incidence in southern Taiwan. Asian J Human Soc Stud. 2014;2(5):661–5.

34. The Ministry of Health and Welfare. Confirmed Cases of Dengue Fever, Taiwan National Infectious Disease Statistics System. http://nidss.cdc.gov.tw. Accessed 30 June 2016.

35. Central Weather Bureau. Daily Precipitation, Climate Statistics. http://www.cwb.gov.tw/V7/climate/monthlyData/mD.htm. Accessed 30 June 2016.

36. Lima-Neto EA, de Carvalho FAT. Centre and range method to fitting a linear regression model on symbolic interval data. Comput Stat Data Anal. 2008;52:1500–15.

37. Lima-Neto EA, de Carvalho FAT. Constrained linear regression models for symbolic interval-valued variables. Comput Stat Data Anal. 2010;54:333–47.

5

Robust IoT-based nursing-care support system with smart bio-objects

Cheng-Fa Chiang[1,2†], Fang-Ming Hsu[1†] and Kuo-Hui Yeh[1*†]

From International Conference on Biomedical Engineering Innovation (ICBEI) 2016 Taichung, Taiwan.

*Correspondence:
khyeh@gms.ndhu.edu.tw
†Cheng-Fa Chiang, Fang-Ming Hsu and Kuo-Hui Yeh contributed equally to this work
[1] Department of Information Management, National Dong Hwa University, Hualien 97401, Taiwan, ROC
Full list of author information is available at the end of the article

Abstract

Background: The significant advancement in the mobile sensing technologies has brought great interests on application development for the Internet-of-Things (IoT). With the advantages of contactlessness data retrieval and efficient data processing of intelligent IoT-based objects, versatile innovative types of on-demand medical relevant services have promptly been developed and deployed. Critical characteristics involved within the data processing and operation must thoroughly be considered. To achieve the efficiency of data retrieval and the robustness of communications among IoT-based objects, sturdy security primitives are required to preserve data confidentiality and entity authentication.

Methods: A robust nursing-care support system is developed for efficient and secure communication among mobile bio-sensors, active intelligent objects, the IoT gateway and the backend nursing-care server in which further data analysis can be performed to provide high-quality and on-demand nursing-care service.

Results: We realize the system implementation with an IoT-based testbed, i.e. the Raspberry PI II platform, to present the practicability of the proposed IoT-oriented nursing-care support system in which a user-friendly computation cost, i.e. 6.33 ms, is required for a normal session of our proposed system. Based on the protocol analysis we conducted, the security robustness of the proposed nursing-care support system is guaranteed.

Conclusions: According to the protocol analysis and performance evaluation, the practicability of the proposed method is demonstrated. In brief, we can claim that our proposed system is very suitable for IoT-based environments and will be a highly competitive candidate for the next generation of nursing-care service systems.

Background

With the rapid growth of information and communications technologies, such as Bluetooth Low Energy (BLE), 3G/4G/5G and NFC/RFID, a comprehensive evolution of the Internet has given rise to a ubiquitous network consisting of mobile intelligent objects, called the Internet of Things (IoT). In IoT-based environments, "contactless data sensing" and "collecting and information analyzing and retrieving" are fundamental components for the provision of human value-added services in a more transparent and faster way than before. Among these services, in particular, the development of IoT-oriented nursing-care service systems are one the most promising and important directions, and are therefore a major focus of government and industry. A nursing-care service system is exploited for data collection, data storing, data retrieval and information display needed in nursing activities via modern information and communication technologies. With the advantages of contactlessness and efficiency brought by the data retrieval on intelligent IoT-objects, innovative types of on-demand nursing-care service systems have promptly been developed in these years. Meanwhile, the issue of system security and patient privacy has been focused by governments and research community. The potential to reveal patient privacy and system security vulnerability may exist wherever personally identifiable information is collected, processed, or stored in a hospital information system. Based on our survey, we present the major principles during patient private data processing: (1) be processed and used for lawful purposes; (2) unauthorized or unlawful processing must be measured; (3) accountability is required; (4) consent for data processing must be guaranteed; (5) be processed with an adequate level of protection and (6) adequate and relevant to the purpose for which it is processed.

In this paper, we present a robust IoT-based nursing-care support system in which fixed environmental sensing objects and intelligent smart objects are deployed in the field and on patients, respectively, to support high-quality nursing-care service. To satisfy the security and privacy requirements, we argue that sturdy cryptographic primitives must be implemented on IoT-objects to construct robust communications among entities. Nevertheless, based on current semiconductor technology, most of IoT-objects cannot afford heavy cryptographic primitives, such as asymmetric cryptography, due to limited computational resources. Therefore, a refinement of the traditional secure communication scheme should be launched in terms of the performance standpoint. That is, we have to thoroughly consider the trade-off between efficiency and the robustness of the adopted cryptographic components to appropriately meet the hardware limitation of IoT-objects and the security requirements we need. According to the analyses, the robust cryptographic module with a reasonable and acceptable computation cost, i.e. SHA-384, SHA-215 and SHA-3 [1, 2], will be good candidate techniques to simultaneously satisfy the security and performance requirements. In conclusion, we would like to demonstrate an efficient IoT-based communication mechanism for nursing-care service systems in which SHA-3 are mainly adopted as the major data protection technique to simultaneously achieve system security and patient privacy during the operation of the proposed system.

In the following, we present the state of the art of IoT application and security. In 2012, Jara et al. [3] proposed an IoT-based knowledge acquisition and management platform. This platform is composed of two parts, i.e. a wireless transmission of continuous vital signs through 6LoWPAN and a patient identification through RFID. The presented system also adopted a data analysis model and pre-processing module for patient health management. Next, Berhanu et al. [4] introduced an e-Health system with IoT devices in which a robust security scheme is included. The authors investigated the impact of antenna orientation on energy consumption to examine the validation of the proposed system. The issue of scalability is also studied through the feasibility of embedding the lightweight security solutions into the ASSET (adaptive security for smart Internet of

Things in e-health) [5]. Later, Torjusen et al. [6] verified that an enabler integrating into the ASSET adaptive security framework and provided an e-healthcare security framework via the IoT. Critical requirements for run-time verification are presented as formal specifications. After that, Bello and Zeadally [7] proposed an intelligent routing cooperation scheme for device-to-device communication. The operation of different network standards in the case of intermittent connection is considered in which the device will be affected by it's limited resources. In 2016, Gope and Hwang [8] proposed an IoT application system for healthcare on body sensor networks (BSN), called BSN-Care, which is able to provide effective real-time monitoring, patient information management and security needs for healthcare. In order to achieve the claimed services, a comprehensive integration of clinical devices and efficient collection of data are demonstrated. Note that the authors also introduced a similar concept for secure communication on IoT in [9].

Yao et al. [10] modified the fast one-way accumulator, proposed by Nyberg [11], to build a lightweight multicast authentication mechanism. However, the proposed method is only applicable to the small-scale IoT networks. It is limited by it's scalability. After that, Ning et al. [12] demonstrated an aggregation-based hierarchical authentication scheme. This method provides a strong security and is applicable to the U2IoT architecture. The main idea is to establish backward and forward anonymous data transmission among multiple targets. In addition, various techniques, i.e. directed path descriptors, homomorphism functions, and Chebyshev chaotic maps, are jointly applied for mutual authentication. Later, Hernández-Ramos et al. [13] proposed a framework for lightweight authentication and authorization. This presented framework is based on the reference model proposed by the EU FP7 IoT-A project. Meanwhile, Kawamoto et al. [14] proposed a location-based authentication system, where ambient information from IoT-based sensors are collected and analyzed as the authentication tokens. To pursue high authentication accuracy, the proposed system automatically and continuously adjusts the system parameters according to surrounding environment situation. Next, Cirani et al. [15] proposed an IoT-OAS architecture with the characteristics of flexible, highly configurable, and easy-to-integrate with existing services. The authors further provide an authorization platform which can invoke an external OAuth-based authorization service. The evaluation of the proposed architecture is based on Contiki OS-based constrained devices. In 2017, Cha et al. [16] demonstrated a privacy-aware mechanism for secure communication and efficient access-control among BLE-based devices. The proposed mechanism is based on elliptic curve cryptography. In addition, the authors presented a framework for the management of security examination reports of BLE-based applications.

Methods

This section introduces the underlying IoT communication architecture and then presents the proposed nursing-care support system consisting of a registration process and an authentication process.

Targeted IoT communication scenario

In this section, we introduce the underlying IoT communication architecture of our proposed nursing-care support system. Figure 1 demonstrates the scenario we target on, i.e. patient management and rehabilitation, in which fixed environmental sensors and medical sensing devices are deployed on field and on the patient, respectively, to support caregivers (such as nurse) in activities of patient care and rehabilitation. Three important components are existed in the identified IoT communication scenario, i.e. a backend nursing-care server, a mobile gateway (usually handheld by the caregiver), and intelligent devices (such as fixed sensor nodes or medical sensing devices). The intelligent devices are utilized for sensing and collecting environmental parameters and patients' bio-data, while the caregiver will operate a handheld smart device as the mobile gateway to communicate with the intelligent devices. Note that the patient's bio-data are such as electrocardiography, electroencephalography, electromyography and blood pressure retrieved from the patient. After that, with the environmental data and the patient's bio-data, the caregiver can identify patient's need in a faster way. More accurate and timely treatment services will then be able to deliver to the patients.

The IoT-based nursing-care support system

In this section, we present the proposed nursing-care support system in which IoT-based intelligent devices are adopted on the patient and the corresponding environment (e.g., Fig. 2). From Fig. 2a, the nurse first utilizes his/her handheld mobile gateway to retrieve the data from smart bio-objects. All of the retrieved data will be forwarded to the backend nursing-care server. At the backend server, a further data analysis for the mining of the patient's needs will be performed. Then, in Fig. 2b, decision support/assistant information will be derived and delivered to the nurse to support better-quality nursing care services on the patient.

In general, the proposed system consists of two phases, i.e. the registration phase and the authentication phase. In the registration phase, the security credentials will be

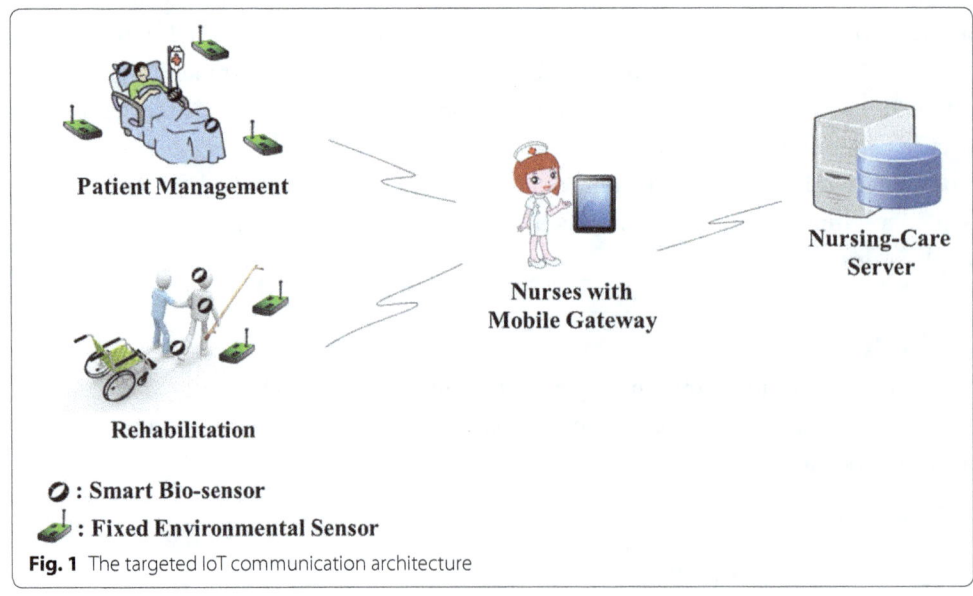

Patient Management

Rehabilitation

Nurses with Mobile Gateway

Nursing-Care Server

⚬ : Smart Bio-sensor

🔲 : Fixed Environmental Sensor

Fig. 1 The targeted IoT communication architecture

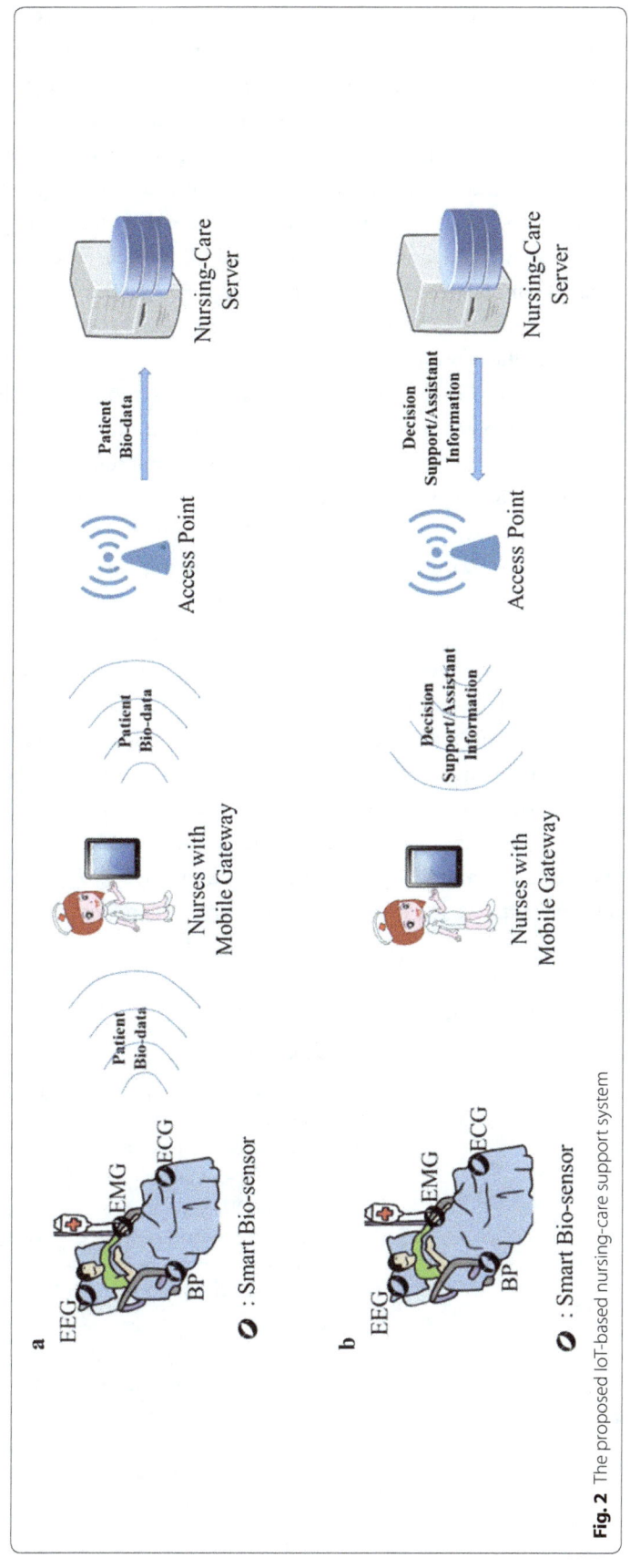

Fig. 2 The proposed IoT-based nursing-care support system

securely agreed and shared among the communication entities, i.e. intelligent devices (smart bio-objects and fixed environmental sensors), the mobile gateway and the nursing-care server, in advance. Next, an authentication phase is operated to secure all of the communication and data exchange among the communication entities. The proposed nursing-care support system is able to achieve the following security requirements: (a) to guarantee mutual authentication among intelligent devices, the mobile gateway and the nursing-care server; (b) to provide anonymity and un-traceability for intelligent devices; (c) to resist against forgery attack and replay attack and (d) to securely establish a robust session key between the mobile gateway and the nursing-care server.

The registration phase of the proposed system

Before introducing our proposed system, we present the symbols and abbreviations throughout this study in Table 1. In the first stage of the registration phase, an intelligent device d_i (i.e. smart bio-sensors or fixed environmental sensors) sends its identity ID_{d_i} to the nursing-care server S as a registration request. On receipt of the request from d_i, the nursing-care server S generates a random number N_{ds} and uses its identity ID_s to compute a secret value $K_{ds} = H(ID_s||N_{ds}||ID_{d_i})$. Next, the nursing-care server S calculates a set of un-linkable shadow identities $SID = \{sid_1, sid_2, ...\}$ for d_i, where each $sid_j \in SID$ and $sid_j = H(ID_{d_i}||N_j||K_{ds})$. Note that N_j is a random number used for deriving each sid_j value. Moreover, a track sequence number Tr_{seq} is generated for fast identification of intelligent device d_i during the authentication process as well as for preventing replay attacks. The Tr_{seq} will be stored and updated at both the nursing-care server S and the device d_i after each authentication session. In that case, the nursing-care server S is able to check the freshness of an incoming request from d_i, and to achieve a fast identification of d_i via Tr_{seq} at the backend database during the authentication session. Finally, the nursing-care server S issues a security credential containing $(ID_{d_i}, K_{ds}, SID, Tr_{seq}, H(.))$ to the intelligent device d_i. At the same time, the nursing-care server will maintain a

Table 1 Notations throughout this study

Symbol	Definition
d_i	Intelligent device (i.e. smart bio-sensors or fixed environmental sensors)
g_j	Mobile gateway (operated by the nurse or the doctor)
S	The nursing-care server
ID_{d_i}	Private identity of d_i
ID_s	Public identity of the nursing-care server S
ID_{g_j}	Public identity of the mobile gateway g_j
AID_{d_i}	One-time-alias identity of d_i
SID	A set of un-linkable shadow identities $SID = \{sid_1, sid_2, ...\}$
K_{ds}	The secret key shared between d_i and S
K_{gs}	The secret key shared between g_j and S
Tr_{seq}	Track sequence number
$N_{ds}, N_j, N_{gs}, N_d, N_g, m$	Random numbers
$H(.)$	Secure one-way hash function, i.e. SHA-3
$\oplus$	Bitwise exclusive-or operation
$\|$	Concatenation operation

record $\left(ID_{d_i}, K_{ds}, SID, Tr_{seq}, H(.) \right)$ corresponding to d_i at the backend database. Note that $H(.)$ denotes a secure one-way hash function such as SHA-3. On the other hand, the registration phase between the mobile gateway g_j and the nursing-care server S are launched in a similar way. The mobile gateway g_j send its identity ID_{g_j} to the nursing-care server as a registration request. Next, the nursing-care server S calculates $K_{gs} = H\left(ID_s || N_{gs} || ID_{g_j} \right)$ with a newly generated random number N_{gs}, and shares a security credential containing the secrets, i.e. ID_{g_j} and K_{gs}, with g_j. The nursing-care server also maintains a tuple $\left(ID_{g_j}, K_{gs}, H(.) \right)$ corresponding to the mobile gateway g_j at the backend database.

The authentication phase of the proposed system

In the authentication phase, we consider that a caregiver (with a mobile gateway) would like to provide on-demand nursing care services to patients via contactless and real-time data collection and analysis mechanisms. Under the insecure IoT communication architecture, an authentication procedure is needed to establish a secure communication channel for data exchange among intelligent devices, the mobile gateway and the nursing-care server. The detailed communication procedures of the authentication phase are as shown in Fig. 3.

- Intelligent Device $d_i \to$ Mobile Gateway g_j : $M_{A_1} = \left\{ AID_{d_i}, N_x, Tr_{seq}(\text{if req.}), ID_{g_j} \right\}$

  First, the intelligent device d_i generates a random number N_d and calculates two values, i.e. $M_x = H\left(K_{ds} || ID_{d_i} \right) \oplus N_d$ and $AID_{d_i} = H\left(ID_{d_i} || K_{ds} || N_d || M_x || ID_{g_j} || Tr_{seq} \right)$.

  Next, d_i constructs a message $M_{A_1} = \left\{ AID_{d_i}, M_x, Tr_{seq}, ID_{g_j} \right\}$ and sends M_{A_1} as an authentication request to the mobile gateway g_j. Note that, if the value Tr_{seq} shared between the intelligent device d_i and the nursing-care server S is out of synchronization, d_i needs to choose a fresh shadow identity sid_j from SID as the value AID_{d_i}. Then, d_i sends $M_{A_1} = \left\{ AID_{d_i}, M_x, ID_{g_j} \right\}$ to the mobile gateway g_j as an authentication request.

- Mobile Gateway $g_j \to$ Nursing-Care Server S: $M_{A_2} = \left\{ M_y, V_1, M_{A_1} \right\}$

  Once the mobile gateway g_j receives the authentication request from the intelligent device d_i, g_j first generates a random number N_g and computes $M_y = K_{gs} \oplus N_g$ and $V_1 = H\left(M_{A_1} || N_g || M_y || K_{gs} \right)$. After that, g_j sends $M_{A_2} = \left\{ M_y, V_1, M_{A_1} \right\}$ to the nursing-care server S.

- Nursing-Care Server $S \to$ Mobile Gateway g_j : $M_{A_3} = \left\{ Tr, ID_s, V_2, V_3 \right\}$

  Upon obtaining the incoming message $M_{A_2} = \left\{ M_y, V_1, M_{A_1} \right\}$, the nursing-care server S first checks whether the track sequence number Tr_{seq} is in the request or not. If it is, S performs step (1). If Tr_{seq} is not included in M_{A_2}, S performs step (2).

  - Step (1): Check the validity of Tr_{seq}. If it holds, look for the corresponding tuple via Tr_{seq} from the backend database. Otherwise, terminate the connection. If Tr_{seq} is valid, S derives $N_d = H\left(K_{ds} || ID_{d_i} \right) \oplus M_x$ and $N_g = K_{gs} \oplus M_y$, and then verifies V_1

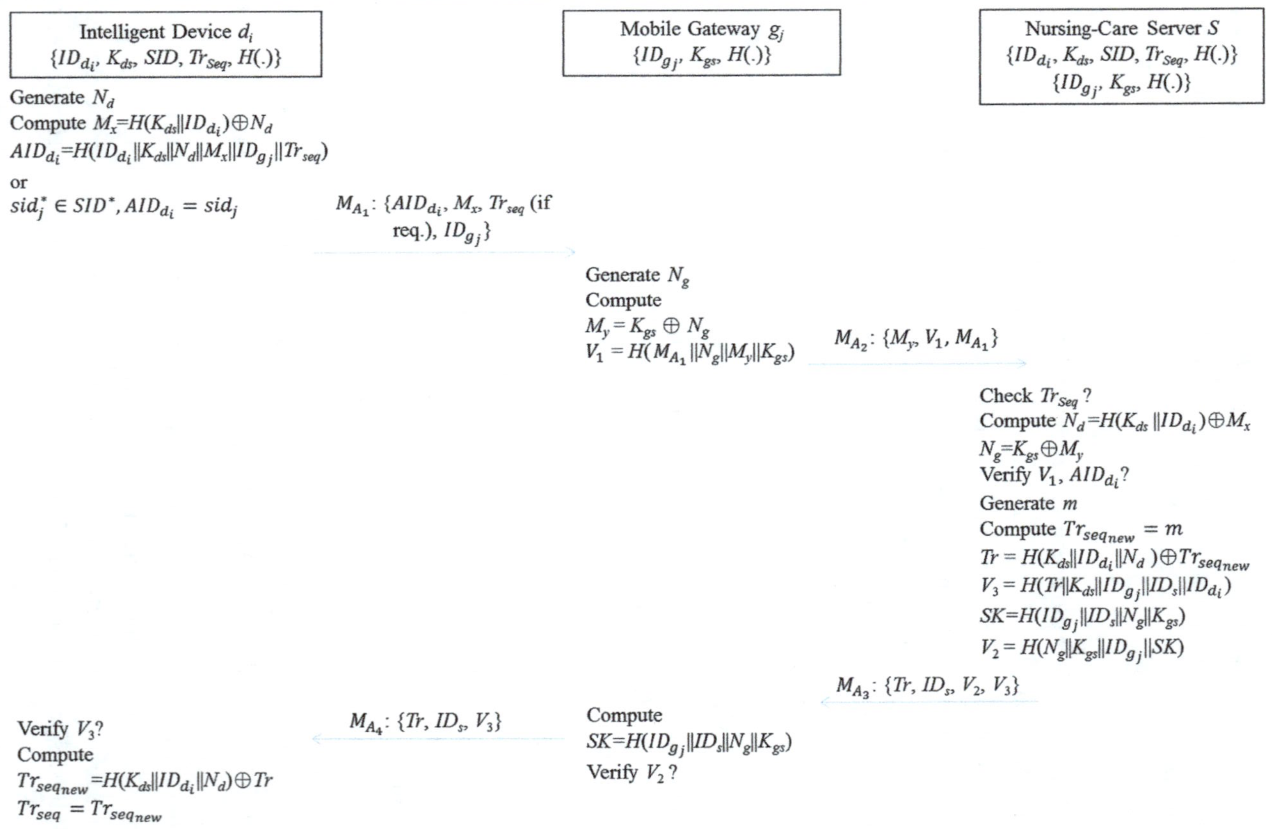

Biomedical Engineering: Integrating Medicine and Technology

Intelligent Device d_i
$\{ID_{d_i}, K_{ds}, SID, Tr_{Seq}, H()\}$

Generate N_d
Compute $M_x = H(K_{ds}||ID_{d_i}) \oplus N_d$
$AID_{d_i} = H(ID_{d_i}||K_{ds}||N_d||M_x||ID_{g_j}||Tr_{seq})$
or
$sid_j^* \in SID^*, AID_{d_i} = sid_j$

$M_{A_1}: \{AID_{d_i}, M_x, Tr_{seq}$ (if req.), $ID_{g_j}\}$

Mobile Gateway g_j
$\{ID_{g_j}, K_{gs}, H()\}$

Generate N_g
Compute
$M_y = K_{gs} \oplus N_g$
$V_1 = H(M_{A_1}||N_g||M_y||K_{gs})$

$M_{A_2}: \{M_y, V_1, M_{A_1}\}$

Nursing-Care Server S
$\{ID_{d_i}, K_{ds}, SID, Tr_{Seq}, H()\}$
$\{ID_{g_j}, K_{gs}, H()\}$

Check Tr_{Seq}?
Compute $N_d = H(K_{ds}||ID_{d_i}) \oplus M_x$
$N_g = K_{gs} \oplus M_y$
Verify V_1, AID_{d_i}?
Generate m
Compute $Tr_{seq_{new}} = m$
$Tr = H(K_{ds}||ID_{d_i}||N_d) \oplus Tr_{seq_{new}}$
$V_3 = H(Tr||K_{ds}||ID_{g_j}||ID_s||ID_{d_i})$
$SK = H(ID_{g_j}||ID_s||N_g||K_{gs})$
$V_2 = H(N_g||K_{gs}||ID_{g_j}||SK)$

$M_{A_3}: \{Tr, ID_s, V_2, V_3\}$

Compute
$SK = H(ID_{g_j}||ID_s||N_g||K_{gs})$
Verify V_2?

$M_{A_4}: \{Tr, ID_s, V_3\}$

Verify V_3?
Compute
$Tr_{seq_{new}} = H(K_{ds}||ID_{d_i}||N_d) \oplus Tr$
$Tr_{seq} = Tr_{seq_{new}}$

Fig. 3 The authentication phase of the proposed IoT-based nursing-care support system

and AID_{d_i}. That is, S will examine (a) whether the received value V_1 and the computed value $H\left(M_{A_1}||N_g||M_y||K_{gs}\right)$ are equal or not, and (b) whether the received value AID_{d_i} and the computed value $H = \left(ID_{d_i}||K_{ds}||N_d||M_x||ID_{g_j}||Tr_{seq}\right)$ are equal or not. Note that if Tr_{seq} in the request does not match the one maintained in the database, the nursing-care server S will reject the request and terminate the connection. A new request from the device d_i will be asked for in which one of the fresh shadow identities sid_j will be picked up from the list SID as an anonymous identity of d_i. In that case, the step (2) will be launched.

- Step (2): The server S will verify the freshness and validity of $AID_{d_i} = sid_j$. If the nursing-care server S cannot identify the sid_j from the backend database, the server will terminate the connection. Next, the S will request the intelligent device d_i to try with another valid shadow identity sid_j.

If one of the above examinations, i.e. step (1) or (2), is passed, the nursing-care server S will generate a random number m, and set $Tr_{seq_{new}} = m$. After that, S calculates $Tr = H\left(K_{ds}||ID_{d_i}||N_d\right) \oplus Tr_{seq_{new}}$, $V_3 = H\left(Tr||K_{ds}||ID_{g_j}||ID_s||ID_{d_i}\right)$, $SK = H\left(ID_{g_j}||ID_s||N_g||K_{gs}\right)$ and $V_2 = H\left(N_g||K_{gs}||ID_{g_j}||SK\right)$, where SK is a session key utilized for the next secure communication between the mobile gateway g_j and the nursing-care server S. Eventually, S sends $M_{A_3} = \{Tr, ID_s, V_2, V_3\}$ to the mobile gateway g_j.

- Mobile Gateway $g_j \rightarrow$ Intelligent Device d_i: $M_{A_4} = \{Tr, ID_s, V_3\}$
  With the incoming message $M_{A_3} = \{Tr, ID_s, V_2, V_3\}$, the mobile gateway g_j computes $SK' = H\left(ID_{g_j}||ID_s||N_g||K_{gs}\right)$ and $H\left(N_g||K_{gs}||ID_{g_j}||SK'\right)$, and then check if the received V_2 is equal to the computed $H\left(N_g||K_{gs}||ID_{g_j}||SK'\right)$. If it holds, it is obvious that a session key SK is securely agreed by g_j and S. After that, the mobile gateway g_j sends $M_{A_4} = \{Tr, ID_s, V_3\}$ to the intelligent device d_i. With M_{A_4}, the device d_i derives $H\left(Tr||K_{ds}||ID_{g_j}||ID_s||ID_{d_i}\right)$ and compares it with the received V_3. If these two values are identical, d_i computes $Tr_{seq_{new}} = H\left(K_{ds}||ID_{d_i}||N_d\right) \oplus Tr$ and sets $Tr_{seq} = Tr_{seq_{new}}$.

Protocol analysis and discussions

In this section, we present a formal security analysis of the communication procedures of the proposed IoT-based nursing-care support system and discuss whether all the proposed security claims can be achieved or not.

- Claim 1: To achieve mutual authentication among communication entities in the proposed nursing-care support system

The mutual authentication via the proposed communication procedures is proven via BAN logic analysis [17, 18]. Basic constructs and logic postulates for the purpose of analysis are first presented, where the symbols P and Q are defined as principals, X and Y are defined as statements, and K ranges over a long-term secret.

Seven constructs is introduced as follows: (1) P believes X means that the principal P believes that X is true; (2) P sees X means that someone has sent a message containing X to P; (3) P said X denotes that P has actually sent a message including statement X at the current session of the protocol or before; (4) P controls X denotes that P has jurisdiction over X; (5) fresh(X) denotes that X has not been sent in a message; (6) $P \xleftrightarrow{K} Q$ means that the secret K is shared between the principals P and Q, and (7) $\{X\}_K$ means that the X is encrypted or protected under the key K. Next, we presented five major rules as logical postulates. First, in the message-meaning rule (referred to rule 1), we believe that if P believes $P \xleftrightarrow{K} Q$ and P sees $\{X\}_K$, then we postulate P believes Q said X. Second, the nonce-verification rule (referred to rule 2) denotes that if P believes fresh (X) and P believes Q said X, then we postulate P believes Q believes X. Third, the jurisdiction rule (referred to rule 3) means that if P believes Q controls X and P believes Q believes X, then we postulate P believes X. Fourth, a rule 4 is identified for that if P sees (X, Y) then P sees X. In addition, if P believes $P \xleftrightarrow{K} Q$ and P sees $\{X\}_K$, then P sees X. Fifth, the final rule 5 denotes that if one part of a formula is fresh, then the entire formula must also be fresh. If P believes fresh (X), then P believes fresh (X, Y). Finally, we demonstrate seven assumptions of our proposed system in the following.

- Assumption 1: d_i, S believe $d_i \xleftrightarrow{ID_{d_i}, K_{ds}, SID, Tr_{seq}} S$
- Assumption 2: g_j, S believe $g_j \xleftrightarrow{ID_{g_i}, K_{gs}} S$
- Assumption 3: d_i, S believe fresh(N_d)
- Assumption 4: g_j, S believe fresh(N_g)
- Assumption 5: d_i believes fresh(m)
- Assumption 6: d_i believes S controls N_d
- Assumption 7: g_j believes S controls N_g

Before the security analysis, we conduct the concrete realization of our proposed communication procedures as follows:

Step 1: $d_i \rightarrow g_j : \left\{ AID_{d_i}, M_x, Tr_{seq} \text{(if req.)}, ID_{g_j} \right\}$, where $AID_{d_i} = H\left(ID_{d_i}||K_{ds}||N_d||M_x||ID_{g_j}||Tr_{seq}\right)$ and $M_x = H\left(K_{ds}||ID_{d_i}\right) \oplus N_d$.

Step 2: $g_j \rightarrow S : \left\{ M_y, V_1, AID_{d_i}, M_x, Tr_{seq} \text{(if req.)}, ID_{g_j} \right\}$, where $M_y = K_{gs} \oplus N_g$ and $V_1 = H\left(M_{A_1}||N_g||M_y||K_{gs}\right)$.

Step 3: $S \rightarrow g_j : \{Tr, ID_s, V_2, V_3\}$, where $Tr = H\left(K_{ds}||ID_{d_i}||N_d\right) \oplus Tr_{seq_{new}}$, $V_2 = H\left(N_g||K_{gs}||ID_{g_j}||SK\right)$, $V_3 = H\left(Tr||K_{ds}||ID_{g_j}||ID_s||ID_{d_i}\right)$ and $SK = H\left(ID_{g_j}||ID_s||N_g||K_{gs}\right)$.

Step 4: $g_j \rightarrow d_i : \{Tr, ID_s, V_3\}$, where $Tr = H\left(K_{ds}||ID_{d_i}||N_d\right) \oplus Tr_{seq_{new}}$ and $V_3 = H\left(Tr||K_{ds}||ID_{g_j}||ID_s||ID_{d_i}\right)$.

After that, the following steps show that the formal analysis of the mutual authentication:

1. g_j sees$\{ID_s, V_2\}$ (Step 3).
2. g_j believe $g_j \xleftrightarrow{ID_{g_i}, K_{gs}} S$ (Assumption 2).
3. g_j believes S said $\{ID_s, V_2\}$ [(1) and (2), inferred by Rule 1].

4. g_j believes fresh(N_g) (Assumption 4).
5. g_j believes S believes $\{ID_s, V_2\}$ [(3) and (4), inferred by Rule 2].
6. g_j believes S controls$\{N_g\}$ (Assumption 7).
7. g_j believes $\{ID_s, V_2\}$ [(5) and (6), Inferred by Rule 3].
8. d_i sees $\{Tr, ID_s, V_3\}$ (Step 4).
9. d_i believes $d_i \overset{ID_{d_i}, K_{ds}, SID, Tr_{seq}}{\longleftrightarrow} S$ (Assumption 1).
10. d_i believes S said $\{Tr, ID_s, V_3\}$ [(8) and (9), Inferred by Rule 1].
11. d_i believes fresh(N_d), fresh(m) (Assumption 3 and 5).
12. d_i believes S believes $\{Tr, ID_s, V_3\}$ [(10) and (11), Inferred by Rule 2].
13. d_i believes S controls $\{N_d\}$ (Assumption 6).
14. d_i believes $\{Tr, ID_s, V_3\}$ [(12) and (13), inferred by Rule 3].

So far, we obtain the following results.

g_j believes S believes $\{ID_s, V_2\}$ [From (5)]

g_j believes $\{ID_s, V_2\}$ [From (7)]

d_i believes S believes $\{Tr, ID_s, V_3\}$ [From (12)]

d_i believes $\{Tr, ID_s, V_3\}$ [From (14)]

Based on the assumption of the trustworthiness of S and the four results (5), (7), (12) and (14), both d_i and g_j can authenticate with each other via S.

- Claim 2: To guarantee anonymity and un-traceability for each intelligent device in the proposed nursing-care support system

During the communication procedures of the proposed system, two random numbers N_d and N_g are utilized to randomize the messages, such as $AID_{d_i}, M_x, M_y, V_1, Tr, V_2$ and V_3, transmitted among the intelligent devices d_i, the mobile gateway g_j and the nursing-care server S. The g_j and S cannot obtain the real identity of d_i. In other words, the identity ID_{d_i} is included in a randomized cipher text during each communication session. Therefore, we can claim that our proposed system can provide the anonymity, and the un-traceability can be guaranteed also. On the other hand, the shadow identity scheme in our system is adopted to deal with the condition of loss of synchronization between the intelligent device and the nursing-care server. Since the shadow identity is randomly chosen, it does not provide any clue for malicious attacks due to the un-linkable property of them.

- Claim 3: To resist against forgery attack and replay attack

Adversaries may counterfeit messages to deceive the legal communication entities, i.e. d_i, g_j and S. However, without N_d, N_g, K_{ds} and K_{gs}, it is hard to create a counterfeited but legitimate messages such as $\left\{M_y, V_1, AID_{d_i}, M_x, Tr_{seq}\text{(if req.)}, ID_{g_j}\right\}$ and $\{Tr, ID_s, V_2, V_3\}$ for spoofing. Even if the adversaries launch a replay attack with a previously eavesdropped message, the previously-used message cannot be successfully verified. This is because the random numbers N_d and N_g must be fresh and on-time valid at each session. As a result, we can claim that the resistance to forgery attack and replay attack can be guaranteed in our proposed system.

- Claim 4: To preserve data confidentiality via a secure transmission channel established between the mobile gateway and the nursing-care server

During the transmission of the proposed system, all of the messages $\left\{M_y, V_1, AID_{d_i}, M_x, Tr_{seq}(\text{if req.}), ID_{g_j}\right\}$ and $\{Tr, ID_s, V_2, V_3\}$ are protected through a secure one-way hash function (e.g., SHA-3), and two robust secrets K_{ds} and K_{gs} chosen by S. Without K_{ds} and K_{gs}, it is difficult to retrieve any useful information from transmitted cipher texts owing to the irreversibility of the one-way hash function. The proposed system thus provides data confidentiality. Moreover, according to the analysis of Claim 1, the mutual authentication of g_j and S is achieved via V_2. It is obvious that a session key SK can securely be agreed by g_j and S during our proposed authentication phase, and this session key SK will then be exploited on the verification of V_2. Therefore, we argue that a robust channel will be established between the mobile gateway and the nursing-care server for secure communication.

Performance evaluation and results

To evaluate the practicability of the proposed IoT-based nursing-care support system, we implement a demo system as a proof-of-concept and evaluate it's performance. The implementation was established under the platform shown in Table 2, where the Raspberry PI II platform is simulated as an intelligent device operating with a smartphone (as the mobile gateway) and a desktop computer (as the nursing-care server). Because a performance bottleneck occurring at the intelligent device is always in a high probability when compared to a smartphone and a desktop computer, in this implementation we focus on the performance evaluation of the intelligent device. A secure one-way hash function, i.e. SHA-3 (512 bits), and bitwise exclusive-or operation are adopted. In addition, in our system implementation, "ID_{d_i}, ID_s, ID_{g_j}, K_{ds} and K_{gs}" are set to 96-bits and "SID, Tr_{seq}, N_{ds}, N_j, N_{gs}, N_d, N_g and m" are set to 64-bits. Each time SID contains 100 sid_j values. The experiments are implemented via Oracle Java 8 and Eclipse 3.8, and we implement the SHA-3 hash function with the support of Bouncy Castle Crypto APIs [19].

Table 3 presents the computation cost required in our proposed IoT-based nursing-care support system. During the registration phase, we need to perform $(2+k)$ times of random number generation and $(2+k)$ times of one-way hash function. Note that k is the size of SID and in our implementation we set k as 100. In that case, in the registration phase we have to execute 102 times the random number generation and 102 times that of the one-way hash function, and we found that the total computation

Table 2 Implementation environment

Environment	Description
Raspberry PI II	Broadcom BCM2836 @ 1 GHz Quad-Core ARM Cortex-A7 Architecture, 1 GB DDR2 RAM and SanDisk 16 GB Class 10 SD Card
Operating system	Raspbian 2016/03
Programming IDE	Eclipse 3.8 with Oracle Java 8 ARM
Crypto API	The Bouncy Castle Crypto APIs
Environment	Description

Table 3 Execution time of the proposed IoT-based nursing-care support system

Phase	Computation cost	Execution time (ms)
Registration	$(2+k)$ RN $+ (2+k)$ H[a]	14.23 ms (i.e. 102RN + 102H)
Authentication	3RN + 6XOR + 14H (with Tr_{seq})	6.33 ms (i.e. 3RN + 6XOR + 14H)
	3RN + 6XOR + 12H (without Tr_{seq})	5.43 ms (i.e. 3RN + 6XOR + 12H)

RN means random number

XOR means bitwise exclusive-or operation

H means the one-way hash function SHA-3 (512 bits)

[a] k is the size of *SID* which contains *ksid$_j$* values

time is around 14.23 ms. Note that 7.09 ms is needed for 102 RN and 7.14 ms is required for 102 H. Next, in the authentication phase we require the execution time of 6.33 ms and 5.43 ms, respectively, to perform all of the cryptographic modules, such as random number generations, exclusive-or operations and the SHA-3 (512-bits) hash function, during a normal communication session. Our implementation presents that the computation cost will be majorly dominated by the SHA-3 hash function as the execution time of the random number generation and exclusive-or operation are comparatively slight. The execution time to perform all of the SHA-3 functions required in our proposed system takes around 94% of the total computation cost. It is obvious that the SHA-3 function may become a bottleneck in terms of the performance points when the scale of the network becomes larger. Note that, in our system implementation, the input bit sequences of SHA-3 function are 192 bits, 288 bits, 800 bits, 896 bits, 992 bits and 1728 bits.

Conclusions

In this paper, we present an efficient IoT-based nursing-care service system to support the caregiver (such as nurse/doctor/administrator) to provide better quality in nursing care activities. In consideration of the trade-off between system security and computation efficiency, we adopt lightweight cryptographic modules as the major data protection technique in the communication procedures of our proposed nursing-care support system. A demo system is implemented as a proof of concept to show the practicability of the proposed method in which a reasonable and user-acceptable computation cost, i.e. at most 6.33 ms, is presented. Moreover, based on the analysis we conducted, the security robustness of the proposed nursing-care support system is guaranteed. In brief, we argue that our proposed system is very suitable for IoT-based environments and will be a highly competitive candidate for the next generation of nursing-care service systems.

Abbreviations

IoT: Internet-of-Things; BLE: Bluetooth Low Energy; NFC: near-field communication; RFID: radio frequency identification; SHA: Secure Hash Algorithm; 6LoWPAN: IPv6 over Low-Power Wireless Personal Area Networks; ASSET: adaptive security for smart Internet of Things in e-health; BSN: body sensor networks.

Declarations
Authors' contributions
CFC and KHY wrote the paper; KHY and FMH conceived and designed the proposed algorithm. All authors read and approved the final manuscript.

Author details
[1] Department of Information Management, National Dong Hwa University, Hualien 97401, Taiwan, ROC. [2] Physical Education Center, National Dong Hwa University, Hualien 97401, Taiwan, ROC.

Acknowledgements
Not applicable

Competing interests
The authors declare that they have no competing interests.

Consent for publication
Not applicable

Funding
Publication of this article was support in part by the Academia Sinica, in part by the Taiwan Information Security Center, and in part by the Ministry of Science and Technology, Taiwan under Grants MOST 105-2221-E-259-014-MY3, MOST 105-2221-E-011-070-MY3, MOST 105-2923-E-182-001-MY3, and MOST 107-2218-E-011-012.

References
1. Dang QH. Secure Hash Standard (SHS). NIST FIPS 180-4; 2015. https://csrc.nist.gov/csrc/media/publications/fips/180/4/final/documents/fips180-4-draft-aug2014.pdf. Accessed 11 Apr 2018.
2. Dworkin MJ. SHA-3 standard: permutation-based hash and extendable-output functions. NIST FIPS-202; 2015. https://csrc.nist.gov/csrc/media/publications/fips/202/final/documents/fips_202_draft.pdf. Accessed 11 Apr 2018.
3. Jara AJ, Zamora MA, Skarmeta AF. Knowledge acquisition and management architecture for mobile and personal Health environments based on the Internet of Things. In: Proceeding of the 11th IEEE international conference on trust, security and privacy in computing and communications; 2012. p. 1811–8.
4. Berhanu Y, Abie H, Hamdi M A test bed for adaptive security for IoT in eHealth. In: Proceeding of the international workshop on adaptive security; 2013. Article No. 5.
5. ASSET—Adaptive security for smart internet of things in eHealth. http://asset.nr.no/asset/index.php/ASSET_-_Adaptive_Security_for_Smart_Internet_of_Things_in_eHealth. Accessed 11 Apr 2018.
6. Torjusen AB, Abie H, Paintsil E, Trcek D, Skomedal Å. Towards run-time verification of adaptive security for IOT in eHealth. In: Proceeding of the 2014 European conference on software architecture workshops; 2014. Article No. 4.
7. Bello O, Zeadally S. Intelligent device-to-device communication in the internet of things. IEEE Syst J. 2016;10(3):1172–82.
8. Gope P, Hwang T. BSN-care: a secure IoT-based modern healthcare system using body sensor network. IEEE Sens J. 2016;16(5):1368–76.
9. Gope P, Hwang T. Untraceable sensor movement in distributed IoT infrastructure. IEEE Sens J. 2015;15(9):5340–8.
10. Yao X, Han X, Du X, Zhou X. A lightweight multicast authentication mechanism for small scale IoT applications. IEEE Sens J. 2013;13(10):3693–701.
11. Nyberg K. Fast accumulated hashing. In: Proceeding of the 3rd fast software encryption workshop; 1996. p. 83–7.
12. Ning H, Liu H, Yang LT. Aggregated-proof based hierarchical authentication scheme for the internet of things. IEEE Trans Parallel Distrib Syst. 2015;26(3):657–67.
13. Hernández-Ramos JL, Pawlowski MP, Jara AJ, Skarmeta AF, Ladid L. Toward a lightweight authentication and authorization framework for smart objects. IEEE J Sel Areas Commun. 2015;33(4):690–702.
14. Kawamoto Y, Nishiyama H, Kato N, Shimizu Y, Takahara A, Jiang T. Effectively collecting data for the location-based authentication in internet of things. IEEE Sens J. 2017;11(3):1403–11.
15. Cirani S, Picone M, Gonizzi P, Veltri L, Ferrari G. IoT-OAS: an OAuth-based authorization service architecture for secure services in IoT scenarios. IEEE Sens J. 2015;15(2):1224–34.
16. Cha SC, Yeh KH, Chen JF. Toward a Robust Security Paradigm for Bluetooth Low Energy-Based Smart Objects in the Internet-of-Things. Sensors. 2017. https://doi.org/10.3390/s17102348.
17. Burrows M, Abadi M, Needham R. A logic of authentication. ACM Trans Comput Syst. 1990;8(1):18–36.
18. Yeh KH, Tsai KY, Hou JL. Analysis and design of a smart card based authentication protocol. J Zhejiang Univ Sci C. 2013;14(12):909–17.
19. The Bouncy Castle Crypto APIs. https://www.bouncycastle.org/. Accessed 11th Apr 2018.

6

"You can tell by the way I use my walk" Predicting the presence of cognitive load with gait measurements

Pritika Dasgupta[1] (ID), Jessie VanSwearingen[2] and Ervin Sejdic[3*]

*Correspondence:
esejdic@ieee.org
[3] Department of Electrical
and Computer Engineering,
Swanson School
of Engineering, University
of Pittsburgh, Pittsburgh,
PA, USA
Full list of author information
is available at the end of the
article

Abstract

Background: There is considerable evidence that a person's gait is affected by cognitive load. Research in this field has implications for understanding the relationship between motor control and neurological conditions in aging and clinical populations. Accordingly, this pilot study evaluates the cognitive load based on gait accelerometry measurements of the walking patterns of ten healthy individuals (18–35 years old).

Methods: Data points were collected using six triaxial accelerometer sensors and treadmill pressure reports. Stride and window extraction methods were used to process these data points and separate into statistical features. A binary classification was created by using logistic regression, support vector machine, random forest, and learning vector quantization to classify cognitive load vs. no cognitive load.

Results: Within and between subjects, a cognitive load was predicted with accuracy values ranged of 0.93–1 by all four models. Various feature selection methods demonstrated that only 2–20 variables could be used to achieve similar levels of accuracies.

Conclusion: Coupling sensors with machine learning algorithms to detect the most minute changes in gait patterns, most of which are too subtle to identify with the human eye, may have a remarkable impact on the potential to detect potential neuromotor illnesses and fall risks. In doing so, we can open a new window to human health and safety prevention.

Keywords: Gait, Cognitive load, Gait classification, Stride extraction, Machine learning, Logistic regression, Support vector machine, Random forest, Learning vector quantization

Background

A person's way of walking, or gait, is a congenital human function, but the significance of the way a person walks is often overlooked [1, 2]. Research has shown that the quality of gait can deteriorate with multi-tasking, illness, and age [1, 2]. In fact, the health status of individuals can be recognized by the way they walk [3, 4]. Since gait is heavily dependent on the brain, nerves, and muscles, cognitive tasks performed in concert with walking can change the gait pattern [5]. Thus, for older adults, maintaining balance and stability while walking often requires additional attention [2, 6].

In addition to motor actions, we often think or solve problems [7]. This behavior is an example of dual tasking, which is defined to be the "concurrent performance of two tasks that can be performed independently, measured separately and have distinct goals" [5, 7–9]. Dual-task walking, particularly when a cognitive task is added during walking, can lead to reduced performance in gait quality and can result in cognitive-motor interference (CMI) [5]. CMI results in a subtle cognitive impact upon the gait of healthy and younger adults. Clinically, we can assess the effect of the addition of a secondary cognitive task to the motor activity via reactions to different stimuli (e.g., colors and sounds), manoeuvering through obstacles, using a cell-phone, or counting backward [7].

Cognitive difficulties have been associated with gait changes. For example, disorders of gait have been related to the cognitive problems among those with mood disorders (e.g., depression), dementia-related illnesses (e.g., Parkinson's and Alzheimer's diseases), and other motor-cognitive disorders. While age-related cognitive dysfunction such as a decreased ability to appropriately allocate attention may be mild, this additional cognitive load has been related to gait problems. In some of these older adults, the cognitive load has been shown to increase fall risk [7, 9, 10]. For older adults in the United States, falls have devastating consequences and are costly, accounting for a significant number of emergency department visits [11].

However, collection and consolidation of the multiple gait characteristics of human gait can be time-consuming and costly to analyze; these are often captured via wearable sensors such as uni-axial gyroscopes and accelerometers, or other equipment such as pressures sensors, video capture, and other equipment [2, 6, 12]. Many of these instruments generate an immense amount of data per observed person [13, 14]. For instance, captured through accelerometer sensors placed on the chest, back or limbs, gait accelerometry data results in thousands of x–y–z kinematic coordinates through time [15–17]. However, gait analysis via wearable sensors is a popular and inexpensive method, with regard to cost and availability of portable sensors [18]. Wearable sensors are a suitable alternative to larger laboratory equipment, because they still provide many benefits to clinical prognosis, diagnosis, and treatment [18].

Data processing methods such as acceleration signal processing and machine learning can help alleviate the stress of manipulating such large datasets. For example, machine learning has been used with different speeds of walking, foot switches, or other forms of walking combined with clinical outcomes (i.e., Parkinson's disease) [13, 19, 20]. However, there has been limited research in machine learning for cognitive load classification using gait observations [7–9]. As a first step, it can be particularly useful to do preliminary studies on healthy and young adults in order to establish a basic level of biomechanical movement [21–23]. For example, Mannini et al. state that understanding human physical activity can have future implications in movement technology (including robotics) that could impact the elderly [21]. Thus, differentiating gait qualities due to cognitive load in healthy and young adults can help us prepare a baseline measurement that can be used to assess gait patterns in a more substantial elderly and ill population.

Through the use of machine learning, researchers can harness this data to predict an individual's gait patterns, cognitive overload, and fall risk [2, 20, 24, 25]. Commonly used algorithms include supervised learning approaches such as logistic regression (LR) [24, 26], support vector machine (SVM) [27, 28], random forest (RF) [29], and k-nearest neighbors (KNN) [30]. However, KNN is negatively affected by increased dimensionality, which is disadvantageous for processing the many features extracted from gait acceleration signal data. An alternative algorithm that bypasses this disadvantage is learning vector quantization (LVQ), a neural network machine learning algorithm, which operates similarly to KNN due to it being a precursor to the nearest neighbor method [31]. There are many studies which have compared SVM with LR, RF, and KNN algorithms; however, there are not many studies that have differentiated between cognitive states using machine learning, particularly with the LR, SVM, RF, and LVQ methods.

The purpose of this study is to demonstrate a range of common machine learning methods that can accurately detect changes in gait with cognitive load in healthy adults. We plan to use a mobile gait data acquisition of stride characteristics of healthy human participants during walking with and without an added cognitive task, and machine learning algorithms, with cross-validation, to describe and classify cognitive status of walking.

Methods

Study design and procedures

This study is a prospective cohort study with a repeated-measurements design, where subjects were their own controls. Participation in the study consisted of two sessions in which each session had five stages: affixing the sensors (10–20 min), first walking trial (10 min), a rest period (10–20 min), second walking trial (10 min), and sensor detachment (10 min). The two sessions of this study for each participant occur with at least 48 h between each session. Each session lasted for about 75–90 min.

The walking trials were performed on a treadmill which rested on top of a flat non-compliant surface. The treadmill captured pressure data [in units of Newton force (N)]. The treadmill was set at a steady pace of 2.2 mph, which is 0.98 m/s which is less than the usual adult walking speed of 1.2–1.3 m/s. This treadmill speed was chosen so that all participants would be at a comfortable, and slower than usual walking speed to be closer to a speed common among community-dwelling older adults and persons with walking difficulties. The first walking trial consisted of walking, while the second walking trial consisted of walking while counting backwards from 10,000 in increments of 7, an arithmetic task that is mentally involved [32, 33]. We will refer to the first walking trial as normal walking, while referring to the second walking trial as walking under cognitive load.

Participants were affixed with six wGT3X-BT triaxial accelerometer sensors (produced by ActiGraph LLC, Ford Walton Beach, Florida, USA) located on their chest, bilateral ankles, wrists, and lower back (Fig. 1). These sensors captured linear accelerations (in units of $\frac{m}{s^2}$) at a frequency of 80 Hz from the x, y, and z directions which correspond to the mediolateral (ML), vertical (V), and anteroposterior (AP) directions. Overall, each sensor relayed 48,000 data points for each walking trial. The sensors were all clinically accepted monitoring devices and presented minimal risk to the subjects.

All recorded data did not include any personal identifying information. This study's data collection and analysis was approved by the University of Pittsburgh Institutional Review Board for ethical conduct and participant safety. All data processing and analysis was done in R (versions 3.3.1–3.4.0) [34]. Machine learning methods were implemented using R's caret package [35].

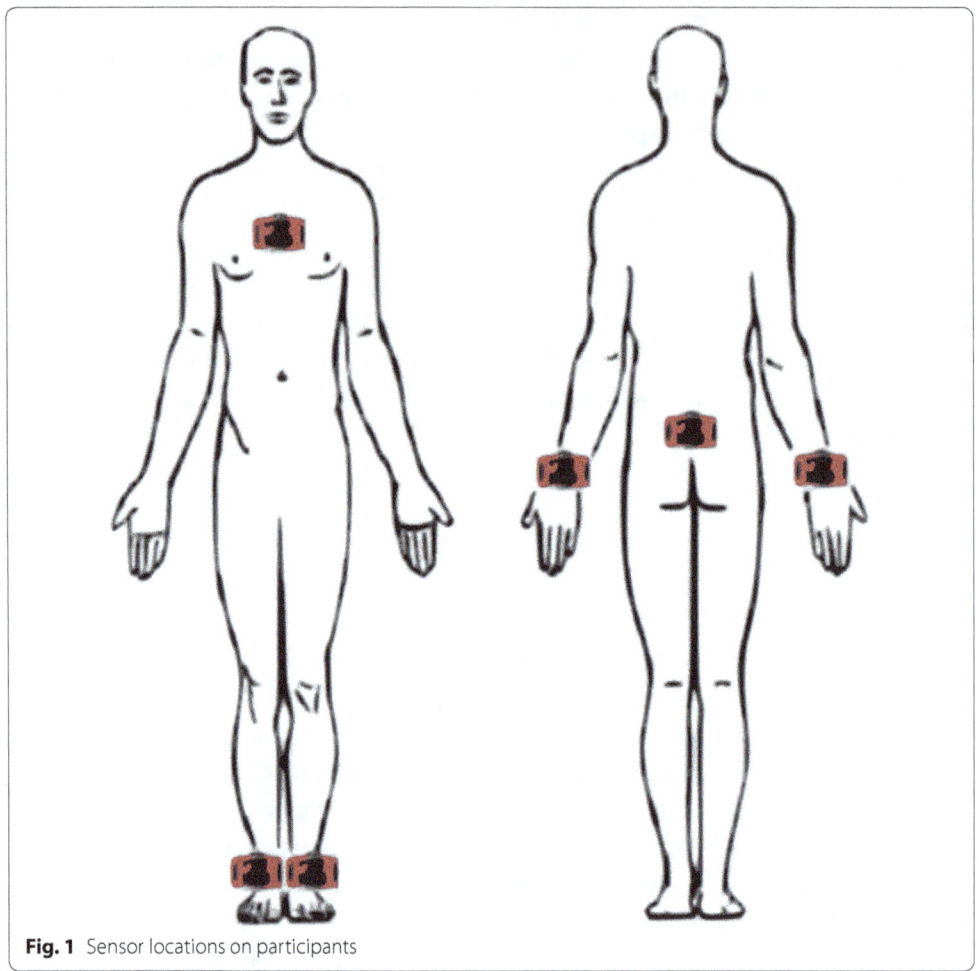

Fig. 1 Sensor locations on participants

Table 1 Demographic characteristics of study participants

Characteristic	Mean (SD) or N (%)
Age	21.40 (4.38)
Gender = male	5 (50 %)
Height (m)	1.72 (0.09)
Weight (kg)	66.36 (8.41)
BMI	22.87 (1.65)

Subjects

Volunteers were recruited from the Pittsburgh area in Pennsylvania, USA. Participation in the study was entirely voluntary and could be discontinued at any time. Acceleration and treadmill pressure data were collected from ten healthy volunteers (18–35 years old) at the University of Pittsburgh. Other than age and a subjective perception of healthiness, no other screening criteria were included in the volunteer recruitment. Demographic characteristics, such as gender, height (m), and weight (kg) were taken at the time of the study. Body mass index (BMI) was also calculated (Table 1).

Age, gender, and BMI have been known to affect biological outcomes; thus, they are included as covariates in our statistical models.

Data processing

For each subject, a total of 24 raw acceleration datasets were obtained; however, only the 12 raw datasets from session 1 were used due to the participants having a significant difference in cognitive task accuracy, the ratio of number of correct responses (of counting backwards from 10,000 in increments of 7) out of all responses, from session 1 to 2 (paired t-statistic $= -2.94$, p-value $= 0.017$), likely due to a training effect. First, a binary (0/1) variable called "cognitive load" was created to represent each trial, where "cognitive load" would be set to 1 (trial 2) and 0 otherwise (trial 1). Second, for each trial, the 6 raw datasets were compiled into one compiled raw dataset. Third, three data transformations were done on these datasets to obtain a stride dataset and two observation window datasets. Fourth, for each transformation, the datasets for trial 1 and trial 2 were combined. This resulted in a total of 30 datasets (3 processed datasets per subject in session 1) (Fig. 2).

Stride extraction

Acceleration signal based stride extraction is a clinically useful tool to evaluate cognitive changes while walking concerning the gait cycle [36]. Successive heel strikes and toe-offs on the same side have been used to define stride, stance, and swing intervals [37]. Heel strike and toe-off events were extracted from the lower back sensor using the algorithm outlined in Sejdić et al. [38]. Local minima points in the V direction correspond to toe-offs, and local minima points in the AP direction are related to heel strikes [38]. The order of which foot initially made the first step was found by taking the average of the first 10 ms of acceleration in the ML direction [38]. If there was positive mean value, the right foot came first; otherwise, the left foot came first [38].

A condensed description of this algorithm consists of the following steps: (1) pre-processing gait accelerometry signals via median filters, (2) determine which foot came first by calculating the average of the mediolateral acceleration signals in the first 10 ms, (3) capture heel strike events via the local minima points in the AP signals, and (4) capture toe-off events via the local minima points in the V signals.

Foot pressure based treadmill reports were used to validate sensor-derived heel strike and toe-off events. When the subject pushes their foot off the ground, the pressure reduces to 0 N; conversely, when the subject heel strikes, the pressure increases. Using a technique from Truong et al. [39], an on/off filter was created to detect heel strike and toe-off events for each foot. This on/off filter produced time points for when the foot was on or off the ground. These time points were matched up to the stride events for validation of Sejdić et al.'s method. An example of stride extraction and validation is shown in Fig. 3.

Window extraction

A more straightforward alternative to stride extraction is the creation of sliding observation windows, by partitioning sensor signals into smaller time segments [40]. Partitioning into sliding observation windows is a conventional technique in activity monitoring

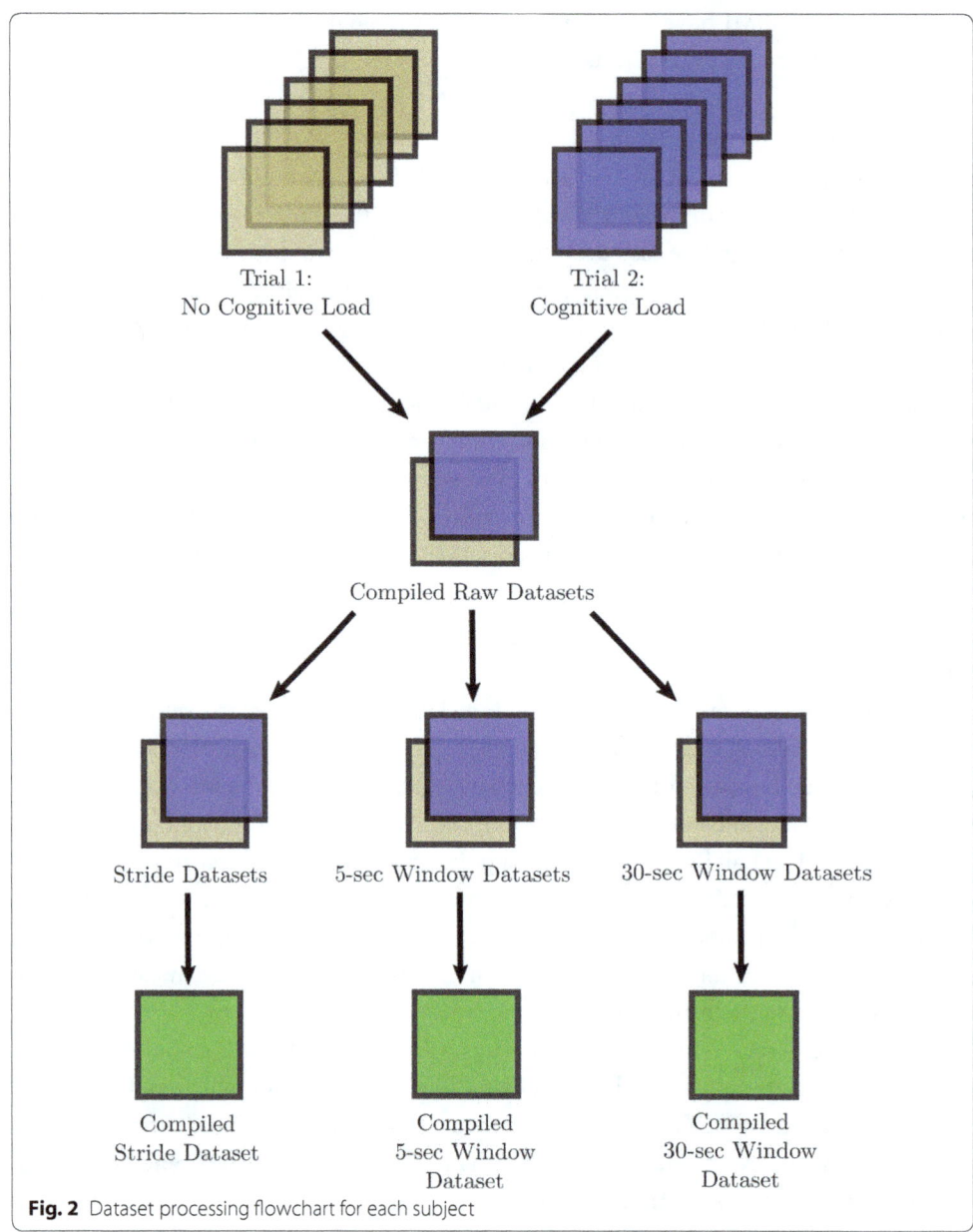

Fig. 2 Dataset processing flowchart for each subject

and machine learning analysis of accelerometer data [21, 40]. For each subject and each sensor, two datasets were formed: (1) acceleration signals were split up in 5-s intervals with 50% overlap, and (2) acceleration signals were split up in 30-s intervals with 50% overlap. An example of how window extraction is done is shown in Fig. 4.

Errant data removal

Data points were removed to eliminate any start-up, pausing, and ending effects to allow the subject to become familiar with getting on and off the treadmill. For the stride datasets, the first fifty strides and the last five strides were removed. Outliers based on average stride time were removed using the interquartile range (IQR) rule [41]. Comparing

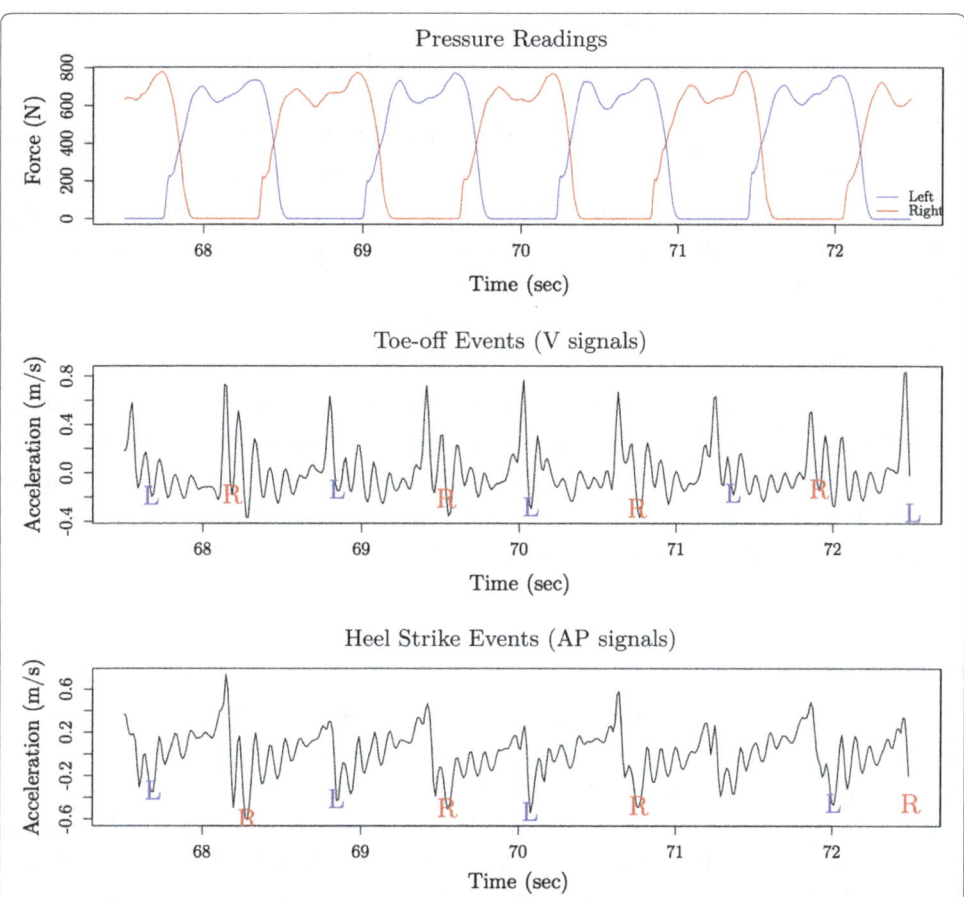

Fig. 3 Stride extraction sample. Subject 1's toe-offs and heel-strikes, which were determined from foot pressure recordings during treadmill walking and acceleration signal recorded using the lower back sensor. The top graph shows the pressure readings from the treadmill, the middle graph are the V acceleration readings, and the bottom graph are the AP acceleration readings. Red and blue colored lines and labels depict the right and left foot respectively

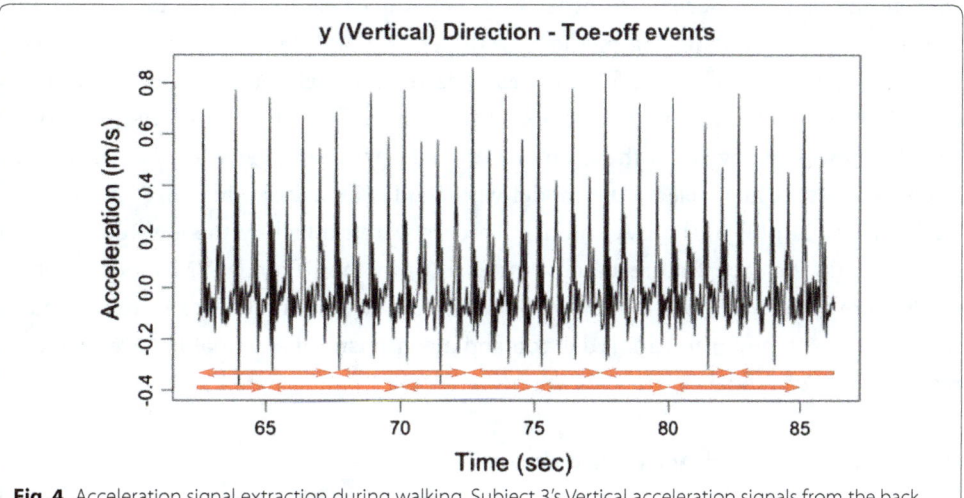

Fig. 4 Acceleration signal extraction during walking. Subject 3's Vertical acceleration signals from the back sensor. 5 s observation windows with 50% overlap are shown

the remaining strides is a statistical challenge because strides are characterized by vectors of unequal lengths, and there are hundreds of strides to compare.

Thus, an ANOVA test [null hypothesis: no statistical difference in means of stride features (see below in "Feature extraction" section) between strides] was done before and after errant data removal and resulted in a p-value of 2.2e−16 and 1, respectively. These results indicate that errant data removal was done correctly because the processed strides were not distinguishable from the other. For the observation window datasets, the first and last minute of the data were removed.

Stride time differences

Additionally, within each participant's stride extracted data (n = 1351 to 1870 strides), an ANOVA test was done between the stride time distributions between the trials. The null hypothesis of this ANOVA test was "no statistical difference in the means of stride times between strides". Typically, a large sample size will lead to statistical significance; in this case, the stride times are expected to be significantly different.

Features

Feature extraction

In order to capture descriptors from each stride and window, it is common in gait studies [24, 42] to calculate descriptive statistics from strides and windows. For each direction (V, AP, and ML), twelve features were extracted from each stride and each observation window in each of the six sensors. These twelve features were the mean, standard deviation, pair-wise correlation, and pair-wise covariances of each of the three directions [43]. From the stride data specifically, times and lengths of the stance phase, swing phase, and overall stride were also extracted. As a result, a 78-feature vector described each stride whereas a 72-feature vector described each observation window.

Feature selection

High dimensionality may lead to over-fitting in machine learning analysis, and it is advantageous to reduce the number of features. First, each of the four models was run with all the features. Second, in order to select which features to include in machine learning models, the following methods were performed: (1) for all four models, a correlation matrix between each feature was constructed; highly correlated feature pairs (r > 0.75) were found and within each pair, the feature with the highest mean absolute correlation was removed; (2) the LR model ran step-wise variable selection, via a likelihood ratio test, which selects the model with the lowest Akaike information criterion (AIC) value [44]; (3) in the LVQ model, features were ranked by the absolute value of the t-statistic for each feature parameter; and (4) in the RF and SVM models, recursive feature elimination was done [45]. Some feature selection methods were utilized so that decreased time for feature reduction and a consensus of essential features could be reached for each model.

Machine learning models and evaluation

The presence of a cognitive load, represented by a binary (0/1) variable, was classified based on gait feature vectors. For binary outcomes, the following four machine learning algorithms were used: LR, LVQ, SVM, and RF. Well-known evaluation metrics were used such as accuracy, sensitivity, specificity, and area under the curve (AUC). In this report, we present the accuracy (Eq. 1) values, with their 95% confidence intervals; sensitivity, specificity, and AUC values are reported in Appendix (Figs. 8, 9, 10, 11, 12).

$$Accuracy = \frac{TP + TN}{TP + TN + FP + FN} \tag{1}$$

where TP is the number of true positives, i.e., the model identifies a cognitively loaded stride/window that was labelled as cognitively loaded; TN is the number of true negatives, i.e., the model identifies a non-cognitively loaded stride/window that was labeled as not cognitively loaded; FP is false "cognitive" load identifications; and FN is false "no cognitive load" identifications.

For each of the machine learning algorithms, we employ three different modelling strategies: within subjects, between subjects, and leave one out. Feature selection was performed after the datasets were processed for each model.

Within and between subjects

For the within-subjects model, each subject's dataset was combined within each of the three data types: strides (n = 1351 to 1870 strides), 5-s observation windows (n = 1280 windows), and 30-s windows (n = 320 windows). For the between subjects model, all subjects' datasets were combined within each of three data types: stride (n = 16,291 strides), 5-s observation windows (n = 12,800 windows), and 30-s windows (n = 3200 windows).

Datasets were split into training and test datasets; datasets were randomly split, using a seed of 7 and the sample function in R, into 80%, which was used for training each machine learning model, and 20%, which was used to evaluate the model's performance. The tenfold cross-validation method was used on each of the training sets, where the training dataset was split into ten subsets. Each subset was held out while the model was trained on all other subsets. This cross-validation process was repeated three times. The final model was chosen by the best accuracy of these runs. Then this model is run on the test dataset to validate the model.

Leave one subject out validation

A leave one subject out model was done to truly capture how well other subjects' features and data can predict an individual's cognitive load. This was done by combining all subjects' datasets for only the stride data type. Each model was trained on nine out of the ten subjects (n = 14,421 to 14,940 strides), and the model was tested on the remaining subject (n = 1351 to 1870 strides). Training and testing were done ten times, and average evaluation metrics were calculated.

Results

An equal number of adult males and females participated. The participants reported good health, which was consistent with the mean BMI derived in the good condition range (Table 1). All participants completed both study sessions.

Stride characteristics

Within each of the ten subjects, the stride time distributions were not significantly different between trials using ANOVA tests with a Tukey's posthoc test (Table 2). Earlier, we indicated that we expected the stride times to be significantly different due to a large sample size of strides per subject. However, the test indicated differences in stride times cannot be differentiated, which suggests that machine learning algorithms will have to be very sensitive when differentiating between cognitive load vs. no cognitive load.

Machine learning results

For each model, all machine learning results reported were derived from the results of the test dataset.

Within each subject, Fig. 5 depicts the accuracy values from all the models. From the strides datasets, over all subjects, the mean (95 % t-distributed CI) of the accuracy of LR is 0.998 (0.995, 1.00), LVQ is 0.999 (0.999, 1.00), RF is 0.998 (0.999, 1.00), and SVM is 0.998 (0.996, 0.999). From the 30 s. window datasets, over all subjects, the mean (95% t-distributed CI) of the accuracy of LR is 1.0 (1.0, 1.0), LVQ is 0.97 (0.93, 1.0), RF is 1.0 (1.0, 1.0), and SVM is 1.0 (1.0, 1.0). From the 5 s. window datasets, over all subjects, the mean (95% t-distributed CI) of the accuracy of LR is 0.997 (0.993, 1.00), LVQ is 0.97 (0.93, 1.0), RF is 0.999 (0.998, 1.00), and SVM is 0.996 (0.994, 0.998).

Common influential features that appear for most participants in these models are the means and standard deviations of the x, y, and z directions for the back sensor and ankle sensors.

Between all the subjects, Fig. 6 depicts the accuracy values from all the models. Common influential features that appear for most participants in these models were similar to the within-subjects models, with the inclusion of the covariances and correlations of the x, y, and z-direction of the chest sensor.

By training on nine out of ten subjects and testing on the remaining one, Fig. 7 depicts the accuracy values from all the models for each tested subject. Over all subjects, the mean (95% t-distributed CI) of the accuracy of LR is 0.59 (0.33, 0.85), LVQ is 0.47 (0.32,

Table 2 Number of strides and stride time comparison between trials 1 and 2, for each subject

Subject	Stride no.	Stride time (s) [mean (SD)]	
		No cognitive load	Cognitive load
1	1511	1.32 (0.23)	1.29 (0.16)
2	1688	1.17 (0.12)	1.20 (0.04)
3	1592	1.28 (0.18)	1.29 (0.07)
4	1705	1.22 (0.17)	1.22 (0.07)
5	1810	1.13 (0.10)	1.12 (0.06)
6	1721	1.18 (0.12)	1.20 (0.09)
7	1870	1.10 (0.11)	1.11 (0.08)
8	1565	1.22 (0.07)	1.25 (0.03)
9	1478	1.24 (0.16)	1.24 (0.04)
10	1351	1.24 (0.07)	1.31 (0.09)

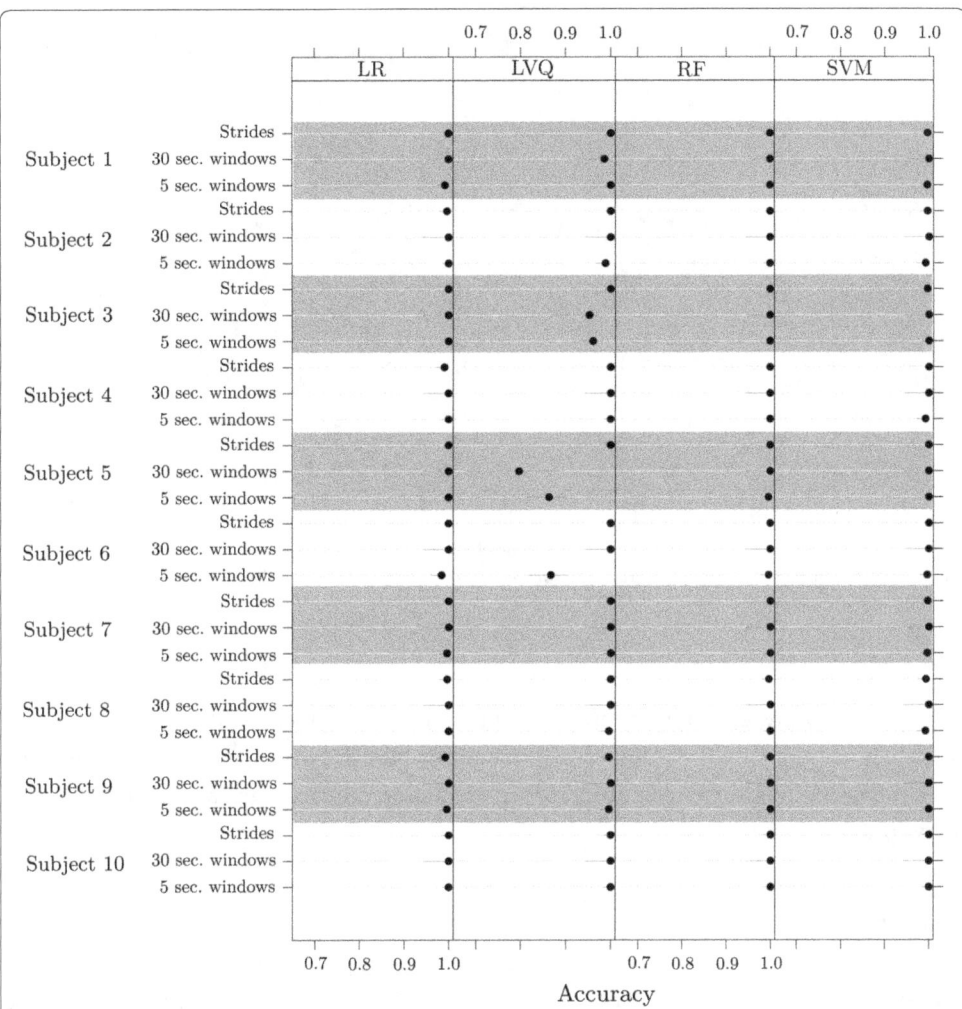

Fig. 5 Within subjects results. Accuracy values (95% confidence intervals in blue) using the "within subjects" model using all three dataset types

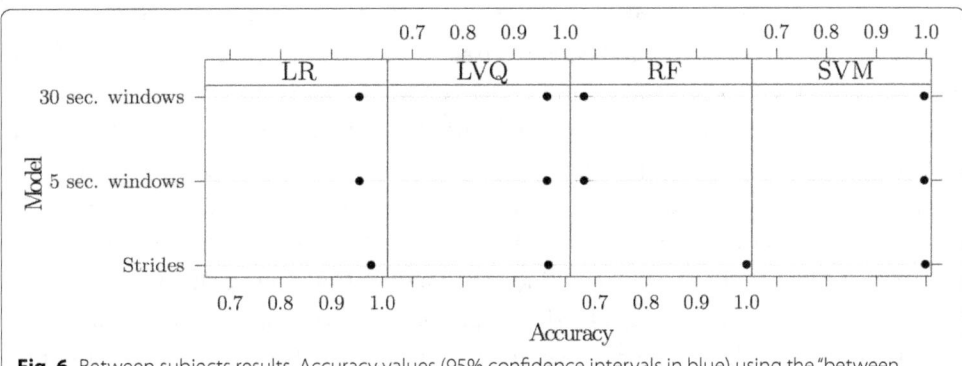

Fig. 6 Between subjects results. Accuracy values (95% confidence intervals in blue) using the "between subjects" model using all three dataset types

0.62), RF is 0.60 (0.40, 0.79), and SVM is 0.49 (0.26, 0.72). Feature reduction varied greatly for these models, likely due to the inconsistency of accuracy results from one subject to the other.

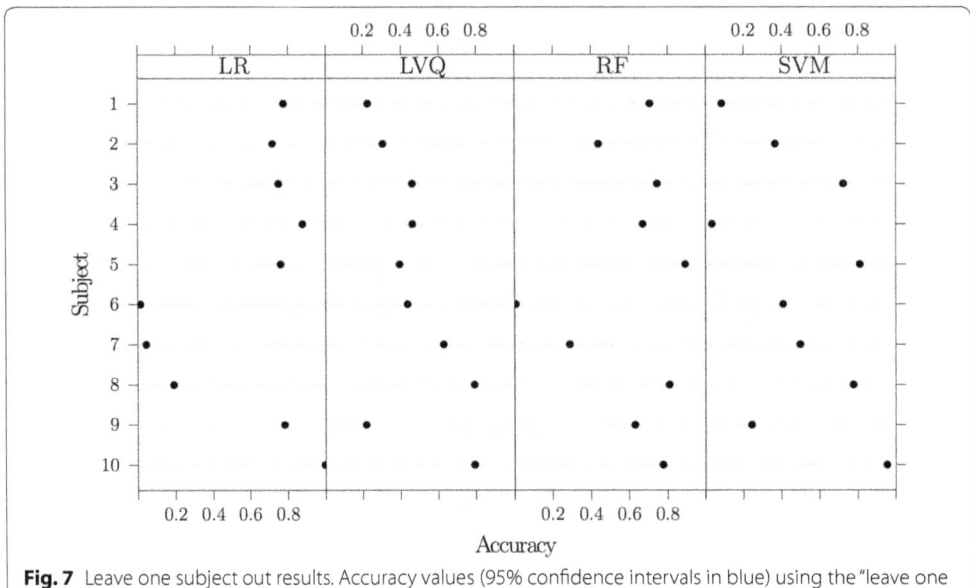

Fig. 7 Leave one subject out results. Accuracy values (95% confidence intervals in blue) using the "leave one subject out" model

We chose to present only accuracies, which is measured by dividing the number of correct predictions by the number of predictions [46]. However, accuracy was not the only evaluation metric; we calculated other evaluation metrics, such as sensitivity, specificity, and AUC values (Figs. 8, 9, 10, 11, 12).

Discussion

This study's purpose was to implement signal processing on raw accelerometry gait data and evaluate the performance of common machine learning methods, LR, RF, LVQ, and SVM, to classify the presence of cognitive load in ten healthy adults.

Our machine learning models consisted of within-subject, between-subject, and leave one subject out classification. Within-subject classification is clinically vital for precision (individual-specific) medicine because it can help health-care practitioners and researchers more accurately predict which treatment strategy will work for a particular person, without regard for differences in individuals [47, 48]. Leave one subject out classification is also relevant for precision medicine, and our attempt with ten individuals is an example of how other people's gait patterns can be used to train a machine learning model to test on an individual. For example, model training on similar patients can be done to help decide a treatment strategy for a patient. Conversely, between-subject classification only assesses an overall baseline of gait patterns over a population.

Not all four machine learning algorithms performed with consistently high accuracy, among all three modelling approaches. In the within subjects model, The results show that the LR, RF, and SVM algorithms recognized cognitive load accurately (with accuracy > 0.93) for both strides and windows. In particular, RF and SVM were consistently strong performers amongst all data transformations, whereas LVQ performance varied. As seen in Fig. 5, subjects 3, 5, and 6 had varied LVQ accuracy values for window datasets. Upon further inspection, feature selection with LVQ for the window datasets was not able to reduce the number of features. Feature selection was able to reduce the number of features in LR, RF, and SVM, but not in LVQ which could mean that there are too many features in the model, which in turn could be picking up random noise instead of the actual trend leading to erratic results. In the between subjects model, LR and SVM algorithms performed well, and LVQ and RF were weak performers. LVQ being a weak performer for both models is a surprising result because LVQ has proven to be a reliable gait predictor in the past [49].

When we assessed the leave one subject out approach, many of the accuracy values were lower than random chance. It was clear that due to gait differences between the individuals, cognitive load prediction varied considerably. Even though the data points for this approach were from the stride dataset and the sample size of these datasets were large when put together for the training set, the result of the machine learning algorithms suffered from having a limited overall sample size of ten individuals. Each individual, albeit being similar in basic demographic characteristics, presumably have different gait patterns. A low accuracy on the leave on out approach in this study may require the following in order to get better results: (1) a high sample size of participants, (2) more detailed demographic and medical characteristics of each participant, (3) different signal processing techniques, and/or (4) varied feature selection approaches by either choosing different features or changing the feature reduction technique. Since this approach has precision medicine implications and can be a bridge between the results of the between subjects approach and the within subjects approach, it is relevant to our discussion. We hope that with a higher sample size, the leave one subject out approach will produce higher accuracy results; however these results do not discount the high accuracy results of the within subjects approach.

Moreover, in machine learning, overfitting can become an issue, when there is low training error and high generalization error. In particular, overfitting can occur when there is a higher model complexity (too many parameters) or a small sample size. To overcome overfitting in our models, we performed feature selection. Feature selection resulted in 2–20 features per model used. Common features included ML signals in back acceleration, stride time and length, V signals in chest acceleration, all signals in both wrist acceleration, and all signals in both ankle acceleration. This suggests that collecting data from one sensor, such as the back, chest, wrists, or ankles, will be sufficient to determine differences in cognitive load. We also evaluated overfitting by including sensitivity, specificity, and AUC values in our Appendix. Specificity and sensitivity values were consistently above 0.90 for all algorithms and for the within and between subjects models.

Overall, these findings are particularly compelling because the treadmill is a "dedicated pacer" and the subjects' were very similar to each other in both demographic and gait characteristics, meaning that the machine learning algorithms, particularly in the within subjects model, were sensitive enough to pick up the cognitive load.

Statistical signal processing is a time-consuming task, but stride extraction was performed successfully and validated by the treadmill pressure reports for each subject (Fig. 3). Even though stride extraction was done for each subject, window extraction was also done to compare how predictive machine learning algorithms can be without respect to gait cycle events. In the within subjects model, accuracy values between windows and strides were easily comparable, with the exception of the LVQ results. Due to the time and memory consuming task of signal processing, window extraction is preferable [50].

The data collection and analysis process had multiple limitations. First, the use of the treadmill is not reflective of overground walking, which could lead to poor generalizability in using this technique and these models for further gait analysis. However, after treadmill familiarization, young and unimpaired individuals, particularly healthy 18–28 year olds, have been shown to have negligible differences or no differences at all between overground and treadmill gait parameters and leg kinematics [51–53]. Second, the low subject sample size (n = 10) may have led to poor accuracy values for the leave one subject out method. However, the stride/window data used for the leave one subject out method had sample sizes that were sufficiently large (please see "Leave one subject out validation" section). Even though there may be low generalizability of the gait data acquired by these subjects, future work in this area has the potential to be remedied by a higher subject sample size and a more extensive set of acceleration signals. Third, the cognitive task accuracy was not included in the machine learning models. Cognitive task accuracy or other measures of how much cognitive load a person had during the walking tasks could have biased the accelerometer data. For example, an individual could have had more cognitive loading than another individual. However, this issue is partially alleviated due to the fact that the machine learning algorithms classified between the presence of cognitive load versus no cognitive load. We can see that this bias could have possibly contributed to the poor leave one subject out results.

These findings contribute to the era of personalized medicine, which aims to improve the potency of therapies at an individual level. To provide personalized medicine in gait research, we must identify appropriate gait features that reliably predict differences in cognitive load. Sensors are relatively inexpensive, and they can be used by healthcare professionals and patients alike to track gait patterns. However, the analysis of raw sensor data can be challenging without an appropriate tool; window separation of these acceleration signals and machine learning analysis fills this gap. The implications of this analysis can lead to the creation of a medical device that can be used in not only the clinic but also in athletics.

Marrying sensors with machine learning algorithms has the potential to be an early indicator of disease or fall risk, especially in older adults. Ageing has been known to contribute to gait difficulties; even more so, gait impairment may be due to an intrinsic disease [54]. In fact, those with gait disorders are more likely to have dementia symptoms than those without gait disorders. We can endeavour to identify these pre-clinical changes in gait that are associated with a cognitive load for diagnosis or aid in other therapies, such as combative cocktail therapies for neuromotor illnesses [55, 56].

Also, there is a quiescent need to identify cognitive status via different types of gait disorders. While this study discriminated between two types of cognitive statuses, this analysis has the potential to describe the overall spectrum of cognitive disorders. Granted, this analysis will heavily depend on the population being studied, but separating the several neurological causes for gait disorders, such as stroke, ataxia, and Parkinson's disease [54], can help add to the stock of knowledge of gait and cognitive faculties.

Thus far, we have determined cognitive status with previously collected gait data. Future work consists of predicting forthcoming continuous clinical outcomes from gait data, predicting the acceleration signals of the next stride using forecasting algorithms such as hidden Markov models and autoregressive integrated moving average, along with other gait parameters, such as cadence.

Conclusion

In our study, we found that just by combining machine learning technology with advanced signal processing methods on sensor data, we were able to detect cognitive states accurately. Our features were derived from gait accelerometry signals from healthy adult subjects. Moreover, we determined that using window extraction methods and selecting gait features from data from only one sensor is satisfactory. These results have an array of clinical implications, namely in personalized medicine and early detection of neuromotor diseases. This successful pilot study provides a clear path for expansion, due to its explicit findings and its potential application towards a much larger population consisting of adults across a broad range of ages. Lastly, this is particularly provocative due to the fusion of machine learning on gait accelerometry data that could lead to the prediction of disease or gait instability in older adults.

Authors' contributions
PD analyzed and interpreted the gait data and was the primary author of the manuscript. JV provided the clinical context of the results, and was a major contributor in writing the manuscript. ES was the primary investigator in this study and edited the manuscript. All authors read and approved the final manuscript.

Author details
[1] Department of Biomedical Informatics, School of Medicine, University of Pittsburgh, Pittsburgh, PA, USA. [2] Department of Physical Therapy, School of Health and Rehabilitation Sciences, University of Pittsburgh, Pittsburgh, PA, USA. [3] Department of Electrical and Computer Engineering, Swanson School of Engineering, University of Pittsburgh, Pittsburgh, PA, USA.

Acknowledgements
Not applicable.

Competing interests
The authors declare that they have no competing interests.

Consent for publication
Not applicable.

Funding
This research is funded by the National Library of Medicine (National Institutes of Health) (Grant Reference Number: 4T15LM007059-30) and, in part, by the Pittsburgh Claude D. Pepper Older Americans Independence Center (NIA P30 AG 024827).

Appendix

See Figures 8, 9, 10, 11, and 12.

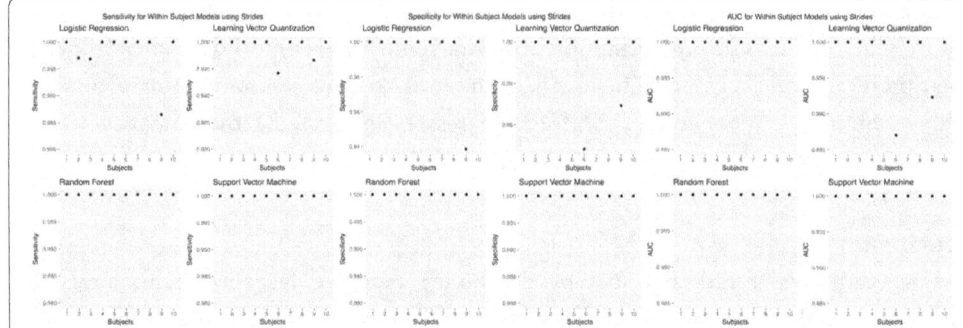

Fig. 8 Sensitivity, specificity, and AUC values for the within subjects (stride dataset) model. Sensitivity values using the "within subjects" model using the stride dataset

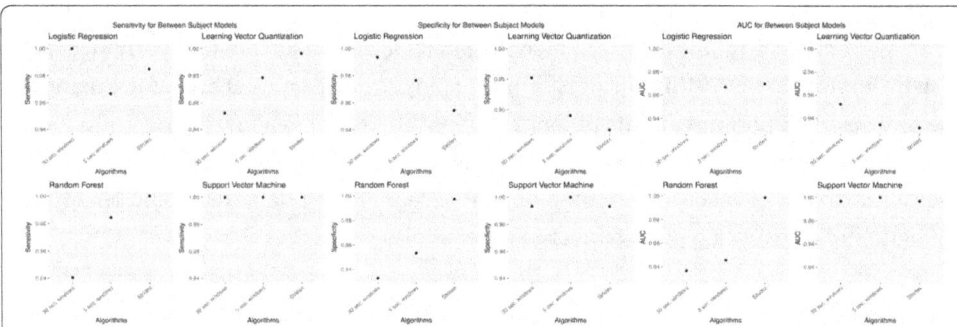

Fig. 9 Sensitivity, specificity, and AUC for the between subjects model. Sensitivity values using the "between subjects" model

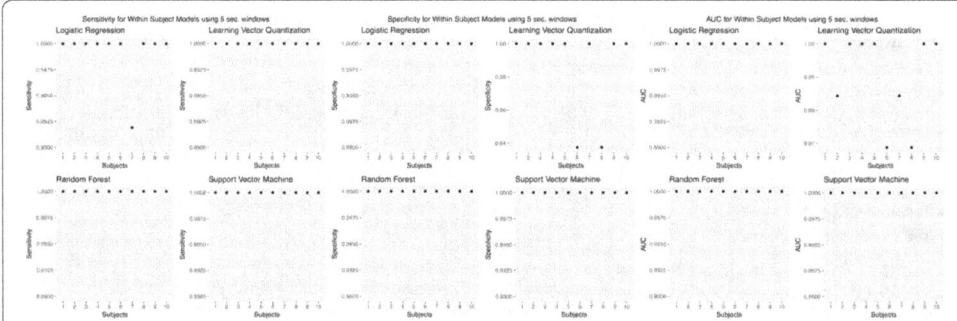

Fig. 10 Sensitivity, specificity, and AUC for the within subjects (5 s windows) model. Sensitivity values using the "within subjects" model using 5 s windows

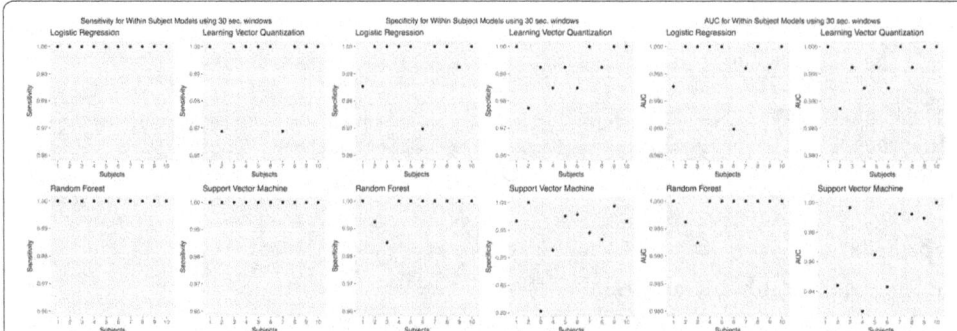

Fig. 11 Sensitivity, specificity, and AUC values for the within subjects (30 s windows) model. Sensitivity values using the "within subjects" model using 30 s windows

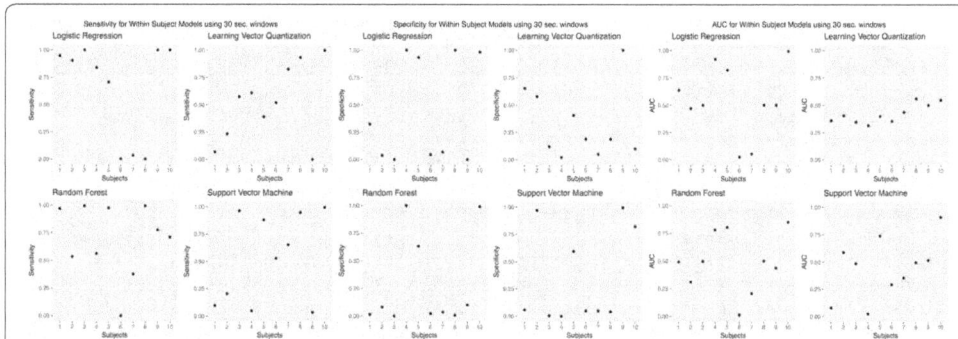

Fig. 12 Sensitivity, specificity, and AUC values for the leave one subject out model. Sensitivity values using the "leave one subject out" model using the stride dataset

References

1. Nutt J, Marsden C, Thompson P. Human walking and higher-level gait disorders, particularly in the elderly. Neurology. 1993;43(2):268–268.
2. Prakash C, Kumar R, Mittal N. Recent developments in human gait research: parameters, approaches, applications, machine learning techniques, datasets and challenges. Artif Intell Rev. 2018;49(1):1–40.
3. Studenski S, Perera S, Patel K, Rosano C, Faulkner K, Inzitari M, Brach J, Chandler J, Cawthon P, Connor EB. Gait speed and survival in older adults. JAMA. 2011;305(1):50–8.
4. Cummings SR, Studenski S, Ferrucci L. A diagnosis of dismobility—giving mobility clinical visibility: a mobility working group recommendation. JAMA. 2014;311(20):2061–2.
5. McIsaac TL, Lamberg EM, Muratori LM. Building a framework for a dual task taxonomy. BioMed Res Int. 2015;2015:1–10. https://doi.org/10.1155/2015/591475.
6. Woollacott M, Shumway-Cook A. Attention and the control of posture and gait: a review of an emerging area of research. Gait Posture. 2002;16(1):1–14.
7. Fraser SA, Li KZ, Berryman N. Desjardins-Crepeau L, Lussier M, Vadaga K, Lehr L, Vu M, Tuong T, Bosquet L, Bherer L. Does combined physical and cognitive training improve dual-task balance and gait outcomes in sedentary older adults? Front Hum Neurosci. 2017;1:1–1. https://doi.org/10.3389/fnhum.2016.00688.
8. Hausdorff JM, Schweiger A, Herman T, Yogev-Seligmann G, Giladi N. Dual-task decrements in gait: contributing factors among healthy older adults. J Gerontol Ser. 2008;63(12):1335–43.
9. Montero-Odasso M, Muir SW, Speechley M. Dual-task complexity affects gait in people with mild cognitive impairment: the interplay between gait variability, dual tasking, and risk of falls. Arch Phys Med Rehab. 2012;93(2):293–9.
10. Montero-Odasso MM, Sarquis-Adamson Y, Speechley M, Borrie MJ, Hachinski VC, Wells J, Riccio PM, Schapira M, Sejdic E, Camicioli RM, Bartha R. Association of dual-task gait with incident dementia in mild cognitive impairment: results from the gait and brain study. JAMA Neurol. 2017;74(7):857–65. https://doi.org/10.1001/jamaneurol.2017.0643.
11. Ellis G, Marshall T, Ritchie C. Comprehensive geriatric assessment in the emergency department. Clin Intervent Aging. 2014;9:2033–43.
12. Tao W, Liu T, Zheng R, Feng H. Gait analysis using wearable sensors. Sensors. 2012;12(2):2255–83.
13. Albert MV, Kording K, Herrmann M, Jayaraman A. Fall classification by machine learning using mobile phones. PLoS ONE. 2012;7(5):36556.
14. Chan H, Yang M, Wang H, Zheng H, McClean S, Sterritt R, Mayagoitia RE. Assessing gait patterns of healthy adults climbing stairs employing machine learning techniques. Int J Intell Syst. 2013;28(3):257–70.
15. Kavanagh JJ, Menz HB. Accelerometry: a technique for quantifying movement patterns during walking. Gait Posture. 2008;28(1):1–15.
16. Bussmann J, Veltink P, Koelma F, Van Lummel R, Stam H. Ambulatory monitoring of mobility-related activities: the initial phase of the development of an activity monitor. Eur J Phys Med Rehab. 1995;5(1):2–7.
17. Mayagoitia RE, Nene AV, Veltink PH. Accelerometer and rate gyroscope measurement of kinematics: an inexpensive alternative to optical motion analysis systems. J Biomech. 2002;35(4):537–42.
18. Chen S, Lach J, Lo B, Yang G-Z. Toward pervasive gait analysis with wearable sensors: a systematic review. IEEE J Biomed Health Inform. 2016;20(6):1521–37.
19. Begg R, Kamruzzaman J. Neural networks for detection and classification of walking pattern changes due to ageing. Aus Phys Eng Sci Med. 2006;29(2):188–95.
20. Tahir NM, Manap HH. Parkinson disease gait classification based on machine learning approach. J Appl Sci. 2012;12(2):180–5.
21. Mannini A, Sabatini AM. Machine learning methods for classifying human physical activity from on-body accelerometers. Sensors. 2010;10(2):1154–75.
22. Bouten CV, Koekkoek KT, Verduin M, Kodde R, Janssen JD. A triaxial accelerometer and portable data processing unit for the assessment of daily physical activity. IEEE Trans Biomed Eng. 1997;44(3):136–47.
23. Meijer GA, Westerterp KR, Verhoeven FM, Koper HB, ten Hoor F. Methods to assess physical activity with special reference to motion sensors and accelerometers. IEEE Trans Biomed Eng. 1991;38(3):221–9.

24. Begg R, Kamruzzaman J. A machine learning approach for automated recognition of movement patterns using basic, kinetic and kinematic gait data. J Biomech. 2005;38(3):401–8.
25. Pogorelc B, Bosnić Z, Gams M. Automatic recognition of gait-related health problems in the elderly using machine learning. Multimedia Tools Appl. 2012;58(2):333–54.
26. Schumacher M, Roßner R, Vach W. Neural networks and logistic regression: Part I. Comput Stat Data Anal. 1996;21(6):661–82.
27. Suykens JA, Vandewalle J. Least squares support vector machine classifiers. Neural Process Lett. 1999;9(3):293–300.
28. Weston J, Mukherjee S, Chapelle O, Pontil M, Poggio T, Vapnik V. Feature selection for support vector machines. In: Advances in neural information processing systems. 2001. p. 668–74.
29. Breiman L. Random forests. Mach Learn. 2001;45(1):5–32.
30. Cover T, Hart P. Nearest neighbor pattern classification. IEEE Trans Inform Theory. 1967;13(1):21–7.
31. Kohonen T. Learning vector quantization. In: Self-organizing maps. Berlin: Springer; 1995. p. 175–189.
32. Logie RH, Baddeley AD. Cognitive processes in counting. J Exp Psychol. 1987;13(2):310.
33. Beauchet O, Dubost V, Gonthier R, Kressig RW. Dual-task-related gait changes in the elderly: does the type of cognitive task matter? Gerontology. 2005;51(1):48–52.
34. R Core Team. R: A language and environment for statistical computing. R Foundation for statistical computing, Vienna: R Foundation for Statistical Computing. 2016. https://www.R-project.org/.
35. Kuhn M. Caret package. J Stat Softw. 2008;28(5):1–26.
36. Brach JS, Swearingen JM, Perera S, Wert DM, Studenski S. Motor learning versus standard walking exercise in older adults with subclinical gait dysfunction: a randomized clinical trial. J Am Geriatr Soc. 2013;61(11):1879–86.
37. Gage JR, Deluca PA, Renshaw TS. Gait analysis: principles and applications. J Bone Joint Surg. 1995;77(10):1607–23.
38. Sejdić E, Lowry KA, Bellanca J, Perera S, Redfern MS, Brach JS. Extraction of stride events from gait accelerometry during treadmill walking. IEEE J Transl Eng Health Med. 2016;4:1–11.
39. Truong PH, Lee J, Kwon A-R, Jeong G-M. Stride counting in human walking and walking distance estimation using insole sensors. Sensors. 2016;16(6):823.
40. Preece SJ, Goulermas JY, Kenney LP, Howard D, Meijer K, Crompton R. Activity identification using body-mounted sensors—a review of classification techniques. Physiol Meas. 2009;30(4):1.
41. Tukey JW. Exploratory data analysis. New York: Wesley; 1977. p. 2–70.
42. Begg RK, Palaniswami M, Owen B. Support vector machines for automated gait classification. IEEE Trans Biomed Eng. 2005;52(5):828–38.
43. Sejdić E, Lowry KA, Bellanca J, Redfern MS, Brach JS. A comprehensive assessment of gait accelerometry signals in time, frequency and time–frequency domains. IEEE Trans Neural Syst Rehab Eng. 2014;22(3):603–12.
44. Akaike H. Factor analysis and AIC. Psychometrika. 1987;52(3):317–32.
45. Guyon I, Weston J, Barnhill S, Vapnik V. Gene selection for cancer classification using support vector machines. Mach Learn. 2002;46(1):389–422.
46. Baratloo A, Hosseini M, Negida A, El Ashal G. Part 1: simple definition and calculation of accuracy, sensitivity and specificity. Emergency. 2015;3(2):48–9.
47. Ashley EA. The precision medicine initiative: a new national effort. JAMA. 2015;313(21):2119–20.
48. Desmond-Hellmann S, Sawyers C, Cox D, Fraser-Liggett C, Galli S, Goldstein D, Hunter D, Kohane I, Lo B, Misteli T. Toward precision medicine: building a knowledge network for biomedical research and a new taxonomy of disease. 2011. pp. 1–142.
49. Kohle M, Merkl D, Kastner J. Clinical gait analysis by neural networks: issues and experiences. In: Proceedings of the tenth IEEE symposium computer-based medical systems. New Jersey: IEEE; 1997. p. 138–43.
50. Wang N, Ambikairajah E, Redmond SJ, Celler BG, Lovell NH. Classification of walking patterns on inclined surfaces from accelerometry data. In: 2009 16th international conference on digital signal processing. New Jersey: IEEE; 2009. p. 1–4.
51. Riley PO, Paolini G, Della Croce U, Paylo KW, Kerrigan DC. A kinematic and kinetic comparison of overground and treadmill walking in healthy subjects. Gait Posture. 2007;26(1):17–24.
52. Matsas A, Taylor N, McBurney H. Knee joint kinematics from familiarised treadmill walking can be generalised to overground walking in young unimpaired subjects. Gait Posture. 2000;11(1):46–53.
53. Lee SJ, Hidler J. Biomechanics of overground vs. treadmill walking in healthy individuals. J Appl Physiol. 2008;104(3):747–55.
54. Snijders AH, Van De Warrenburg BP, Giladi N, Bloem BR. Neurological gait disorders in elderly people: clinical approach and classification. Lancet Neurol. 2007;6(1):63–74.
55. Pantelopoulos A, Bourbakis NG. A survey on wearable sensor-based systems for health monitoring and prognosis. IEEE Trans Syst Man Cybern. 2010;40(1):1–2.
56. Muro-De-La-Herran A, Garcia-Zapirain B, Mendez-Zorrilla A. Gait analysis methods: an overview of wearable and non-wearable systems, highlighting clinical applications. Sensors. 2014;14(2):3362–94.

7

Regressing grasping using force myography: an exploratory study

Rana Sadeghi Chegani⊕ and Carlo Menon*

*Correspondence:
cmenon@sfu.ca
Menrva Research Group,
School of Mechatronic
Systems Engineering
and Engineering Science,
Simon Fraser University,
250-13450-102 Avenue,
Surrey, BC V3T 0A3, Canada

Abstract

Background: Partial hand amputation forms more than 90% of all upper limb amputations. This amputation has a notable effect on the amputee's life. To improve the quality of life for partial hand amputees different prosthesis options, including externally-powered prosthesis, have been investigated. The focus of this work is to explore force myography (FMG) as a technique for regressing grasping movement accompanied by wrist position variations. This study can lay the groundwork for a future investigation of FMG as a technique for controlling externally-powered prostheses continuously.

Methods: Ten able-bodied participants performed three hand movements while their wrist was fixed in one of six predefined positions. The angle between Thumb and Index finger (θ_{TI}), and Thumb and Middle finger (θ_{TM}) were calculated as measures of grasping movements. Two approaches were examined for estimating each angle: (i) one regression model, trained on data from all wrist positions and hand movements; (ii) a classifier that identified the wrist position followed by a separate regression model for each wrist position. The possibility of training the system using a limited number of wrist positions and testing it on all positions was also investigated.

Results: The first approach had a correlation of determination (R^2) of 0.871 for θ_{TI} and $R^2_{\theta_{TM}} = 0.941$. Using the second approach $R^2_{\theta_{TI}} = 0.874$ and $R^2_{\theta_{TM}} = 0.942$ were obtained. The first approach is over two times faster than the second approach while having similar performance; thus the first approach was selected to investigate the effect of the wrist position variations. Training with 6 or 5 wrist positions yielded results which were not statistically significant. A statistically significant decrease in performance resulted when less than five wrist positions were used for training.

Conclusions: The results indicate the potential of FMG to regress grasping movement, accompanied by wrist position variations, with a regression model for each angle. Also, it is necessary to include more than one wrist position in the training phase.

Keywords: Force myography, Random forest, Finger movement prediction, Continuous grasping predication, Partial hand prosthesis

Background

3.6 million individuals are expected to live with an amputation by 2050 [1]. Historical data show that approximately 35% of amputations are related to upper extremities and the majority of these (i.e. over 90%) are partial hand amputation [1]. Partial hand amputation (amputation of digits or hands), which is also known as minor upper-limb loss, can impose a notable influence on the person's life. It can have an adverse effect on their self-image, can cause loss of job and emotional distress [2]. Epidemiological studies show that the number of partial hand amputations per year is one every 18,000–20,000 residents [3]. A 2018 report indicates that there were 209,053 hospital visits specifically related to hand and digit injuries between 2002 and 2010, and 5000 work-related finger amputations in 2010 in the United States alone [4]. Although the minor amputation is a common injury, due to the vulnerability of the fingers, the research in the field of partial hand amputation has moderately progressed compared to the research in major upper limb amputation filed [3, 5]. One solution to resolve some of the problems regarding the amputation is to fit partial hand amputees with a prosthesis [3]. Several prosthesis options have been investigated and marketed by researchers and companies. Some of these options include passive, body-powered, activity-specific, and externally-powered prostheses [6]. The externally-powered prostheses use myosignals (signal from muscles activation and function) from the user to predict their intention to control the hand. In the commercially available externally-powered prosthesis, surface electromyography (sEMG) is more frequently used due to its portability and non-invasiveness.[1].

The commercial prostheses provide a set of pre-programmed hand gestures for the user. With correct sensor placement and training, the user can control the hand to shape one of the pre-programmed hand gestures. Although this classification approach can achieve a high classification accuracy, it limits the user to some specific hand gestures [7].

The user of an externally powered prosthesis will not be limited to a set of hand gestures if they can control each finger continuously. To provide such control for the users, researchers have investigated different regression methods to predict continuous finger movements and grasping actions using sEMG [8–10]. The majority of the researchers asked the participant to keep their hand in a specific position and did not consider the effect of wrist movement on the performance of their technique. As a result, Pan et al. [11] collected data in different wrist positions from the Middle finger of six able-bodied participants and the Index finger of two partial-hand amputees. The result of their study showed the possibility of predicting the finger movement in the presence of wrist movement with sEMG signal.

Although the sEMG signal is a common signal to control prostheses, it has some disadvantages. The sEMG signal can be unstable due to the environmental factors, like electrical noise, and user-based factors, such as sweating. This disadvantages resulted in marginal progress despite extensive research on the signal [12]. A low-cost alternative to sEMG to control a prosthesis is FMG. FMG is defined as tracking the volumetric changes in a muscle associated with the muscles contraction or relaxation during the functional movement of the limb [13]. The FMG technique presents some advantages over sEMG technique. Unlike the sEMG technique the FMG technique does not need skin preparation; the signal is not sensitive to the users sweating; and the electrical noise does not affect the signal [14]. Researchers have investigated FMG for hand gesture classification and continuous force prediction during the past decade [13, 15–18]. In two separate researches Connan et al. [19] and Jiang et al. [18] studied the FMG and sEMG signals that were collected from the participants at the same time while they were hold-ing specific wrist positions or hand gesture. They have shown that FMG has a compara-ble performance to sEMG, during hand and wrist movement classification.

Cho et al. [20] used FMG signal collected from the residual and intact limb of tran-sradial amputees to classify different hand gestures. They compare the classification accuracy between the signal collected from the residual limb and the intact limb. Their investigation indicate that the residual limb has a lower accuracy during the classifica-tion, due to the degraded muscle tone. However, they were able to classify six hand ges-tures with good accuracy in the residual limb. The work of Cho et al. [20] along with other works [21, 22], have showed FMG has the potential to yield promising result when used to control a prosthetic arm in transradial amputee. During the experiment Cho et al. [20] asked the participants hold their elbow in a 90° angle. To investigate the pos-sibility of resolving this constrain, Radmand et al. [17] introduced a high density FMG. They were able to classify hand and wrist motions with a low classification error. The authors also suggested to include data from eight different static positions in the train-ing dataset. This positions were defined as specific locations in 3D space in front of the participant to cover a person's workspace and ensure proper performance with limb position variation. Using different static positions resulted in classifying hand and wrist gestures in 3D space with a high classification accuracy.

As FMG signal demonstrates promising results for hand and wrist classification, Kad-khodayan et al. [23] used the signal for continuous finger movement prediction. They recorded the relative displacement of the tip of the Index finger, the Middle finger and the Thumb with respect to a reference point on the hand, during three hand move-ments. The authors asked the participants to keep their hand in a strictly fixed position. They used a support vector regression (SVR) with a Radial bases kernel function as their regression algorithm. The authors used ten-fold cross-validation to validate the model's performance. The work of Kadkhodayan et al. [23] can indicate the potential use of FMG signal for continuous grasping movement prediction.

One point about the work of Kadkhodayan et al. [23] is worth mentioning. Kadk-hodayan et al. [23] asked participants to keep their hand in a stationary position. In par-tial hand amputees with a functional wrist, the attempt to perform a hand grasp, either with the remaining digits or to control a prosthesis, is accompanied by wrist movement. Pan et al. [11] mentioned since the wrist movement affects the myosignal it has a nega-tive effect on the system's performance during continuous movement prediction. In other words, the system trained in a fixed hand position may not be able to perform as well with different hand positioning [5]. The same effect might be observed in prediction using FMG signal.

In our study, the performance of FMG for continuous grasping prediction is investi-gated. To build upon the work of Kadkhodayan et al. [23], we included different wrist positions in the study. To the authors' best knowledge the effect of the wrist positioning on the FMG performance for predicting continuous grasping movement has not been investigated. In this paper we propose to use random forest (RF) algorithm [24] as our prediction algorithm, due to its potential advantages to the conventional algorithms. Some of the advantages are as follows. The RF algorithm is robust to overfitting; it has only two parameters to optimize (the number of variables in the random subset at each node to split and the number of trees in the forest), while it is not very sensitive to these parameters [24, 25]. The features of the RF algorithm can make it a potential alternative to the conventional algorithms, for the goal of this study. Four regression methods, linear regression (LR), SVR, neural network regression (NNR), RF, and three classification methods, linear discriminate analysis (LDA), support vector machine (SVM), and RF had been investigated and compared in the present work. After determining the model to predict the continuous grasping movement, the effect of the variation of the wrist positioning on the performance of the model was investigated. The investigation was done to determine if it is possible to train on a fixed wrist position and predict the grasping movement in different wrist positions. In addition to that, the minimum number of the wrist positions that need to be included in the training phase was examined. The result of this work can create the groundwork for a further investigation of the potential of the FMG technique to be used as a controlling technique for continuously controlling a prosthesis device.

This paper is organized as follows: next section explains the proposed experimental protocol and processing of the experimental data; after that, an overview of experimental results is provided. Discussion and concluding remarks are presented at the end.

Methods

This section describes the setup and methods that were used for the data collection. It also specifies the different grasp types and wrist positions that were used during the data collection. In addition to that the processing of the data and the machine learning algorithms are defined. The last part presents the outcome measures to evaluate the performance of the models.

Experimental setup

An array of 18 force sensing resistors (FSR® 400, Short, Interlink Electronics, Westlake Village, CA) were placed in a flexible band, cut from 2 mm thick foam. Figure 1a shows

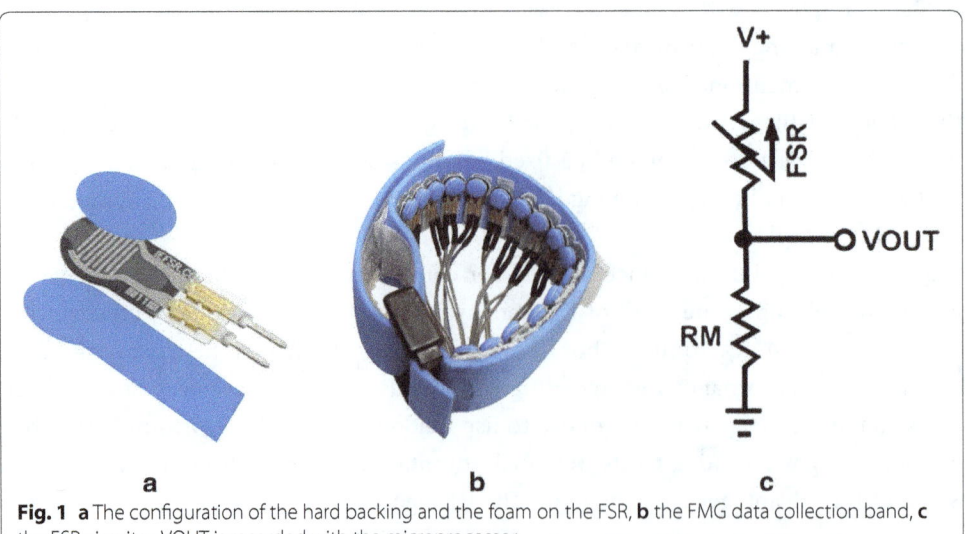

Fig. 1 **a** The configuration of the hard backing and the foam on the FSR, **b** the FMG data collection band, **c** the FSR circuitry. VOUT is recorded with the microprocessor

a close view of a FSR. FSRs are polymer thick film (PTF) sensors that show a decrease in resistance by increasing the applied force on their surfaces. This specific model can be activated with 0.2 N force, and its sensitivity range is up to 20 N [26]. As the FSRs are flexible devices, by wrapping the band around the participant's wrist, each FSR will bend. To stop the FSRs from bending, each FSR was supported with a piece of 1.5 mm thick acrylic sheet which was cut to the FSR's exact shape and dimensions. The acrylic piece provided a hard backing for the FSR which stopped it from bending. A piece of 1.5 mm thick foam was placed on each FSR, to concentrate the pressure from the wrist on the FSR's sensing area. The sensors were placed on the flexible band, 4 mm apart from each other. Figure 1b illustrates the FMG data collection band. As mentioned in Interlink Electronics [27] to have a simple force reading the sensor was connected to a resistor in a voltage divider circuit (Fig. 1c). The value of the output voltage read from the resistor (VOUT in Fig. 1c) was used as a measure of the applied pressure. An Atmega 32 microprocessor was used to read this value. The microprocessor was connected to an on-site computer with a USB cable, to record the data. A velcro strap and a buckle were used to wrap the band around the participants' limb. The band placement was kept uniform among the participants. It was placed on the upper limb above the head of Ulna bone. The buckle of the band was kept on the Radius bone. The reasoning behind this placement is that, the muscles and tendons that control hand digits are mostly deep muscles. The nature of FMG is to pick up the effect of limb volume changes, from muscle and tendon movements, on the skin surface. On the forearm, moving from the elbow to the wrist, the digit controlling muscles get closer to the skin surface, and the changes would be more localized for the FMG band. It is worth mentioning that these changes are visible on forearm belly muscle as well, but they are more localized near the wrist. As the changes regarding the digit movements were of great interest in this research this specific placement was selected.

The participants sat behind a desk with an adjustable height. The height of the desk and their chair was adjusted in a way that participants could rest their elbow on the table. Since the goal of this study was to investigate the effect of the wrist position on the FMG, to eliminate any other unwanted movement, participants' hand and forearm were fixed in a brace (Fig. 2). To maintain participants' comfort pieces of the polystyrene foam were used as support for forearm, wrist, and hand in the brace. The brace included a joint under the wrist. For moving to each wrist position the nut and screw of the wrist joint were unlocked. Then the participant was asked to move their

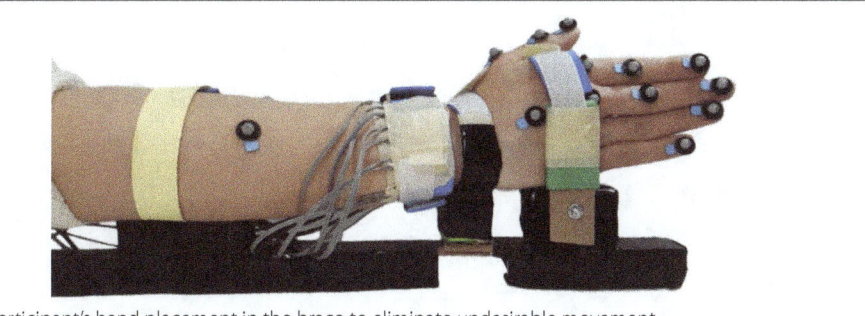

Fig. 2 Participant's hand placement in the brace to eliminate undesirable movement

hand to the specific position. The nut and screw were locked after hand positioning. With this consideration, the elbow was kept relaxed, and the wrist was fixed in a specific position during each data collection session.

A motion capture system (Qualisys AB, Gothenburg, Sweden) was used to track the finger movements. The system consisted of eight cameras, 15 markers, and the motion capture software which is called QTM. The markers were placed on the participants' hand, wrist and forearm (Fig. 3). On the index and the Middle finger, markers were placed on fingertips, proximal interphalangeal (PIP) joint, and metacarpophalangeal (MCP) joint. On the Thumb, the markers were placed on the Thumb's tip, interphalangeal (IP) joint, and MCP joint. On the wrist, the marker was placed on the Radiocarpal joint. On the forearm, the markers were placed on the Radius bone in line with the wrist marker and between the Ulna and Radius perpendicular to the Radius's marker. Two markers were placed on the Little and Ring fingers' tips as well. The data collected from the markers using the cameras was used to create a 3D model of the hand, fingers, and forearm to track their movements.

Each marker on the participant's hand should be visible to at least three cameras during data collection, for the QTM software to track the position of the marker in 3D space. Figure 4 shows the camera placement and the participant. Cameras were placed around the participant, to cover approximately 250° of the area around them.

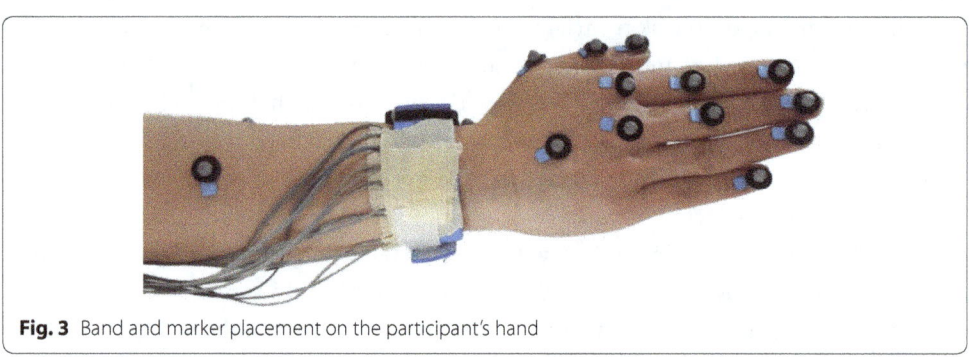

Fig. 3 Band and marker placement on the participant's hand

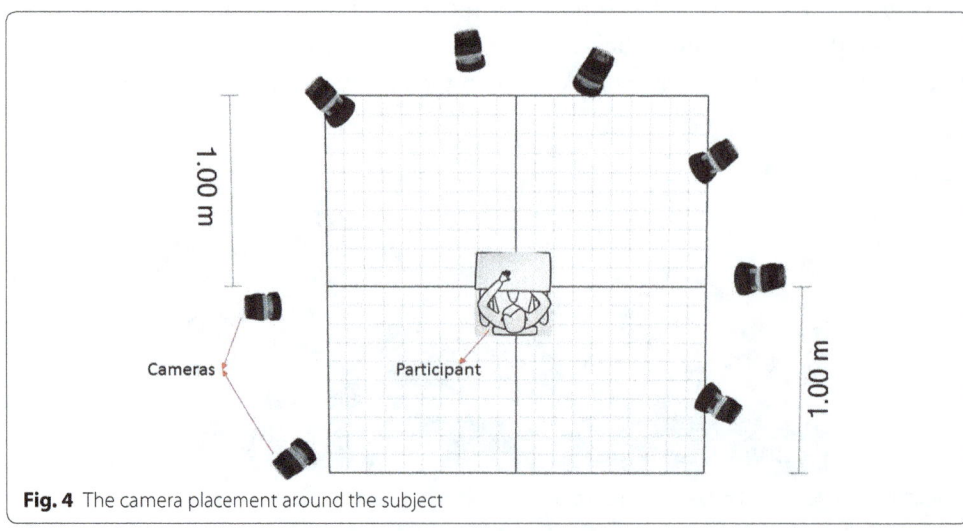

Fig. 4 The camera placement around the subject

The camera placement guaranteed that all the markers were at least visible to three cameras, during data collection.

A program in C# was coded, to collect data from the FMG band and the motion capture system at the same time. The program consisted of two threads, main thread and background thread. The band was read in the background thread and the motion capture data was read in the main thread. Every time the program received a data point from the band, the corresponding 3D location data of the markers from the motion capture system was read. The sampling frequency of the motion capture system was set to 100 Hz. The sampling rate of the FMG band was set to 15 Hz. This was sufficient to track finger movements in this study, which is slower than 1 Hz [23].

Grasp types and wrist positions

Three hand grasps were selected, to investigate the effect of the continuous grasping movements on the FMG. In an attempt to estimate the continuous grasping movement Kadkhodayan et al. [23] selected three grasp types, as a subcategory from Cutkosky's grasp taxonomy [28]. To build upon the work of Kadkhodayan et al. [23] the same grasp types were selected, and the investigation of the effect of the wrist movement variations was added. The three grasps are opposed Thumb-Index Finger grip, opposed Thumb-two Finger grip, and Heavy wrap-Large Diameter. Figure 5 shows the snapshots of the grasp types during flexing and extending fingers. The first two

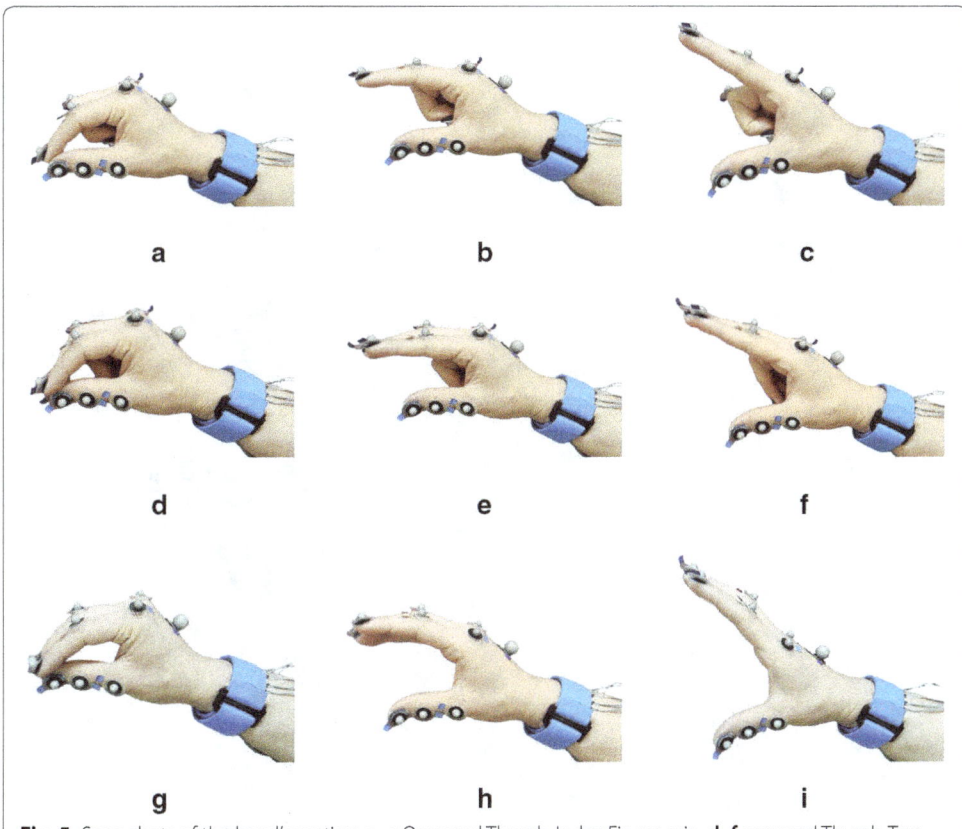

Fig. 5 Snap shots of the hand's motion. **a–c** Opposed Thumb-Index Finger grip, **d–f** opposed Thumb-Two Finger grip. **g–i** Heavy wrap-Large Diameter

grasps are precision grasps which mainly are used for manipulating a small object with fingertips. The third grasp is a power grasp and is mainly used for grasping large cylindrical objects. For simplicity the grasps were called "Index Finger-Thumb", "Two Fingers-Thumb" and "Large Diameter" grasp, for the rest of the paper.

Since three-finger robotic hands and prosthetic hands are commonly used [29–31], the movement of the Index finger, Middle finger, and Thumb were studied in this work. The participants did the grasps repeatedly, which provided data from three dynamic hand movements.

In addition to the grasping movement, we looked at the effect of the position of the wrist on the FMG. To do this six static wrist positions were used to collect data. These positions included keeping the wrist in Extension, Flexion, Neutral, Pronation, Radial deviation, and Ulnar deviation. Figure 6 illustrates the wrist positions. This study is a feasibility study regarding the investigation of the effect of the wrist position on the FMG signal, and is a first step toward removing the wrist position effect. Thus, only the six static wrist position following the work of Pan et al. [11] were included in this study, and wrist transition between the static positions was not considered. As the motion capture system was not able to see the markers on the fingers during the hand movements in Supination wrist position, this position was not included in

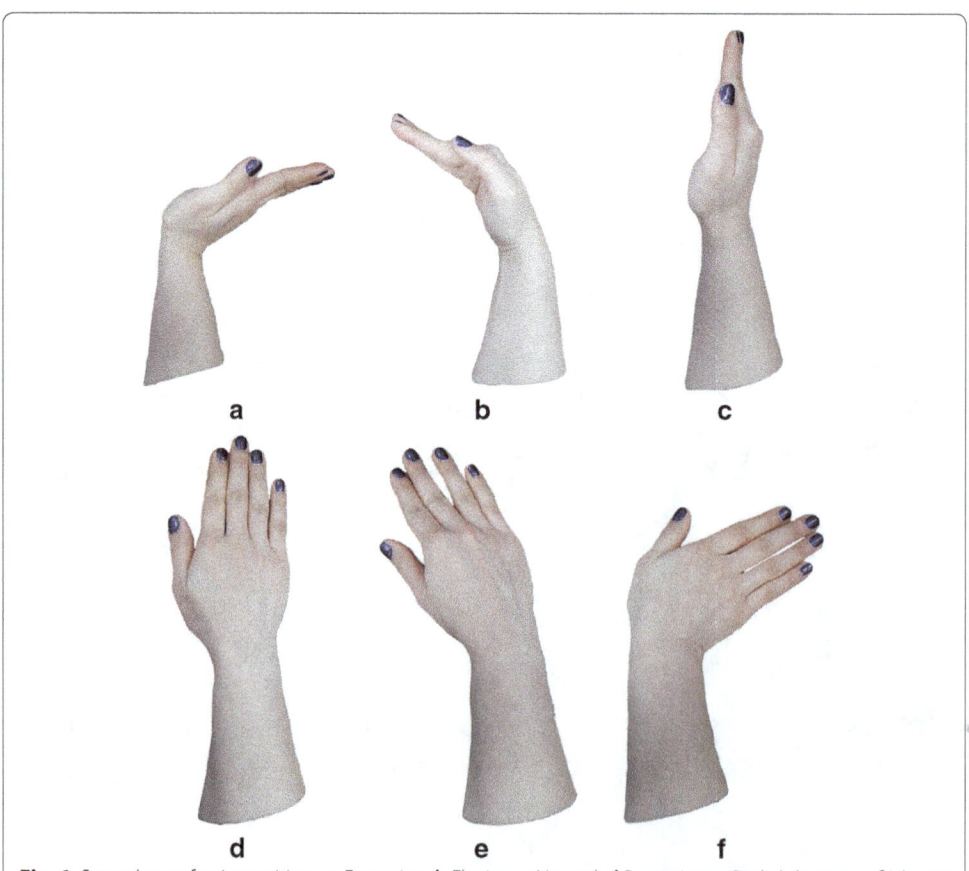

Fig. 6 Snap shots of wrist positions. **a** Extension, **b** Flexion, **c** Neutral, **d** Pronation, **e** Radial deviation, **f** Ulnar deviation

the study. The mentioned grasp types and wrist positions were demonstrated for the participants. After fixing the participant's elbow and wrist position, they were asked to flex and extend their fingers and Thumb as if they were grasping objects of different sizes. The participants did not hold or squeeze any object. The effect of applying force on an object or surface is not investigated in this study, since it is a feasibility study on the wrist movement effect on the signal. The participants were asked to do each grasp for 10 s. As there are three hand movements this adds up to 30 s in total. The participants repeat the hand movements 30 times. Since the sampling frequency of the band was set to 15 Hz, the repetitions provided approximately 13500 data point for each wrist position. The dataset for each data collection session nearly included 81000 data points. Participants could rest between the repetitions. Participants were asked if they need a rest every 10 min. In addition, there was a mandatory rest half way through the experiment for 15 min. There were no limitations on the speed, and participants performed the motion at their natural pace. With the considerations above, the data collection session for each participant, including setup, rest between repetitions, and recording data, was approximately 3 h long. It is worth mentioning that the fatigue has a lower influence on FMG signal than sEMG signal [19].

In the partial prosthetic hands usually the individual control of each finger joint is not provided, and the hand does not have the same kinematic of the real hand. As a result, if the finger joints are predicted, to control a prosthesis they need to be converted to the kinematics of the prosthesis to be able to control it. This motivated us to select measures different than the finger joint angles for the grasping movement. Kadkhodayan et al. [23] used the length of a vector connecting the fingertips to a reference point on the hand. As this vector can be affected by the size of the participant's hand, we selected another measure independent of the hand's dimensions. The angle between Index finger and Thumb and Middle finger and Thumb were selected (Fig. 7).

The processing of the motion capture data

The collected data from the location of the markers were used to calculate three vectors. The first one was defined as the vector starting from the MCP joint of the Thumb and ending at the Thumb's tip (V_{MT}). The second vector started at the MCP joint of the Thumb and ended at the Index finger's tip (V_{MI}). The last vector started at the MCP joint of the Thumb and ended at the Middle finger's tip (V_{MM}). These vectors were used to

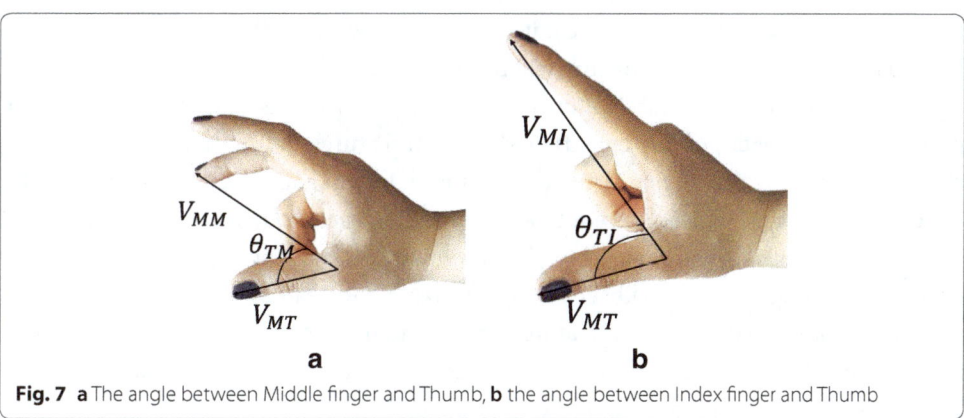

Fig. 7 **a** The angle between Middle finger and Thumb, **b** the angle between Index finger and Thumb

calculate the angle between the Index finger and the Thumb (θ_{TI}), and the angle between the Middle finger and the Thumb (θ_{TM}). Figure 7 illustrates the angles. Formulas 1 and 2 show the calculation of each angle.

$$\theta_{TI} = \tan^{-1}\left(\frac{|\vec{V}_{MT} \times \vec{V}_{MI}|}{\vec{V}_{MT} \cdot \vec{V}_{MI}}\right) \tag{1}$$

$$\theta_{TM} = \tan^{-1}\left(\frac{|\vec{V}_{MT} \times \vec{V}_{MM}|}{\vec{V}_{MT} \cdot \vec{V}_{MM}}\right) \tag{2}$$

Considering the six wrist positions and three hand movements, the dataset included 18 different states (3 hand movements in 6 wrist positions) for θ_{TI}. As the Middle finger was not moving in Index Finger-Thumb grasp, the data from this movement was not included in the dataset of θ_{TM}. This provided 12 states (2 hand movements in 6 wrist positions) for the investigation of θ_{TM}.

Classification and regression

For predicting grasp movements, two approaches were explored and compared. The approaches are one-step regression and two-step regression. The changes in the value of θ_{TI} and θ_{TM} were used as the target variable in training and testing phases, and the FMG data was used as the predictor variable. All data processing and analysis were done in an offline setting. The training and testing were performed on a server running Cenots ver. 6.9 equipped with four twelve-core (2 threads per core) Intel(R) Xeon(R) processors (E7-4860 v2 @ 2.60GHz) and 1000 GB RAM. To have a fair comparison between different models and approaches, all programs were run with a single core.

One-step regression

In the first approach, one-step regression, a single regression model was used for training the model and testing it, regardless of the hand movement or the position of the wrist. Both training and testing datasets included data from all static wrist positions and dynamic hand movements. Four regression models, namely LR, SVR, NNR, and RF regression were trained, tested and compared to identify which one performs better toward the goal of this paper. To the authors' best knowledge, RF has not been used on FMG data before. Thus the algorithm will be explained in short.

Random forest is an ensemble of un-pruned trees. Consider a dataset that has N data points, and each data point is constructed from M features. In the random forest algorithm developed by Breiman [24], to construct each tree a random subset of the samples including N' data points is selected. In a standard tree each node is split based on all M features, but in the algorithm developed by Breiman [24], each node is divided using the best guess among a random subset of the features. So instead of using all M feature to make a decision at each node, M' features are used to make a decision. The prediction of a new sample is made by aggregating (the majority of votes for classification and averaging for regression) the prediction of the whole forest.

In this paper, the random forest package of Matlab (introduced in R2009a) was used. The number of trees was set to 100. Matlab's default was used for the number of random features to use for splitting at each node (M'), which was the square root of the number of features for classification ($M' = \sqrt{M}$) and one-third of the number of features for regression ($M' = \frac{M}{3}$). Matlab's default was also used to determine the size of the random subset of the data (N') to construct each tree, which was the same as the training dataset ($N' = N$). In our case, we have 18 sensors which provided 18 channels; thus four random features were used for classification and six random features were used for splitting at each node for regression.

Two-step regression

Previous works on sEMG signal have shown that the movement of the wrist degrades the signal and there is a need to define separate models for different wrist positions [5, 11]. Pan et al. [11] indicated when sEMG is used with different models for different wrist positions, it is possible to estimate the continuous finger movements. The two-step regression approach was designed with an inspiration of their work. Comparing this approach and the one-step regression approach can indicate that whether FMG needs different models for different wrist position, similar to sEMG. In addition, by including the two approaches we can roughly compare sEMG and FMG for the same application.

This approach consisted of a six-class classifier and six regression models corresponding to six wrist positions. At the first step, the data points belonging to each wrist position were gathered together. For each group, a regression model was trained. Then a classification model was trained to classify the wrist positions. To test a new data point, first, the classifier found the wrist position that the point was in. Then based on the predicted wrist position the corresponding regression model was selected and used to predict the angle.

Three classification methods were explored, to choose the appropriate classification method for the second approach. As LDA and SVM with a Gaussian kernel have been used in the literature to classify wrist and hand movements using FMG, these two classifiers were tested [14, 18, 32]. In addition to those algorithms RF [24, 25] is introduced for classification using the FMG signal.

Analysis of the effect of the wrist position variation

As it was mention in the "Background" section, the movement of the wrist can affect the performance of the prediction. To analyze how the wrist position affects the performance first, one wrist position was included in the training phase, while all six positions were included in the testing set. The result of this part can confirm whether it is necessary to train on different wrist positions or not. One-way ANOVA and Tukey HSD test for post-hoc analysis were used to compare including six wrist positions, and including just one of the wrist positions in the training dataset. After that, the possibility of including less than six positions in the training phase, and the inclusion of all six positions in the testing phase was investigated.

To do the investigation, the data of one or a set of wrist positions were left out of the training dataset, while their data were included in the testing dataset. The reasoning behind it was if for instance, the data in the Neutral wrist position was similar to the data in the

Pronation wrist position, it would be possible to include just the data from one of those wrist positions in training data while including both of them in the testing phase.

Five cases were considered. In each case, one, two, three, four or five static wrist positions were removed from the training set, while their data was included in the test dataset. In case of removing two to five wrist positions, all combinations of removing wrist positions from the training data were considered. For instance, in case of removing two wrist positions, 15 combinations needed to be considered. Considering all combinations provides 62 different options in total. The performance of each option was calculated and compared to the performance of the system when all six positions were used for the training phase. One-way ANOVA and Tukey HSD test were used to determine which combinations do not have a statistically significant difference with including all six positions in the training dataset.

In all statistical analysis a confidence interval of 95% was considered ($\alpha = 0.05$). The statistical significance test was run on θ_{TI} and θ_{TM} independently. For each angle, the statistic tests were run on two outcome measures, namely R^2, and the percentage root mean square error ($RMSE\%$). The combinations that did not demonstrate a statistically significant difference with including six wrist positions in the training test were identified for each angle, and each outcome measure. The identified combinations that were overlapped between two angles, and within the two outcome measures, were selected as the combinations that can be used instead of including all wrist positions, without influencing the performance of the prediction.

Outcome measures

Each of the approaches was investigated separately for θ_{TI} and θ_{TM}. The collected dataset included thirty repetitions of each hand movement in each wrist position. In this work, a cross-repetition method was used for evaluating the performance. In a cross-repetition method, the data collected from a set of repetitions were left out as the testing dataset and were not used for the model optimization. To shape training and testing datasets, for each approach, six repetitions out of thirty repetitions were randomly selected, without replacement and set out for testing and other twenty-four repetitions were used for training. Which means 20% of the data were used in the testing phase and 80% were used in the training phase. The selection of test data set was made five times, to make sure that each repetition has been in the test set at least once. The result is reported as an average among repetitions and participants. R^2, $RMSE\%$, the average training time, and the estimated prediction time for a new data point were measured as ways to evaluate the performance of the modeling.

$$R^2 = 1 - \frac{\sum_{i=1}^{N}(y_i - y_{i'})^2}{\sum_{i=1}^{N}(y_i - \bar{y}_i)^2} \tag{3}$$

$$RMSE\% = \frac{\sqrt{\frac{1}{N}\sum_{i=1}^{N}(y_i - y_{i'})^2}}{y_{max} - y_{min}} \times 100 \tag{4}$$

Formula 3 shows the calculation of R^2, and Formula 4 shows the calculation of $RMSE\%$ value. The expected value is shown with y, and $y\prime$ shows the predicted value, $\bar{y}$ indicates the average of y in the test set and N indicates the number of data points in the test set. To measure the estimated prediction time for a new data point, the prediction time for the testing phase was measured and divided by the number of data points in the test dataset.

Concerning the two-step regression approach, it is essential to keep in mind that the evaluation measurements were calculated using the final output, which is the predicted value of θ.

Before training the algorithms, the dataset was scanned. If an angle data point was missing the corresponding FMG data was ignored, and the data point was removed. No other pre-processing was done on the FMG signal, and the raw FMG data was used for classification and regression. Train and test datasets were the same in the one-step regression and two-step regression. The same sets were also used for analyzing the effect of the wrist position variation.

Results

Ten able-bodied subjects, six males and four females, age 23–41 participated in this study. The experiment received ethics approval from Simon Fraser University, and participants gave their written informed consent. The result of each approach is presented separately. The comparison between two approaches helps to identify the appropriate approach, among two for estimating grasping movement. At last, the effect of the wrist position variation on the performance of the model is investigated.

One-step regression

Table 1 shows the results for θ_{TI}. The $R^2_{\theta_{TI}}$ values for LR, SVR, NNR and RF were 0.694, 0.868, 0.813 and 0.872 respectively. The $R^2_{\theta_{TI}}$ value indicated that RF performed slightly better than SVR, and NNR, while it significantly outperformed LR. The RF algorithm was about four times faster than SVR and more than 49 times faster than NNR during the training phase. The RF algorithm is also significantly faster than SVR and NNR to predict the output for a new data point.

Table 2 indicates the results for θ_{TM}. The $R^2_{\theta_{TM}}$ values for LR, SVR, NNR and RF were 0.873, 0.941, 0.911 and 0.941 respectively. Looking at the result of θ_{TM} regression, it showed the similarity between SVR and RF, while RF was over two times faster than SVR during the training, and more than seven times faster during testing. RF also performs slightly better than NNR, and the NNR is about 33 times slower than the RF algorithm

Table 1 Regression algorithms comparison for θ_{TI}, one-step regression approach

	R^2	RMSE%	Training time (min)	Estimated prediction time (ms) for each sample point
LR	0.6945 ± 0.07	14.59 ± 1.71	0.011 ± 0.012	0.0006
SVR	0.8680 ± 0.03	9.55 ± 1.07	6.172 ± 0.946	1.097
NNR	0.8129 ± 0.06	11.29 ± 1.51	72.925 ± 7.40	3.730
RF	0.8718 ± 0.03	9.40 ± 1.05	1.487 ± 0.14	0.076

Table 2 Regression algorithms comparison for θ_{TM}, one-step regression

	R^2	RMSE%	Training time (min)	Estimated prediction time (ms) for each sample point
LR	0.8734 ± 0.05	8.98 ± 1.27	0.008 ± 0.01	0.0006
SVR	0.9415 ± 0.02	6.12 ± 0.83	2.35 ± 0.45	0.585
NNR	0.9110 ± 0.03	7.53 ± 0.97	32.00 ± 3.38	2.449
RF	0.9411 ± 0.02	6.13 ± 0.84	0.95 ± 0.08	0.076

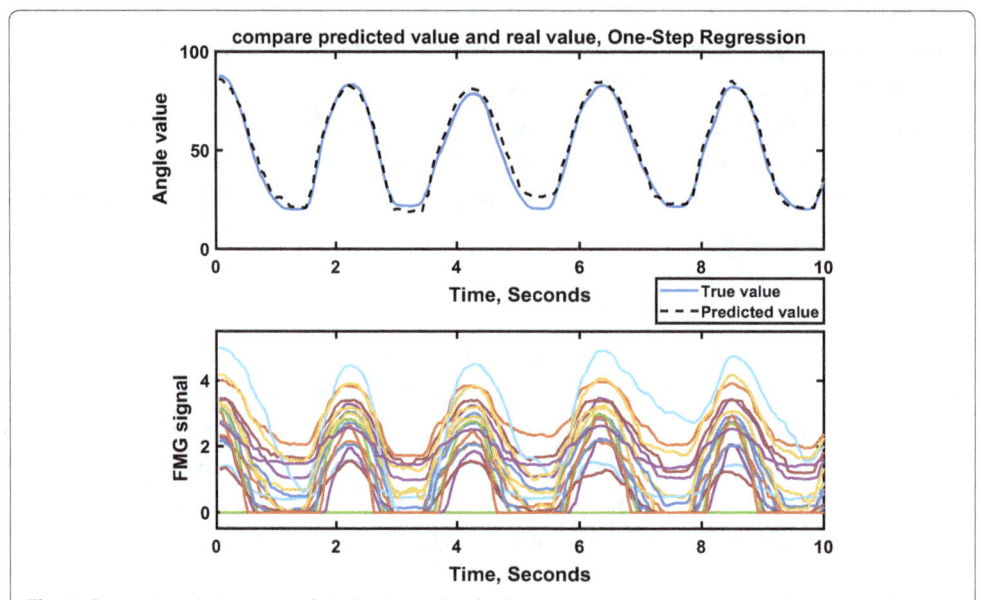

Fig. 8 Comparison between predicted value and real value, one-step regression approach using random forest, for θ_{TM}, subject #10, repetition #19, Large Diameter grasp, in Pronation wrist position

in both training and testing. The result confirms that the RF algorithm is a good alternative to the other three conventional algorithms for this research.

Figure 8 illustrates an example of predicting θ_{TM} using one-step regression approach with RF regression algorithm. Using RF regression R^2 value of 0.872 and 0.941 were obtained for θ_{TI} and θ_{TM} respectively.

Two-step regression

Based on the result of the first approach, RF was selected as the regression model. Table 3 indicates the result of the classifiers comparison. The accuracy of 85.01%, 95.55%, and 95.76% were obtained for LDA, SVM, and RF on average. RF algorithm was able to do the classification marginally better than SVM, while it was 1.5 times faster, and it

Table 3 Classification algorithms comparison for classifying six wrist positions

	Accuracy (%)	Sensitivity (%)	Specificity (%)	Time (s)
LDA	85.01 ± 9.55	90.65 ± 11.06	97.87 ± 3.22	0.10 ± 0.07
SVM	95.55 ± 4.41	97.46 ± 4.71	99.52 ± 0.78	58.82 ± 33.26
RF	95.76 ± 4.37	97.42 ± 4.48	99.45 ± 0.92	38.77 ± 3.91

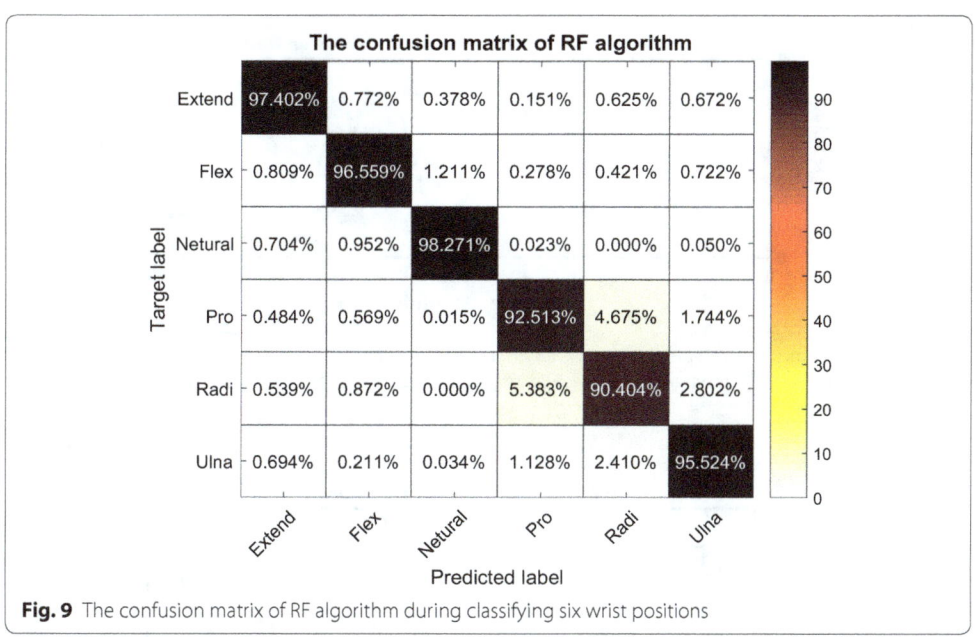

Fig. 9 The confusion matrix of RF algorithm during classifying six wrist positions

outperforms the LDA classifier. The result and training time confirm that RF algorithm can perform as an alternative to LDA and SVM to classify wrist position using FMG signal. Figure 9 illustrates the RF algorithm's confusion matrix. The confusion matrix indicates that the Pronation and Radial wrist positions are more likely to be miss-classified as each other compared to the other wrist positions. This miss-classification can be a result of the small deviation angle during the Radial deviation, which was an average of 18° for the participants.

Table 4 indicates the average results of the second approach. The $R^2_{\theta_{TI}} = 0.874$ and $R^2_{\theta_{TM}} = 0.942$ were obtained. Figure 10 shows a sample of the target and predicted value using the two-step regression approach. The training and testing dataset were the same as Fig. 8.

Figure 11 shows the average R^2 of two-step regression in six static wrist positions. The data were grouped into the wrist positions based on the result of the classifier.

Two one-sided t-test (TOST) [33] to test the equivalence was run on the $R^2_{\theta_{TI}}$ and $R^2_{\theta_{TM}}$ to compare the two approaches. The value of level of significance was set to 0.05 ($\alpha = 0.05$), the lower limit of the equivalence interval was set to -0.015 and the upper limit of the equivalence interval was set to 0.015. The null hypothesis was set as: $(\mu_{R^2_{one-step}} - \mu_{R^2_{two-step}}) > 0.015$ or $(\mu_{R^2_{one-step}} - \mu_{R^2_{two-step}}) < -0.015$, and the alternative hypothesis was set as: $-0.015 < (\mu_{R^2_{one-step}} - \mu_{R^2_{two-step}}) < 0.015$. The test resulted

Table 4 The results of two-step regression approach

	Classification accuracy (%)	R^2	RMSE%	Training time (min)	Estimated prediction time (ms) for each sample point
θ_{TI}	95.76 ± 4.38	0.8744 ± 0.03	9.31 ± 1.07	2.498 ± 0.179	68.174
θ_{TM}	96.45 ± 3.91	0.9424 ± 0.02	6.06 ± 0.87	1.665 ± 0.117	52.682

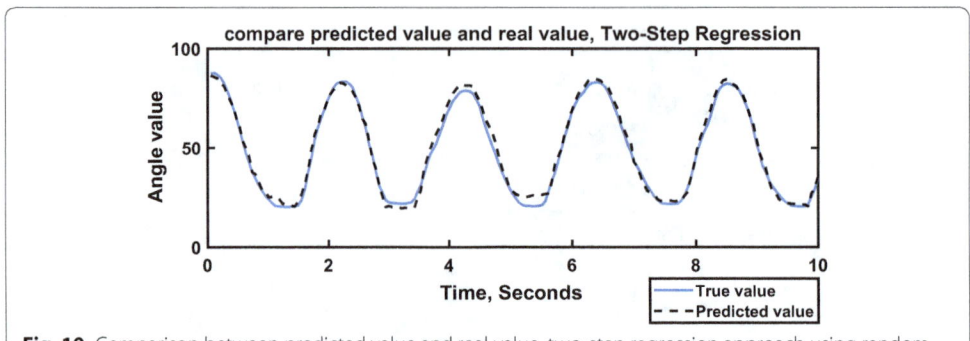

Fig. 10 Comparison between predicted value and real value, two-step regression approach using random forest, for θ_{TM}, subject #10, repetition #19, Large Diameter grasp, in Pronation wrist position

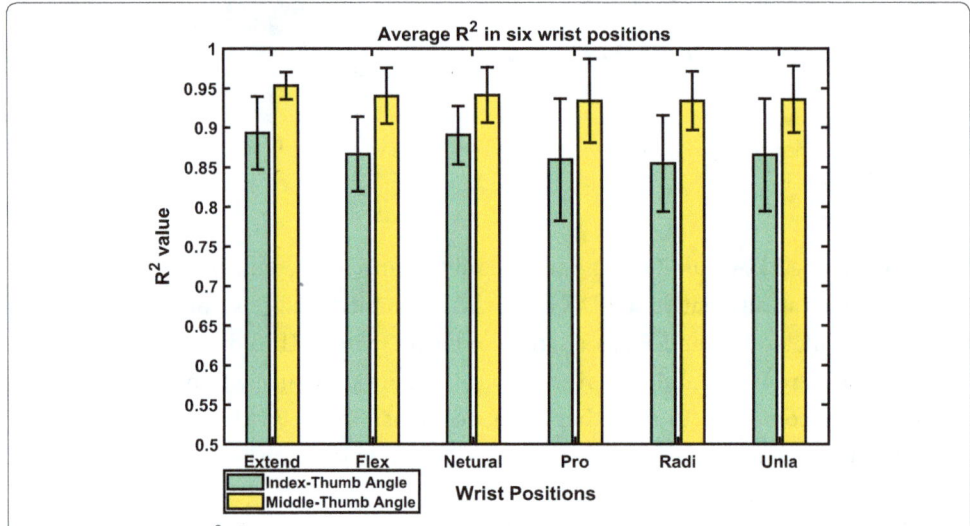

Fig. 11 The average R^2 of two-step regression in six static wrist positions, the data are grouped to the classes based on the result of the classifier

in $p < \alpha$ for both $R^2_{\theta_{TI}}$ and $R^2_{\theta_{TM}}$, which rejects the null hypothesis. Furthermore the confidence interval of the difference falls entirely inside the equivalence limits. This indicates the results from the two approaches are equivalent and they are not different from each other. The result indicates that the second approach had a similar performance to the first approach while it was more than 1.5 times slower than the one-step regression approach in training phase and significantly slower during the prediction of a new data point. The results suggest that one regression model for each angle (θ_{TI} and θ_{TM}) can predict the angle value in the presence of variations of the position of the wrist. Subsequently, the first approach was selected to investigate the effect of the wrist position variation on the result.

Analysis of the effect of the wrist position variation

All the possible combinations of removing one or more wrist positions from the training dataset were explored. Each combination was compared to the case of including all the wrist positions in the training dataset.

Table 5 indicate the result of including only one wrist position in the training dataset. As the result indicates, training on only one wrist position can notably decrease the R^2 value and increase the $RMSE\%$. In all cases for both θ_{TI} and θ_{TM} the statistical analysis for R^2 and $RMSE\%$ resulted in, $p \ll \alpha$. The p-value means we cannot conclude that there was no statistically significant difference between including all six positions in the training dataset or just including one wrist position. The result can confirm that including more than one wrist position in the training dataset is necessary, to predict continuous grasping movement in the presence of wrist position variations.

After that by comparing all different options for removing wrist positions from the training dataset, the positions that can be removed, without significant influence on the result of the prediction are identified. In case of removing one of the wrist positions from the training dataset, four options resulted in a p-value of more than 0.05. The $p > 0.05$ indicates that no statistically significant difference can be established between that specific option and including all six wrist position in the training dataset.

Table 6 presents the result of the wrist positions that need to be included in the training dataset and their corresponding R^2 and $RMSE\%$ value. The results indicated that the Extension, Pronation, Radial, or Ulnar wrist position could be removed from the training dataset. By removing any of these four wrist positions, no statistically significant difference can be established between the calculated R^2 and $RMSE\%$ with the values calculated with including all the six positions in the training dataset. The full result of the statistical analysis of each angle, with R^2 and $RMSE\%$, is provided in supplementary

Table 5 **The result of including only one wrist position in the training dataset, using one-step regression**

Included wrist positions in training dataset	θ_{TI}		θ_{TM}	
	R^2	$RMSE\%$	R^2	$RMSE\%$
Extension, Flexion, Neutral, Pronation, Radial, Ulnar	0.872 ± 0.034	9.402 ± 1.052	0.941 ± 0.021	6.133 ± 0.843
Extension	0.412 ± 0.230	19.90 ± 2.79	0.566 ± 0.304	15.740 ± 4.460
Flexion	0.553 ± 0.345	22.64 ± 4.16	0.436 ± 0.339	18.359 ± 4.007
Neutral	0.550 ± 0.214	20.70 ± 3.50	0.593 ± 0.163	16.067 ± 2.818
Pronation	0.372 ± 0.160	17.56 ± 2.73	0.700 ± 0.170	13.533 ± 2.924
Radial	0.231 ± 0.157	17.45 ± 2.60	0.631 ± 0.203	15.027 ± 3.305
Ulnar	0.412 ± 0.277	18.10 ± 3.66	0.641 ± 0.277	14.120 ± 3.550

Table 6 **The result of the algorithm using five wrist positions in the training dataset, one-step regression**

Included wrist position in training dataset	θ_{TI}		θ_{TM}	
	R^2	$RMSE\%$	R^2	$RMSE\%$
Extension, Flexion, Neutral, Pronation, Radial, Ulnar	0.872 ± 0.034	9.402 ± 1.052	0.941 ± 0.021	6.133 ± 0.843
Flexion, Neutral, Pronation, Radial, Ulnar	0.821 ± 0.056	11.072 ± 1.375	0.901 ± 0.051	7.810 ± 1.575
Extension, Flexion, Neutral, Radial, Ulnar	0.842 ± 0.046	10.432 ± 1.362	0.927 ± 0.034	6.748 ± 1.221
Extension, Flexion, Neutral, Pronation, Ulnar	0.845 ± 0.046	10.320 ± 1.343	0.925 ± 0.039	6.797 ± 1.187
Extension, Flexion, Neutral, Pronation, Radial	0.832 ± 0.051	10.722 ± 1.238	0.919 ± 0.039	7.239 ± 1.070

material (Additional file 1), which is the pair-wise comparison between all the 62 combinations and including all six wrist position in the training dataset. In addition, a sample of the FMG signal in different hand movements and wrist positions is provided in the supplementary material. Additional file 2 indicates that despite notable differences between some wrist positions in a hand movement, for example Extension and Pronation wrist positions, there are similarities in others, like Pronation and Ulnar deviation.

Discussion

The FMG signal was investigated for continuous grasping movements estimation, with ten able-bodied participants. Concerning the goal of the study, two different methods were explored. The first was using a regression algorithm to predict the grasping movements; and the second was using a classification prior to the regression algorithm and six regression models corresponding to the wrist positions. The angle between Index finger and Thumb (θ_{TI}) and Middle finger and Thumb (θ_{TM}) were selected as the measures of the grasping movement.

Based on the presented result the two approaches had similar performance. However, the second approach is slower than the first approach, during both training and prediction phases. One regression model for the θ_{TI} angle and a regression model for θ_{TM} were able to estimate the continues grasping movement with an average R^2 of 0.906 and $RMSE\%$ of 7.77%. Pan et al. [11] were able to predict continuous finger movement in static wrist positions using sEMG with an average R^2 about 0.8. They used a switching regime approach and trained different regression models for different wrist positions. In addition, Pan et al. [11] estimates the testing time for a new data point to be 53.011 ms. Our work shows that FMG can predict a new data point in approximately 0.076 ms during the first approach. The present study implies that the FMG signal demonstrates comparable performance to sEMG in the same application, while there is no need to have different models for different wrist positions, as long as the data from different wrist positions are provided during the training.

This study demonstrated a more comprehensive approach compared to the work of Kadkhodayan et al. [23], and took one step closer to a practical case, as the effect of the wrist variation was considered in this work. Adewuyi et al. [5] investigate the possibility of training the system on one wrist position and use it with variations in the position of the wrist, using sEMG signal. Their result indicated that the variation of the wrist position can have a negative effect on the performance of the model. Our result shows that the same effect can be observed in FMG signal. If the system was only trained on one wrist positing, it was not able to predict the continuous grasping movement, with wrist position variation, with an average R^2 more than 0.6. Our further investigations revealed that it is possible to include five wrist positions in the training dataset and include all six positions in the testing phase. In other words, it is possible to remove Extension, Pronation, Radial, or Ulnar wrist position from the training dataset, without any statistically significant effect on the performance of the model. This provides four options for the training dataset, that include five wrist positions instead of six wrist positions (Table 6, rows 2–5). The Tukey HSD test was run to do a pairwise analysis between the four options, which provided six combinations. The results showed $p > 0.05$ in all six combinations. The p-value indicated no statistically significant differences can be established between the four options. In conclusion, it is possible to use any of them without degrading the performance of the system.

With respect to one step regression, the RF algorithm was able to have comparable results to SVR, and NNR while it was faster. In addition to that, the time of the training using RF algorithm was not sensitive to the size of the dataset. As mentioned in "Motion Data Processing" the θ_{TM} included 12 states and θ_{TI} included 18 states. The time of the training using the RF algorithm was 1.49 and 0.95 min for θ_{TI} and θ_{TM} respectively, and the training time of SVR algorithm was 6.17 and 2.35 for the angles respectively. This training time indicated that in RF algorithm by increasing the size of the dataset the training time did not increase drastically. The training time using SVR algorithm for θ_{TI} is over 2.5 times more than the training time for θ_{TM} with fewer data points. Regarding the NNR algorithm, the training time of the θ_{TI} is about twice the training time of the θ_{TM}. The training times indicated the sensitivity of the SVR and NNR models to the size of the training dataset. The result for θ_{TM} dataset also indicates that the RF regression was a good alternative for the other commonly used regression models for the goal of this paper.

The results indicated, independent of the approach, θ_{TI} has a lower R^2 and higher $RMSE\%$ than θ_{TM}. This result can be justified by looking at the hand movements that are included in the dataset for each angle. As it was mentioned in the "Methods" section, since the Middle finger is not moving in the Index Finger-Thumb movement, this movement is not included in the dataset of θ_{TM}. The Index Finger-Thumb movement is more challenging to predict than the other two hand movements. The FMG band is placed on the forearm near the wrist. The tendons and muscles responsible for the movement of fingers, Thumb, and wrist are passing from this location. Keeping three hand movements in mind, in the Index Finger-Thumb movement, only Index and Thumb were moving. This means only the tendons and muscles that are controlling the Thumb and Index fingers were affecting the sensors that collect FMG, in a fixed wrist position. While in Two Fingers-Thumb movements the tendons and muscles responsible for moving Middle finger were influencing the signal as well. In the Large Diameter grip, all the fingers were moving, and there were more changes in each sensor's reading in this movement. Every one of the sensors in the band can provide information for machine learning algorithm. Since in the Two Finger-Thumb movement and Large Diameter grip more sensors are influenced, more information is provided for the machine learning algorithm. This results in a better performance in estimating the angle in these two hand movements. Despite the facts mentioned above, it is worth noticing that both methods were able to estimate θ_{TI} with a R^2 more than 0.8.

Conclusion

In this study, FMG signal was introduced and explored to provide continuous grasping movement prediction toward controlling partial hand prosthesis. With having the importance of the effect of the wrist movement on the performance of the grasping movement prediction in mind, the effect of the wrist movement on the FMG signals performance during predicting continuous grasping movement was explored. To the authors' best knowledge this is the first work that considers the effect of the wrist movement on the performance of the FMG during continuous grasping movement prediction.

Two approaches were defined and examined for predicting continuous grasping movement in the presence of different wrist positioning. The first approach was to use one regression model for grasping movement prediction, irrespective of hand movements and wrist positions. During the second approach first, a classifier determined the wrist position that participant was in and after that, a regression model, trained for that wrist position, was used to predict the measures of the grasping movement. The two approaches had a similar performance, while the first approach was faster than the second approach. This indicates one regression model with adequate training data was able to predict the grasping movement.

The effect of the variations in the wrist position was investigated. The result concluded that it is necessary to include at least five wrist positions in the training dataset to be able to predict continuous grasping movement in different wrist positions.

In short, the presented study indicates the potential of FMG signal to be used as a control technology to provide continuous grasping movement control for partial hand amputees. Future works can investigate the possibility of improving the performance by adding feature extraction from the FMG signal and studying deep learning algorithms for regression. Additionally, more grasping movements can be included in the study as the number of the grasping movements covered in this study were limited to three grasp movements. To continue the work in the future, it is essential to test the ability of the signal to perform grasping movement regression, while the wrist is not constrained to specific angles and include a continuous wrist movement estimation in the algorithm if it is needed. It is also necessary to test the signal to control a simulated hand or a robotic hand gripper in real-time. In addition, the effect of the FMG band placement on the limb as well as the effect of the fatigue on performance of the system should be investigated. Also, the effect of object handling and the resistance from holding and squeezing an object should be investigated. At last, it is important to test the result of the study with partial hand amputee participants and test the potential and limitations of the signal and the algorithm.

Abbreviations
FMG: force myography; R^2: correlation of determination; RF: random forest; sEMG: surface electromyography; MCP: metacarpophalangeal; PIP: proximal interphalangeal; FSR: force sensing resistor; PTF: polymer thick film; IP: interphalangeal; LR: linear regression; SVR: support vector regression; NNR: neural network regression; LDA: linear discriminate analysis; $RMSE\%$: percentage root mean square error; SVM: support vector machine.

Authors' contributions
RSC wrote the content of the paper, carried out the data collection, and analysis. CM identified the application, coordinated the research, and edited the manuscript. Both authors read and approved the final manuscript.

Acknowledgements
The authors would like to thank Dr. Carolyn Sparrey, for her cooperation and assistance during the research.

Competing interests
The authors declare that they have no competing interests.

Funding
Research supported by the Natural Sciences and Engineering Research Council of Canada (NSERC), the Canadian Institutes of Health Research (CIHR), the Canada Research Chair (CRC) program and Canada Foundation for Innovation Leader's Opportunity Fund (CFI IOF 2013-26913).

References

1. Ziegler-Graham K, MacKenzie EJ, Ephraim PL, Travison TG, Brookmeyer R. Estimating the prevalence of limb loss in the united states: 2005 to 2050. Arch Phys Med Rehabil. 2008;89(3):422–9.
2. Gavrilova N, Harijan A, Schiro S, Hultman CS, Lee C. Patterns of finger amputation and replantation in the setting of a rapidly growing immigrant population. Ann Plast Surg. 2010;64(5):534–6.
3. Imbinto I, Peccia C, Controzzi M, Cutti AG, Davalli A, Sacchetti R, Cipriani C. Treatment of the partial hand amputation: an engineering perspective. IEEE Rev Biomed Eng. 2016;9:32–48.
4. Reavey PL, Stranix JT, Muresan H, Soares M, Thanik V. Disappearing digits: analysis of national trends in amputation and replantation in the United States. Plast Reconstr Surg. 2018;141(6):857–67.
5. Adewuyi AA, Hargrove LJ, Kuiken TA. Towards improved partial-hand prostheses: the effect of wrist kinematics on pattern-recognition-based control. In: 2013 6th international IEEE/EMBS conference on neural engineering (NER). New York: IEEE; 2013. p. 1489–92.
6. Whelan LR, Farley J. Functional outcomes with externally powered partial hand prostheses. J Prosthet Orthot. 2018;30:69–73.
7. Krasoulis A, Member S, Vijayakumar S. Evaluation of regression methods for the continuous decoding of finger movement from surface EMG and accelerometry. In: 7th annual international IEEE EMBS conference on neural engineering, Montpellier. 2015. p. 631–4.
8. Smith RJ, Tenore F, Huberdeau D, Etienne-Cummings R, Thakor NV. Continuous decoding of finger position from surface EMG signals for the control of powered prostheses. In: Engineering in Medicine and Biology Society, 2008. EMBS 2008. 30th annual international conference of the IEEE; New York: IEEE; 2008. p. 197–200.
9. Ngeo J, Tamei T, Shibata T. Continuous estimation of finger joint angles using muscle activation inputs from surface EMG signals. In: Engineering in Medicine and Biology Society (EMBC), 2012 annual international conference of the IEEE. New York: IEEE; 2012. p. 2756–9.
10. Pan L, Sheng X, Zhang D, Zhu X. Simultaneous and proportional estimation of finger joint angles from surface EMG signals during mirrored bilateral movements. In: International conference on intelligent robotics and applications. Berlin: Springer; 2013. p. 493–9.
11. Pan L, Zhang D, Liu J, Sheng X, Zhu X. Continuous estimation of finger joint angles under different static wrist motions from surface EMG signals. Biomed Signal Process Control. 2014;14:265–71.
12. Castellini C, Artemiadis P, Wininger M, Ajoudani A, Alimusaj M, Bicchi A, Caputo B, Craelius W, Dosen S, Englehart K. Proceedings of the first workshop on peripheral machine interfaces: going beyond traditional surface electromyography. Front Neurorobotics. 2014;8:22.
13. Wininger M, Kim N-H, Craelius W. Pressure signature of forearm as predictor of grip force. J Rehabil Res Dev. 2008;45(6):883.
14. Xiao ZG, Menon C. Counting grasping action using force myography: an exploratory study with healthy individuals. JMIR Rehabil Assistive Technol. 2017;4(1)
15. Li N, Yang D, Jiang L, Liu H, Cai H. Combined use of FSR sensor array and SVM classifier for finger motion recognition based on pressure distribution map. J Bionic Eng. 2012;9(1):39–47.
16. Castellini C, Ravindra V. A wearable low-cost device based upon force-sensing resistors to detect single-finger forces. In: 2014 5th IEEE RAS & EMBS international conference on biomedical robotics and biomechatronics. New York: IEEE; 2014. p. 199–203.
17. Radmand A, Scheme E, Englehart K. High-density force myography: a possible alternative for upper-limb prosthetic control. J Rehabil Res Dev. 2016;53(4).
18. Jiang X, Merhi L-K, Xiao ZG, Menon C. Exploration of force myography and surface electromyography in hand gesture classification. Med Eng Phys. 2017;41:63–73.
19. Connan M, Ruiz Ramírez E, Vodermayer B, Castellini C. Assessment of a wearable force-and electromyography device and comparison of the related signals for myocontrol. Front Neurorobotics. 2016;10:17.
20. Cho E, Chen R, Merhi L-K, Xiao Z, Pousett B, Menon C, et al. Force myography to control robotic upper extremity prostheses: a feasibility study. Front Bioeng Biotechnol. 2016;4:18.
21. Ahmadizadeh C, Merhi L-K, Pousett B, Sangha S, Menon C. Toward intuitive prosthetic control: solving common issues using force myography, surface electromyography, and pattern recognition in a pilot case study. IEEE Robot Autom Mag. 2017;24(4):102–11.
22. Phillips SL, Craelius W. Residual kinetic imaging: a versatile interface for prosthetic control. Robotica. 2005;23(3):277–82.
23. Kadkhodayan A, Jiang X, Menon C. Continuous prediction of finger movements using force myography. J Med Biol Eng. 2016;36(4):594–604. https://doi.org/10.1007/s40846-016-0151-y.
24. Breiman L. Random forests. Mach Learn. 2001;45(1):5–32.
25. Liaw A, Wiener M. Classification and regression by randomforest. R News. 2002;2(3):18–22.
26. Interlink Electronics: FSR® 400 Series datasheet. Interlink Electronics.
27. Interlink Electronics: FSR® Integration Guide. Interlink Electronics. Rev. C
28. Cutkosky MR. On grasp choice, grasp models, and the design of hands for manufacturing tasks. IEEE Trans Robot Autom. 1989;5(3):269–79.
29. Zollo L, Roccella S, Guglielmelli E, Carrozza MC, Dario P. Biomechatronic design and control of an anthropomorphic artificial hand for prosthetic and robotic applications. IEEE/ASME Trans Mechatron. 2007;12(4):418–29.
30. Manti M, Hassan T, Passetti G, D'Elia N, Laschi C, Cianchetti M. A bioinspired soft robotic gripper for adaptable and effective grasping. Soft Robot. 2015;2(3):107–16. https://doi.org/10.1089/soro.2015.0009.
31. Jiang H, Wachs JP, Duerstock BS. Integrated vision-based robotic arm interface for operators with upper limb mobility impairments. In: 2013 IEEE 13th international conference on rehabilitation robotics (ICORR), Seattle. 2013. p. 1–6. https://doi.org/10.1109/ICORR.2013.6650447

8

Comfort level discussion for prosthetic sockets with different fabricating processing conditions

Cheung-Hwa Hsu[*], Chao-Hui Ou, Wei-Lun Hong and Yu-Han Gao

From International Conference on Biomedical Engineering Innovation (ICBEI) 2016 Taichung, Taiwan.

*Correspondence:
chhsu@nkust.edu.tw
Department of Mold and Die
Engineering, National
Kaohsiung University
of Science and Technology,
Kaohsiung, Taiwan, ROC

Abstract

Background: In the past, manufacture of prosthetic socket by using traditional hand-made method not only consumed research time but also required a special assembly approach. Recently, reverse engineering and rapid prototype technology have grown up explosively, and thus, provide a choice to fabricate prosthetic socket.

Methods: Application 3D computer aided design and manufacturing (computer-aided design/computer-aided engineering) tools approach the surface shape stump data is digitized and can be easily modified and reused. Collocation investigates gait parameters of prosthetic socket, and interface stress between stump and socket with different processing conditions. Meanwhile, questionnaire was utilized to survey satisfaction rating scale, comfort level, of subjects using this kind of artificial device.

Results: The main outcome of current research including gait parameters, stress interface and satisfaction rating scale those would be an informative reference for further studies in design and manufacture as well as clinical applications of prosthetic sockets.

Conclusions: This study found that, regardless of the method used for socket fabrication, most stress was concentrated in tibia end pressure-relief area. This caused discomfort in the area of tibia end to the participant wearing prosthesis. This discomfort was most evident in case when the prosthetic socket was fabricated using RE and RP.

Keywords: Prosthetic socket, Interface pressures, Rapid prototyping, Gait analysis

Background

Limb asymmetries in the physically disabled cause gait abnormalities, which is why the average gait speed in below-knee amputees (only 64 m/min) is lower than the normal gait speed of 91 m/min [1]. More comfortable walking can be achieved in unilateral amputees by increasing the prosthetic ankle angle [2]. Prostheses must be custom made for each amputee stump based on the individual needs of a patient. Much time should be allowed for adjusting a prosthetic socket and the alignment of prosthesis because the highest levels of patient satisfaction can be achieved only through multiple fitting trials, modifications, and adjustments. Traditionally, prosthetic sockets were produced via a manual and complicated manufacturing process conducted by experienced and technically skilled prosthesis. Computer software and hardware advances brought about a fast development of 3D computer aided design (CAD) and manufacturing (CAM) tools [3]. This study applied 3D CAD, reverse engineering (RE), and rapid prototyping (RP) techniques to develop a new technology to manufacture prosthetic sockets and avoid the inconveniences of the traditional handmade method. Gait analysis data of patients was recorded using a motion analysis system. Knee joint stresses and moments were computed through inverse dynamics. Indentation tests of soft tissues were used to measure pressure discomfort and pain thresholds. A special scale was developed for pain assessment such that combined a numeric rating scale (NRS) and visual analogue scale (VAS). In the 11-point NRS, 0 and 10 points represented the lowest and highest comfort levels for the prosthetic sockets, respectively. The VAS used a straight line to represent pain tolerance levels, with the two ends of the line indicating opposite extremes of pain [4].

Methods

Figure 1 shows the process of RP-based fabrication of prosthetic sockets. First, a plaster cast of the stump is taken by means of vacuum forming. Next, the stump model created using a computed tomography (CT) scan is imported into a drawing software to construct a 3D stump model and the stump surface is modified with respect to pressure-tolerant and pressure-relief areas. After these modifications and verification of the model

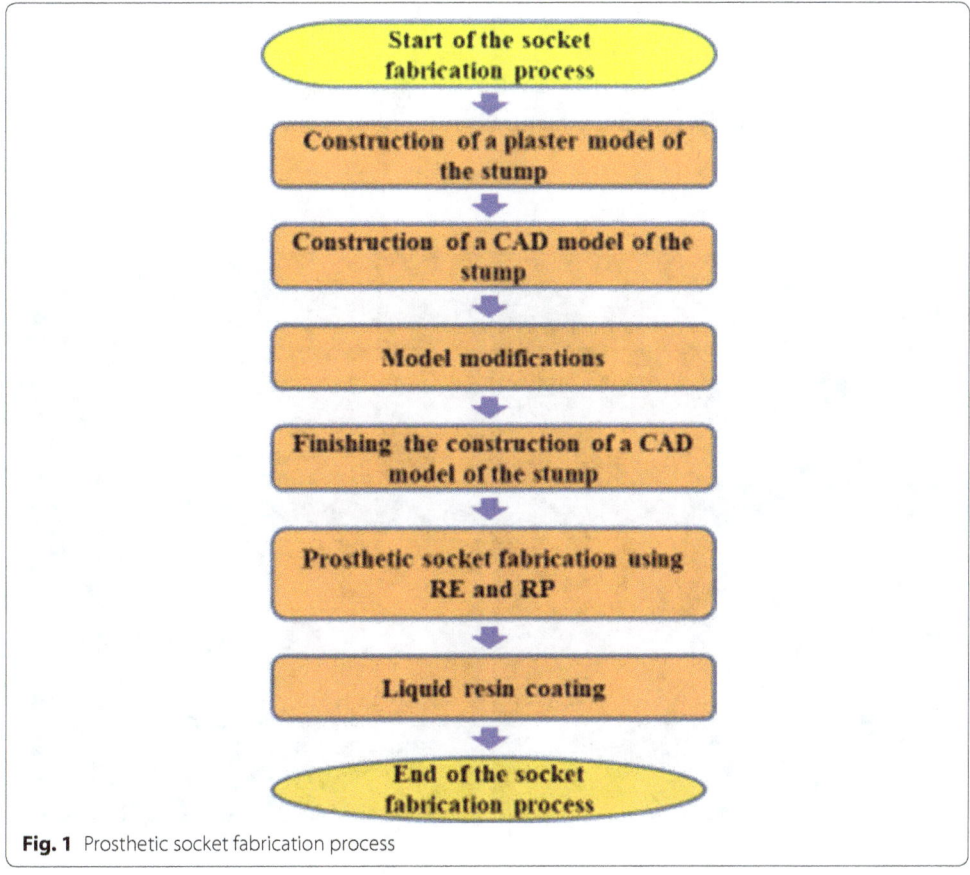

Fig. 1 Prosthetic socket fabrication process

by a specialist in prosthetics, it is processed in an RP machine and prosthetic sockets are fabricated (see Fig. 2) and coated with the liquid resin for greater strength and, thus, an amputee's comfort. The internal liner serves as a stump–socket buffer. The inner surface of sockets is covered with a 4 mm thick material using RP and sockets are placed into plastic bags. Plaster is injected and, once dries and hard, the resulting plaster model is used as a model of the internal liner. Due to the use of rigid plastics in most prosthetic sockets, elastic and skin-like material should be selected for the internal liner, such as silica gel or polyurethane. The plaster cast model is then coated with the internal liner the material of which is heated to fit the curved surface. Finally, prosthesis modifies the liner shape to fit the plaster cast. The entire procedure of RP-based fabrication of a prosthetic socket is presented in Fig. 3. Insertion of the liner into the socket is the last step in such a manufacturing process.

Results and discussion

Prosthetic sockets fabricated using the traditional handmade method and those fabricated using RE and RP were compared based on the differences in their gait analysis results. In gait analysis, a 3D space measurement system was applied to examine joint angle and time at each joint motion. Three types of parameters were studied, namely spatial and temporal parameters, kinematics parameters, and dynamics parameters. Instrumentation plans were designed according to the lower limb data and different positions of the patient were measured. Table 1 provides the participants body measurement data that was further used for dynamic analysis of their gait.

Gait analysis

Spatial and temporal parameters (see Tables 2, 3) present spatial and temporal parameters of the prosthetic sockets manufactured with different methods. Their comparison

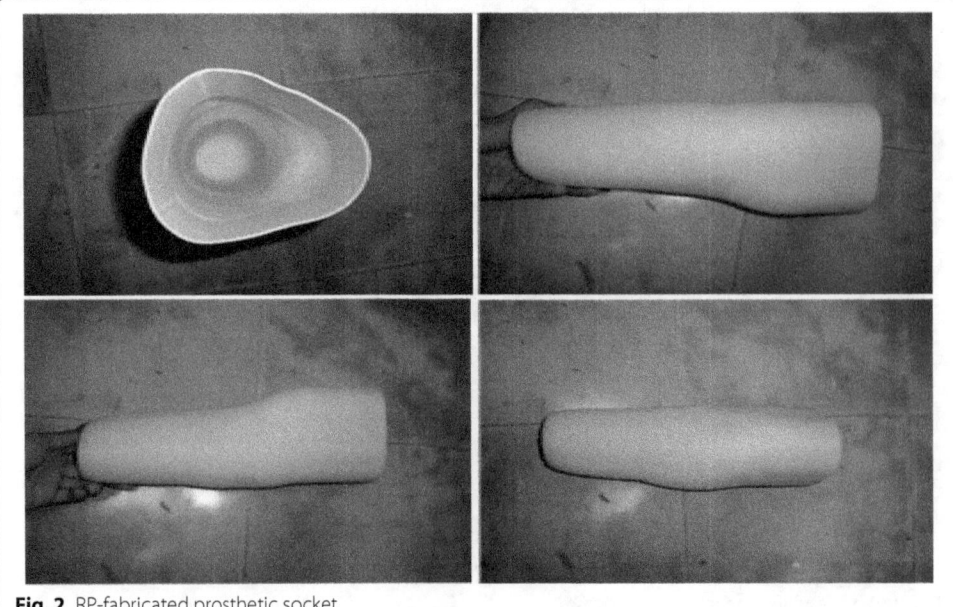

Fig. 2 RP-fabricated prosthetic socket

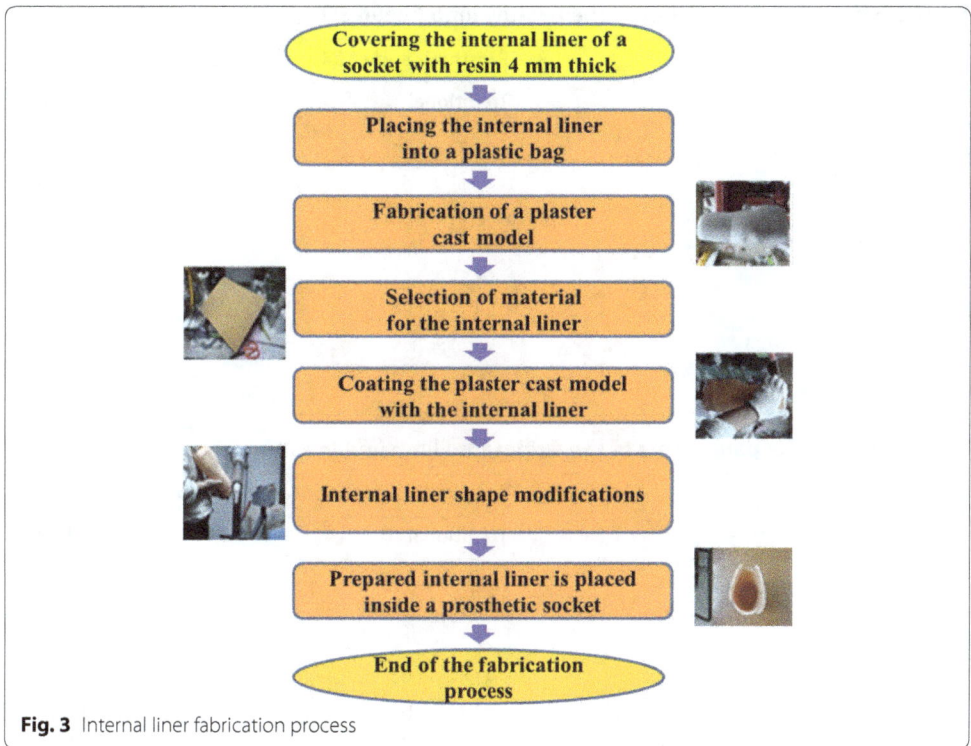

Fig. 3 Internal liner fabrication process

Table 1 Body measurement data of participants

Participants	Uni-lateral below-knee prosthesis users
Amputation side	Left
Sex	Male
Age	26
Height (cm)	175.0
Weight (kg)	84.1
Thigh length (cm)	Healthy = 48.7; amputated = 48.7
Thigh circumference (cm)	Healthy = 46.5; amputated = 46.5
Calf length (cm)	Healthy = 27.0; amputated = 19.2
Calf circumference (cm)	Healthy = 41.3; amputated = 33.5
Foot length (cm)	Healthy = 26.5; amputated = 25.0
Foot width (cm)	Healthy = 10.4; amputated = 9.5
Prosthesis weight (kg)	2.0

showed that velocity and cadence were higher for the participant who wore the prosthetic socket fabricated with the traditional handmade method rather than that fabricated using RE and RP. This indicated that the handmade prosthetic socket better corresponded to the participant's habits in terms of mobility and operation. In terms of stride length, larger values were observed when the prosthetic socket fabricated using RE and RP, rather than the traditional handmade method, was used. However, the former also demonstrated a greater difference in step length between an amputated and healthy limb, meaning that asymmetry between two limbs is a serious issue in the use of prosthetic sockets fabricated using RE and RP. The percentage of gait cycle spent in

Table 2 Kinematics parameters of the healthy limb (mean SD)

Gait parameters	Prosthetic socket fabrication method	
	Traditional	RE and RP
Forward velocity (cm/s)	107.73 ± 5.34	98.25 ± 1.77
Cadence (steps/min)	89.11 ± 1.81	79.13 ± 0.87
Stride length (cm)	142.64 ± 4.67	149.29 ± 0.77
Step width (cm)	31.52 ± 1.94	30.78 ± 1.18
Step length (cm)	68.55 ± 2.43	68.11 ± 1.78
Stance phase (%)	64.29 ± 1.84	65.66 ± 0.10
Swing phase (%)	35.71 ± 1.84	34.34 ± 0.10

Table 3 Kinematics parameters of the amputated limb (mean SD)

Gait parameters	Prosthetic socket fabrication method	
	Traditional	RE and RP
Forward velocity (cm/s)	107.67 ± 4.29	96.91 ± 2.76
Cadence (steps/min)	90.35 ± 3.76	80.49 ± 1.80
Stride length (cm)	144.25 ± 2.15	145.88 ± 0.20
Step width (cm)	31.52 ± 1.94	30.78 ± 1.18
Step length (cm)	74.89 ± 2.95	79.47 ± 0.14
Stance phase (%)	63.44 ± 0.97	59.77 ± 0.34
Swing phase (%)	36.56 ± 0.97	40.23 ± 0.34

stance phase was found to be higher in the healthy side than the amputated side. This indicated that the participants tended to put a greater load on the healthy limb, while not willing to load the amputated limb, which was more evident in case of the prosthetic socket fabricated using RE and RP.

Kinematics parameters

Figures 4, 5, and 6 show data on angular changes in the hip, knee, and ankle joints of the healthy and amputated limbs in sagittal, coronal, and transverse planes when wearing different prosthetic sockets. Angular changes in hip joints are presented in Fig. 4. In the sagittal and transverse planes, angular changes in the hip joint of the healthy limb did not differ greatly; in the coronal plane, when the prosthetic socket fabricated with the traditional handmade method was used, the hip joint angle of the healthy limb was larger than that for the prosthetic socket fabricated using RE and RP. With regard to the amputated limb, a relatively large extension angle in the sagittal plane in case of the prosthetic socket fabricated with the traditional handmade method indicated that it was better able to support the body weight and maintain balance than the prosthetic socket fabricated using RE and RP. Angular changes in knee joints are presented in Fig. 5. While no large differences were observed in the coronal and transverse planes, in the sagittal plane, healthy knee flexion during stance phase was more evident in the participant who wore the prosthetic socket fabricated with the traditional handmade method; it is inferred that low gait velocity decreased impact forces. Moreover, in the participant who wore the prosthetic socket fabricated using RE and RP, the amputated limb was not able

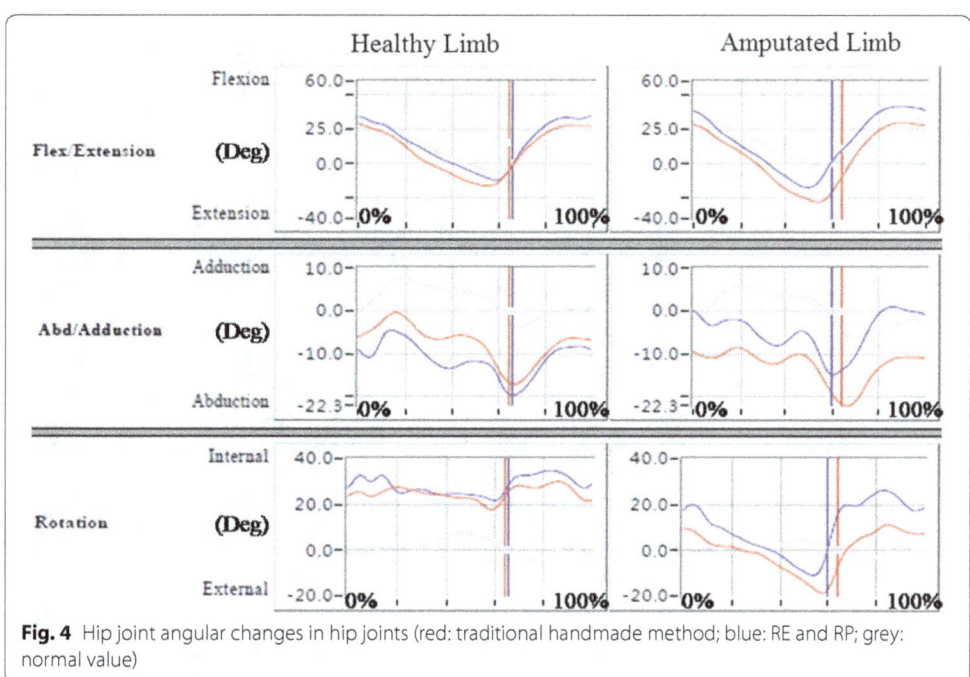

Fig. 4 Hip joint angular changes in hip joints (red: traditional handmade method; blue: RE and RP; grey: normal value)

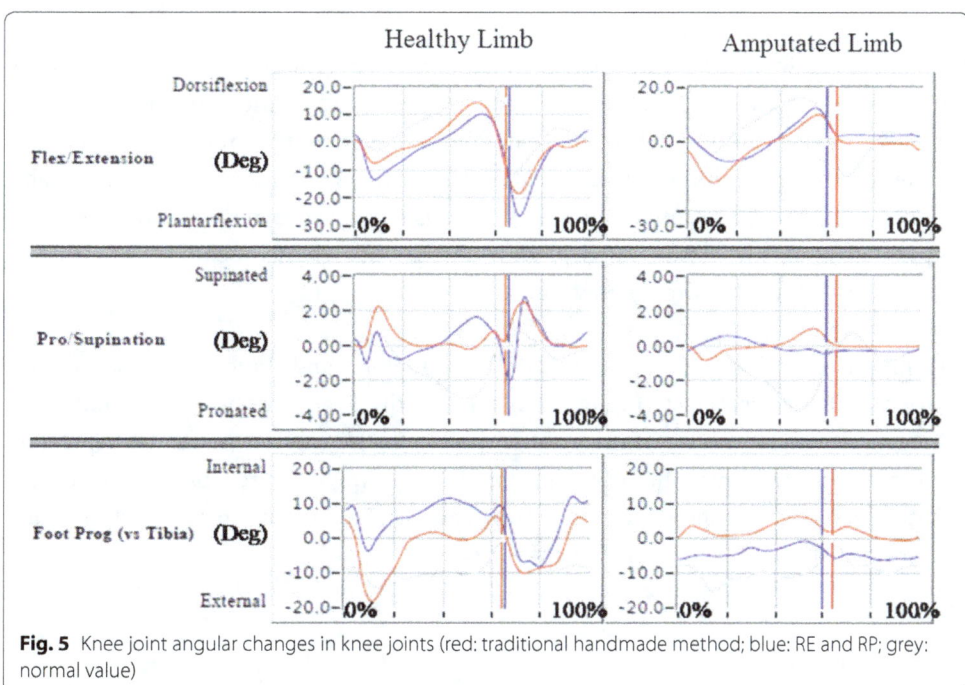

Fig. 5 Knee joint angular changes in knee joints (red: traditional handmade method; blue: RE and RP; grey: normal value)

to extend fully due to limited knee extension, lack of peak values, and socket locking inefficiency.

Angular changes in ankle joints are presented in Fig. 6. In the sagittal plane, greater ankle dorsiflexion range of motion and better performance in the healthy limb during the entire stance phase was observed when the prosthetic socket was used that

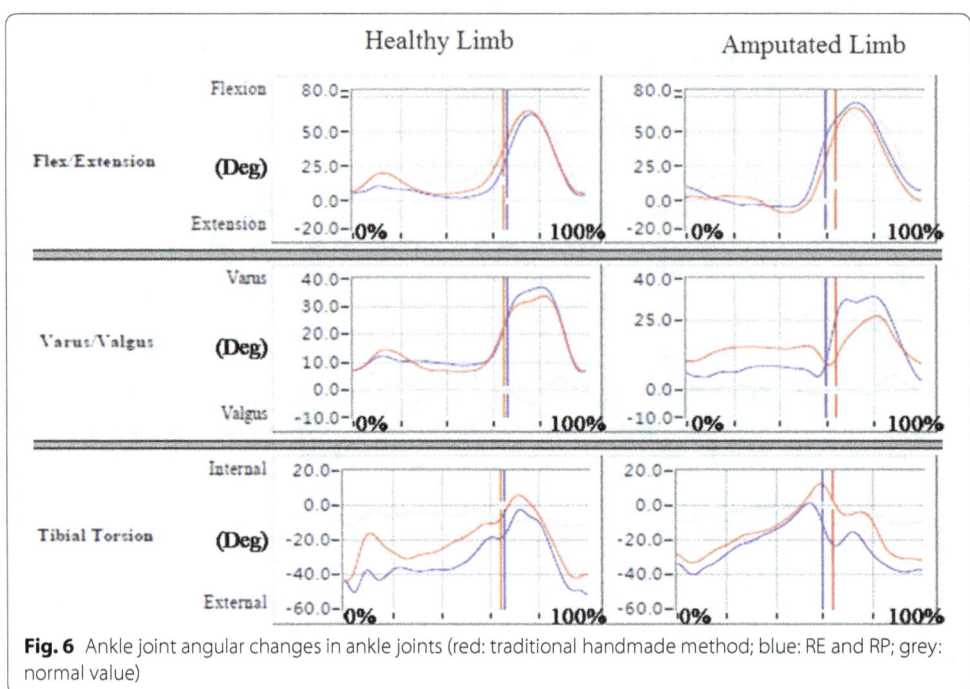

Fig. 6 Ankle joint angular changes in ankle joints (red: traditional handmade method; blue: RE and RP; grey: normal value)

was fabricated with the traditional handmade method. Due to the prosthetic socket fabricated using RE and RP being firmer in terms of adjustment to the amputated limb, the range of motion in the amputated limb in its case was not large. Depending on the socket fabrication method, peak values for the ankle joint angle considerably differed in the healthy limb in coronal and transverse planes and were higher in the amputated limb in the transverse plane when the prosthetic socket was fabricated with the traditional handmade method.

Comparison of spatial and temporal parameters and kinematics parameters showed that for greater step lengths in the healthy limb, the hip, knee, and ankle joint angles varied greatly in the coronal plane. In the sagittal plane, higher values for the amputated limb hip, knee, and ankle joint angles were observed when the prosthetic socket was fabricated using RE and RP. This is due to larger step length and shorter stance phase in case of such sockets.

Dynamics analysis

Measurements of force in the amputated limb obtained using a force plate are given for prosthetic sockets fabricated using the traditional handmade method (see Fig. 7) and RE and RP (see Fig. 8). Two stress peaks occur in stance phases of 25% and 75%, particularly, when two-limb support changes into one-limb support with one limb leaving the ground and when one-limb support changes again into two-limb support with the leg stepping on the ground. Due to the lighter weight of the prosthetic socket fabricated using RE and RP, its force values were lower.

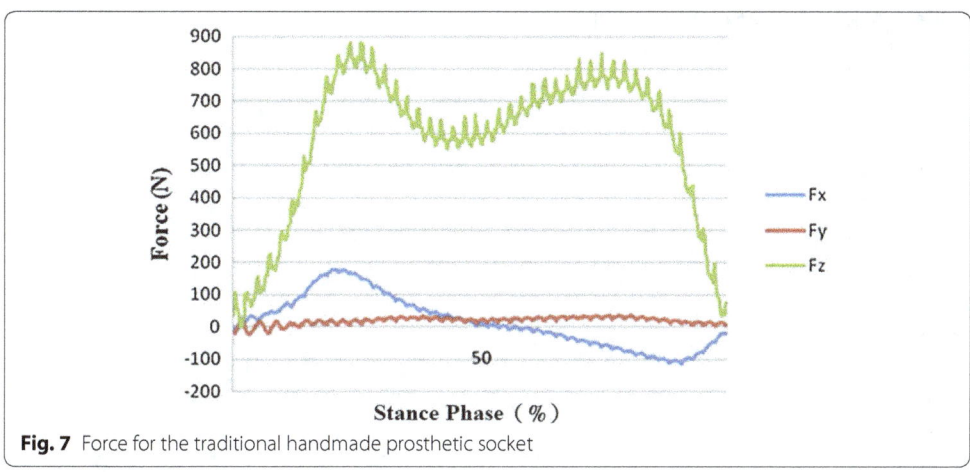

Fig. 7 Force for the traditional handmade prosthetic socket

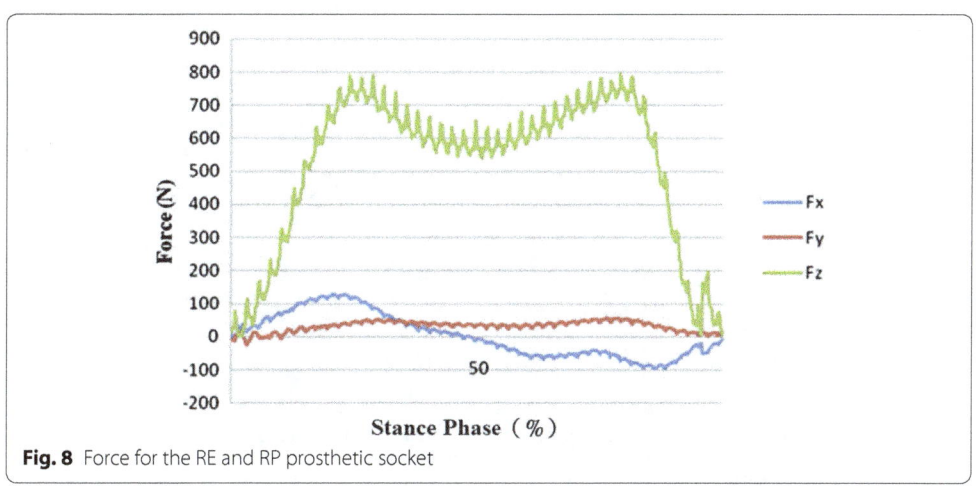

Fig. 8 Force for the RE and RP prosthetic socket

Pressure pain measurement results

A pressure pain detector was used to measure pressure pain in the participants. A specialist made measurements in pressure-tolerant (patella tendon, anterior tibia muscle, medial tibia flare, medial tibia angle, and calf muscle) and pressure-relief (tibia crest, tibia end, fibular head, and fibular end) areas of the amputated limb to compare pressure pain tolerance in these areas. Empirical analysis results indicated that, on average, pressure pain tolerance in pressure-tolerant areas was higher than that in pressure-relief areas, with the highest and lowest values observed in the areas of patella tendon and fibular end, respectively, in both participants. Pressure pain measurement results for pressure-tolerant and pressure-relief areas are provided in Table 4, Figs. 9, and 10.

Stump–socket interface stress results

This study sought to examine stump–socket interface stress distribution in participants wearing prosthetic sockets fabricated using the traditional handmade method and RE and RP while walking at their normal gait velocity. Additionally, based on the dynamics data obtained from the force plate, boundary conditions were set for the quasi-static

Table 4 Measurement results (unit N)

Measurement area	Case 1	Case 2
Pressure-tolerant areas (unit N)		
Tibias anterior muscle	12	11
Medial tibia flare	9.5	11
Calf muscle	12.5	9
Medial tibial angle	11	9.5
Patella tendon	14.5	19
Pressure-relief areas (unit N)		
Tibia end	7.5	9.3
Tibia crest	13.2	9.5
Fibular head	9.2	10.2
Fibular end	7.7	8.5

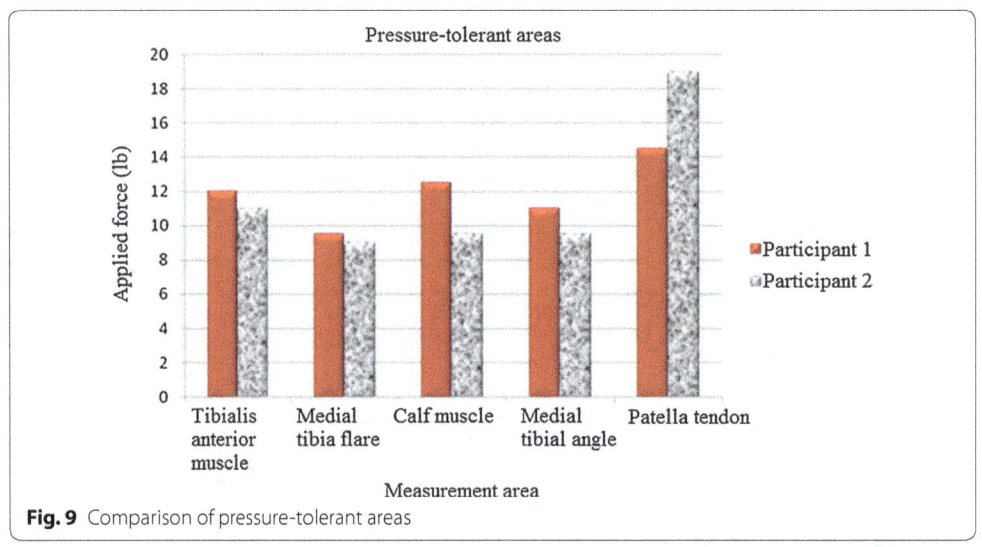

Fig. 9 Comparison of pressure-tolerant areas

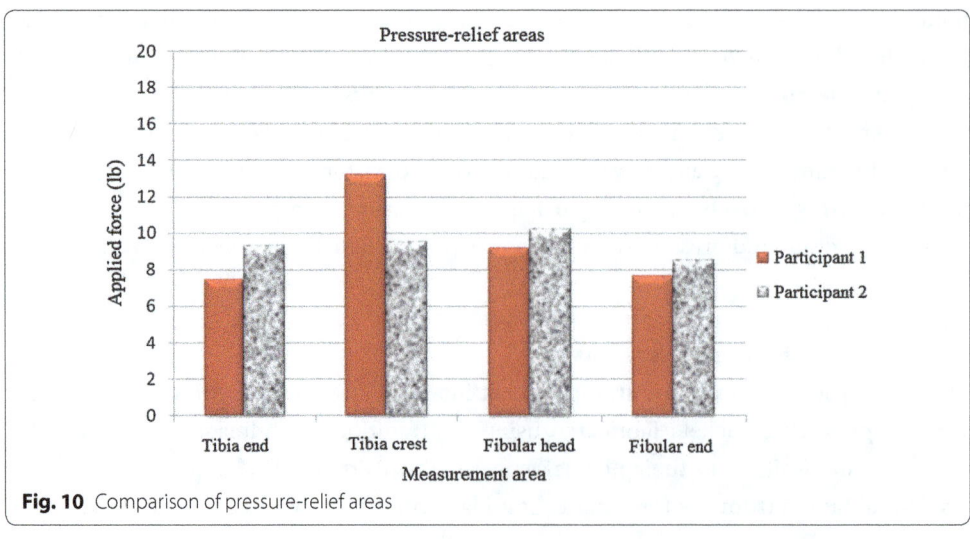

Fig. 10 Comparison of pressure-relief areas

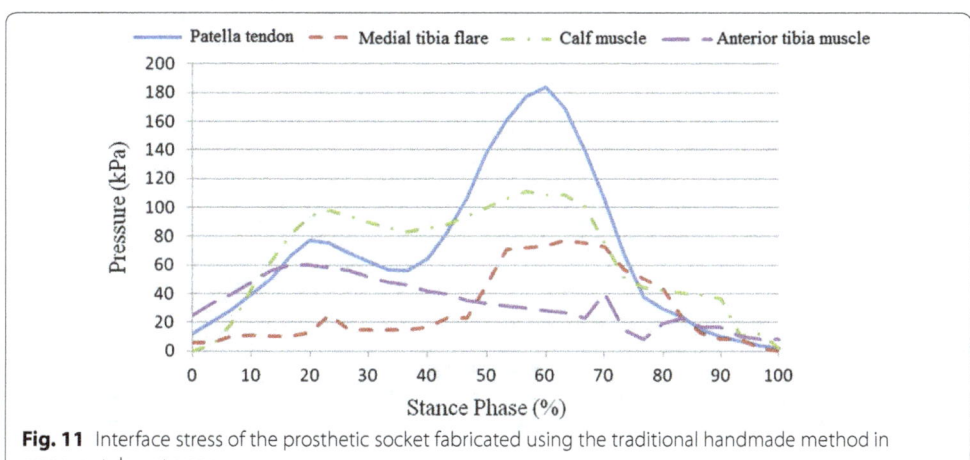

Fig. 11 Interface stress of the prosthetic socket fabricated using the traditional handmade method in pressure-tolerant areas

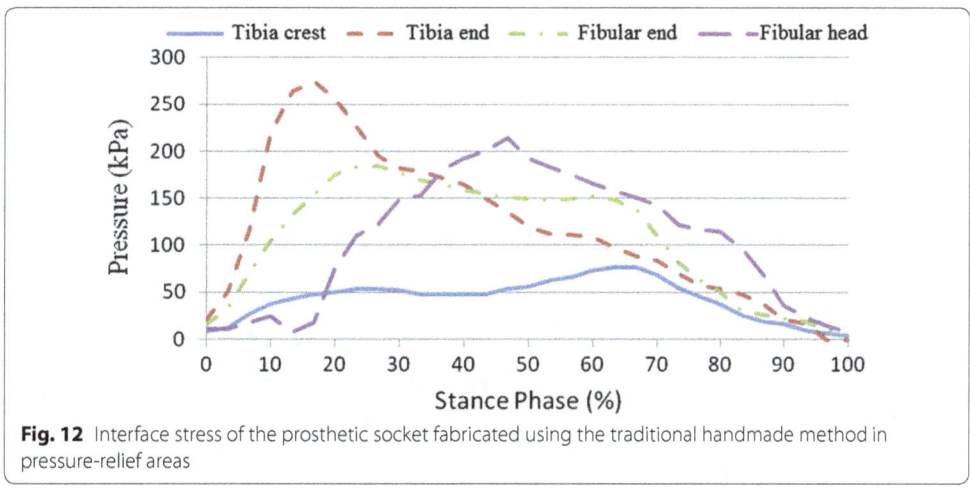

Fig. 12 Interface stress of the prosthetic socket fabricated using the traditional handmade method in pressure-relief areas

finite element model used for interface stress analysis, followed by comparison of empirical results. Interface stress was examined in a total of eight areas, including pressure-tolerant (patella tendon, medial tibia flare, calf muscle, and anterior tibia muscle) and pressure-relief (tibia crest, tibia end, fibular end, and fibular head) areas. Table 5 shows the peak values for stump–socket interface stress in different areas for the participants

Table 5 Interface stress of different prosthetic sockets (unit kPa)

Measurement area	Traditional handmade method	RE and RP
Patella tendon	183.85	113.02
Medial tibia flare	72.52	107.81
Calf muscle	110.93	107.77
Anterior tibia muscle	60.41	95.48
Tibia crest	77.08	77.67
Tibia end	274.79	372.10
Fibular end	183.85	173.14
Fibular head	214.65	192.12

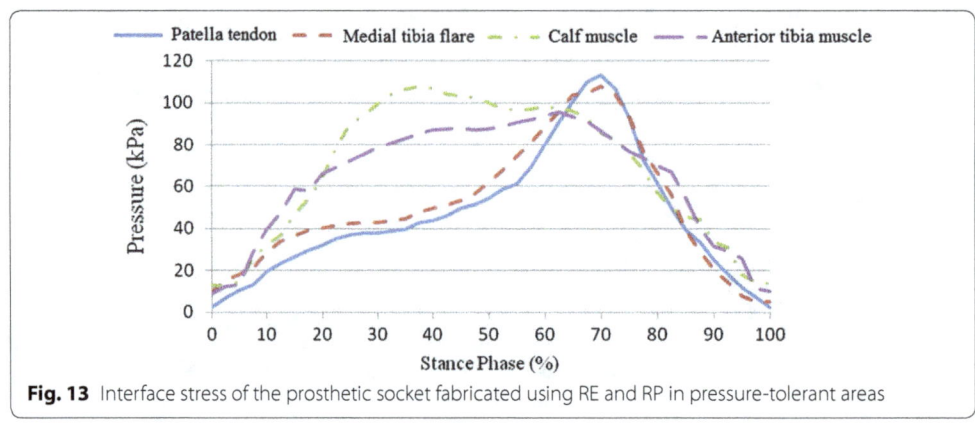

Fig. 13 Interface stress of the prosthetic socket fabricated using RE and RP in pressure-tolerant areas

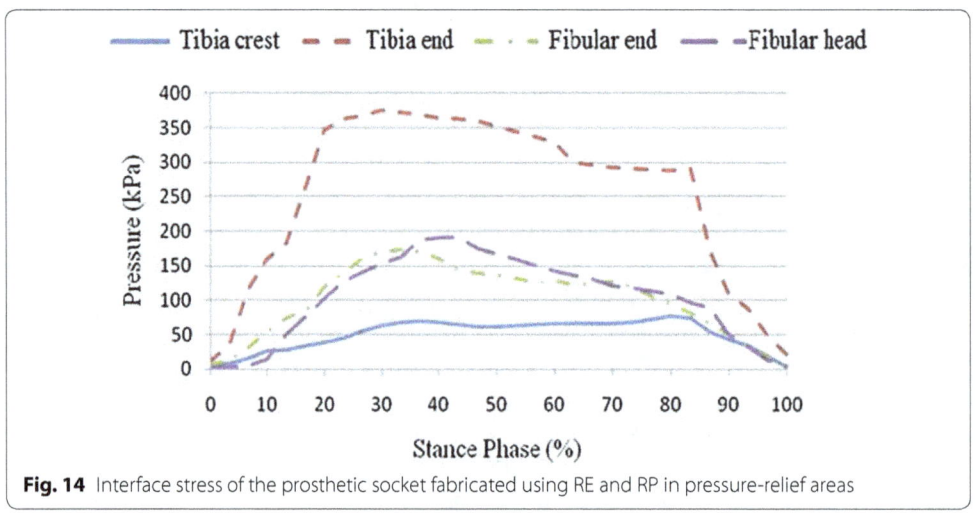

Fig. 14 Interface stress of the prosthetic socket fabricated using RE and RP in pressure-relief areas

wearing the different prosthetic sockets. Figures 11 and 12 show the measurement results for the interface stress between a stump and the socket fabricated using the traditional handmade method in pressure-tolerant and pressure-relief areas, respectively. Figures 13 and 14 show the measurement results for the interface stress between a stump and the socket fabricated RE and RP in pressure-tolerant and pressure-relief areas, respectively.

As seen from Figs. 11, 12, 13, and 14, regardless of the socket fabrication method, the highest interface stress values were observed in the patella tendon among pressure-tolerant areas and the tibia end among pressure-relief areas. The interface stress in the patella tendon area was greater for the participant who wore the prosthetic socket fabricated with the traditional handmade method, meaning that such a socket received more loading than that which was fabricated using RE and RP, thus, prolonging the stance phase of the amputated limb. The interface stress in the medial tibia flare area was smaller for the participant who wore the prosthetic socket fabricated using the traditional handmade method, meaning that such a socket was larger than that which was fabricated using RE and RP, thus, imposing less stress on the enclosed stump. No substantial difference was found in interface stress values in the areas of the calf muscle and anterior tibia muscle for the two participants. In the fibular end area, stump–socket interface stress was larger for the participant who wore the prosthetic socket fabricated with the traditional handmade method due to its larger size; whereas, the interface stress in the tibia end was relatively high in both cases, while being more evident in case of the socket made using the RE and RP fabrication method.

Satisfaction rating scale analysis results

The satisfaction rating scale was used to analyze amputees' satisfaction with the prosthetic sockets fabricated with the traditional handmade method and RE and RP in terms of their comfort level and mobility. The comfort dimension included items regarding the ease of taking on and off the prosthesis and factors affecting the amputee's subjective perception, such as aesthetic appearance, smell, and sound of the prosthesis (Table 6). Comfort-related satisfaction results are presented in Fig. 15. The mobility dimension was related to the participants' walking and included stability when walking up or down stairs or slopes, convenience when getting in and out of vehicles, and the ability to walk on slippery surfaces or in narrow places (Table 7). Mobility-related satisfaction results are presented in Fig. 16. The results showed that, for comfort, the prosthetic sockets fabricated using RE and RP showed better results than those fabricated with the traditional handmade method in terms of prosthesis weight and the amount of required prosthetic socks; whereas, no significant difference was observed for other items. In the mobility dimension, the prosthetic sockets fabricated with the traditional handmade method outperformed those fabricated using RE and RP. The results showed that, at the time of the survey, the participants were more satisfied with the prosthetic sockets fabricated with the traditional handmade method.

Conclusions

This study sought to compare amputees' satisfaction with prosthetic sockets fabricated under different processing conditions, particularly, using RE and RP and the traditional handmade method. The main factors affecting their satisfaction with the prostheses were gait normality and stump–socket interface stress distribution. For the sake of accuracy, this study conducted an empirical analysis of gait and interface stress and measured the prosthesis users' satisfaction levels. The following conclusions were made based on the results:

1. With regards to gait analysis, a greater difference in step length between the amputated and healthy limbs was observed in the participant who wore the prosthetic socket fabricated using RE and RP, meaning that asymmetry between two limbs is a serious issue in the use of such prosthetic sockets. Observation of the gait cycles found that the healthy limbs had greater stance phases than the amputated limb. This indicated that the participant tended to put a greater load on the healthy limb, while not willing to load the amputated limb, which was more evident in case of the pros-

Table 6 Comfort level dimension content

Item	Score	Subjective perception content	Patient's subjective perception
The prosthesis feels comfortable when standing up and sitting down	1	Absolutely uncomfortable	
	2	A bit uncomfortable	
	3	Satisfactory	
	4	Comfortable	
	5	Very comfortable	
The prosthesis is easy to take on and off	1	Impossible	
	2	Very difficult	
	3	Moderately difficult	
	4	A bit difficult	
	5	Easy	
The prosthesis feels light	1	Very heavy	
	2	A bit heavy	
	3	Satisfactory	
	4	Light	
	5	Very light	
After wearing the prosthesis for how long does it feel hot?	1	1 h	
	2	1–4 h	
	3	4–7 h	
	4	7–11 h	
	5	12 h and more	
How many prosthetic socks are needed?	1	Silk socks × 1; silica gel socks × 1; thick quilted socks × 3	
	2	Silk socks × 1; silica gel socks × 1; thick quilted socks × 2	
	3	Silk socks × 1; silica gel socks × 1; thick quilted socks × 1	
	4	Silk socks × 1; silica gel socks × 1; thin quilted socks × 1	
	5	Silk socks × 1; thin quilted socks × 2	
The prosthesis makes weird sounds	1	Always	
	2	Frequently	
	3	Sometimes	
	4	Rarely	
	5	Never	
When on, the prosthesis causes an allergic reaction to skin	1	Always	
	2	Frequently	
	3	Sometimes	
	4	Rarely	
	5	Never	
It is possible to wear any type of shoes	1	Impossible	
	2	Usually possible	
	3	Sometimes possible	
	4	Usually possible	
	5	Possible	
The prosthesis is resistant to dirt and easy to clean	1	Impossible to clean	
	2	Usually impossible to clean	
	3	Sometimes possible to clean	
	4	Usually possible to clean	
	5	Possible to clean	

Table 6 (continued)

Item	Score	Subjective perception content	Patient's subjective perception
The prosthesis makes it difficult to dress	1	Always	
	2	Frequently	
	3	Sometimes	
	4	Rarely	
	5	Never	

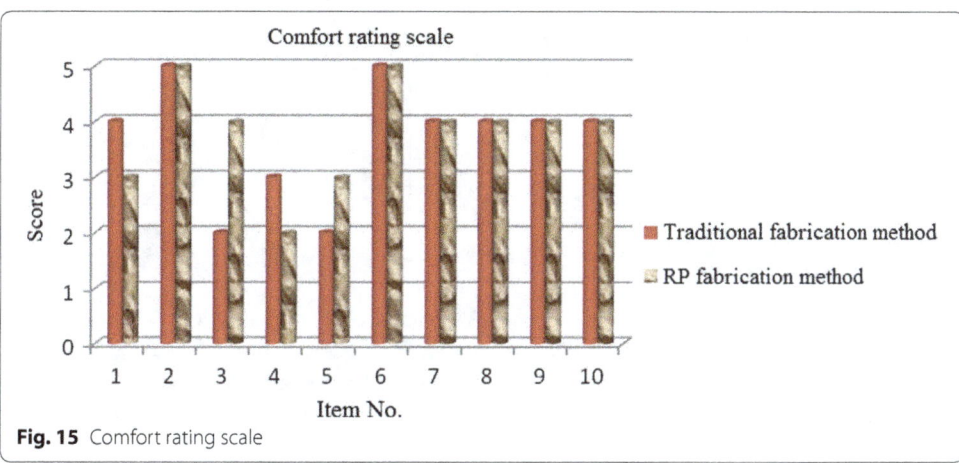

Fig. 15 Comfort rating scale

thetic socket fabricated using RE and RP. Gait analysis data demonstrated that the prosthetic socket fabricated using the traditional handmade method had better characteristics.

2. With regards to stump–socket interface stress, this study found that, regardless of the method used for socket fabrication, most stress was concentrated in tibia end pressure-relief area. This caused discomfort in the area of tibia end to the participant wearing prosthesis. This discomfort was most evident in case when the prosthetic socket was fabricated using RE and RP.

3. Analysis of the survey data obtained with the satisfaction rating scale showed a higher level of satisfaction toward prosthetic sockets fabricated with the traditional handmade method rather than those fabricated using RE and RP. The latter outperformed the former in terms of comfort level, while being inferior in terms of user satisfaction towards the prosthesis mobility. These results imply that mobility is an important consideration when introducing new prosthetic sockets to an amputee.

4. Survey data analysis showed that important factors for prosthesis users included the sensation of heat, ease of taking on and off the prosthesis, and its weight and mobility. Another important issue is reducing back and lower back pain in prosthesis users. It is hoped that the survey results can serve as a reference for the design and fabrication of prosthetic sockets.

Table 7 Mobility dimension content

Item	Score	Subjective perception content	Patient's subjective perception
How long can you walk while wearing the prosthesis?	1	Less than 1 h	
	2	1–4 h	
	3	4–7 h	
	4	7–11 h	
	5	More than 12 h	
Do you feel stable when walking on uneven surfaces while wearing the prosthesis?	1	Absolutely unstable	
	2	Very unstable	
	3	Moderately unstable	
	4	A bit unstable	
	5	Stable	
Is it difficult to walk fast with the prosthesis for half an hour?	1	Impossible	
	2	Very difficult	
	3	Moderately difficult	
	4	A bit difficult	
	5	Easy	
It is difficult to walk up and down stairs while wearing the prosthesis?	1	Impossible	
	2	Very difficult	
	3	A bit difficult	
	4	Able to keep pace with a healthy person	
	5	Walking as a healthy person	
How much physical strength is required to use the prosthesis?	1	Unable to walk	
	2	Requires much effort	
	3	Requires some effort	
	4	Requires little effort	
	5	Does not require any effort	
Can you walk stably up and down slopes while wearing the prosthesis?	1	Impossible	
	2	Very difficult	
	3	A bit difficult	
	4	Able to keep pace with a healthy person	
	5	Walking as a healthy person	
Can you get in and out of a vehicle while wearing the prosthesis?	1	Impossible	
	2	Very difficult	
	3	Moderately difficult	
	4	A bit difficult	
	5	Easy	
Can you walk on slippery surfaces while wearing the prosthesis?	1	Impossible	
	2	Very difficult	
	3	A bit difficult	
	4	Able to keep pace with a healthy person	
	5	Walking as a healthy person	
Can you walk stably on narrow streets while wearing the prosthesis?	1	Absolutely unstable	
	2	Very unstable	
	3	Moderately unstable	
	4	A bit unstable	
	5	Stable	

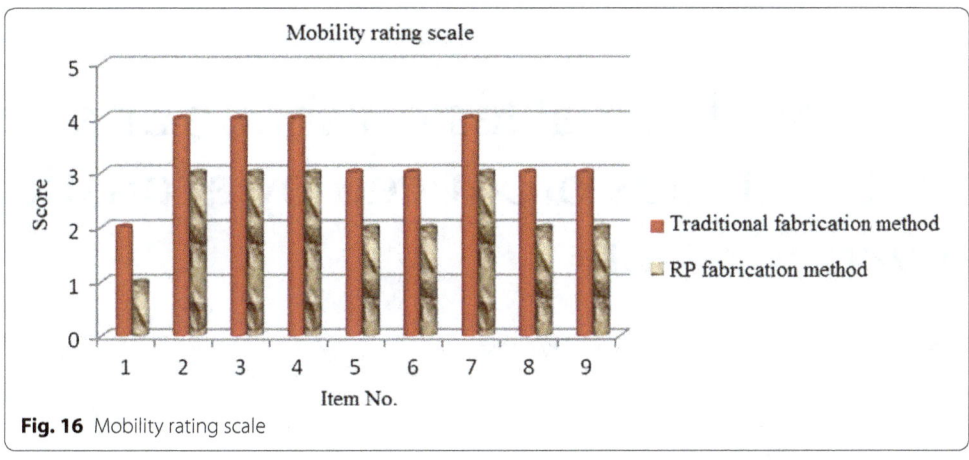

Fig. 16 Mobility rating scale

Abbreviations
CAD: computer-aided design; CAE: computer-aided engineering; RE: reverse engineering; RP: rapid prototyping; NRS: numeric rating scale; VAS: visual analogue scale; CT: computed tomography.

Declarations
Author's contributions
C-HH, C-HO, W-LH and Y-HG have designed the study protocol and written the paper. They measurement and analyze interpret the data. C-HH has given final approval and decided to submit the paper for publication. C-HO designed the study. W-LH and Y-HG construct of a plaster model and CAD of the stump analysis and collect data. C-HH designed and supervises statistical analysis. C-HO carry out gait analysis, W-LH and Y-HG collect and analysis kinematics parameters. All authors read and approved the final manuscript.

Acknowledgements
Not applicable.

Competing interests
The authors declare that they have no competing interests.

Funding
The study is funded by the Department of Mold and Die Engineering, National Kaohsiung University of Science and Technology, research project.

References
1. Robinson JL, Smidt GL, Arora JS. Accelerographic, temporal and distance gait factors in below-knee amputees. Phys Ther. 1977;57:898–904.
2. Goh JCH, Solomonidis SE, Spence WD, Paul JP. Biomechanical evaluation of SACH and uniaxial feet. Prosthetic Orthotic Int. 1984;8:47–154.
3. Zhang M, Mak AFT, Roberts VC. Finite element modeling of a residual lower-limb in a prosthetic socket: a survey of the development in the first decade. Med Eng Phys. 1998;20(5):360–73.
4. Downie WW, Leatham PA, Rhind VM, Wright V, Branco JA, Anderson JA. Studies with pain rating scales. Ann Rheum Dis. 1978;37:378–81.

9

The effects on thermal lesion shape and size from bubble clouds produced by acoustic droplet vaporization

Ying Xin[1], Aili Zhang[1]* ⓘ, Lisa X. Xu[1] and J. Brian Fowlkes[2]*

*Correspondence:
zhangaili@sjtu.edu.cn;
fowlkes@umich.edu
[1] School of Biomedical
Engineering, 400 Med-X
Research Institute, Shanghai
Jiao Tong University, 1954
Huashan Rd, Shanghai, China
[2] Department of Radiology,
University of Michigan Health
System, 3226C Medical
Sciences Building I, 1301
Catherine Street, Ann Arbor,
MI, USA

Abstract

Background: Bubbles formed by acoustic droplet vaporization (ADV) have proven to be an effective method for significant enlargement of the thermal lesions produced by high intensity focused ultrasound (HIFU). We investigated the influences of bubble cloud shape and droplet concentration on HIFU thermal lesions, as these relate to the ADV technique.

Methods: Unlike previous studies where the droplets were simultaneously vaporized with the HIFU exposure for thermal lesion formation, droplets were vaporized by pulse wave (PW) ultrasound prior to continuous wave (CW) ultrasound heating in this experimental study. Under different experimental conditions, we recorded and quantified by the image processing methods the morphology and size of the bubble clouds created and the corresponding thermal lesions formed.

Results: The results demonstrated that different ADV droplet concentrations produced a variety of thermal lesion shapes and sizes. The lesion volume could be increased using PW ultrasound followed by CW exposure, especially for higher droplet concentrations, e.g. 3.41×10^6/mL yielded a tenfold increase over that seen using CW alone.

Conclusion: These findings could lead to optimization of HIFU therapy by selecting a bubble forming strategy and droplet concentrations, especially using lower ultrasound powers which is desirable in clinical applications.

Keywords: Acoustic droplet vaporization (ADV), Bubble cloud, Thermal lesions, HIFU ablation

Introduction

High intensity focused ultrasound (HIFU) treatment is gaining popularity due to its noninvasiveness. By focusing ultrasound energy in a small area inside the body, the temperature can be elevated to over 56 °C in seconds and cause irreversible necrosis and damage of the tumor tissue [1]. However, due to the need to protect normal tissue and the small lesion size from one single application of HIFU, the process is time consuming, especially for large tumors which may require a treatment time of several hours [1, 2].

To shorten the treatment time, researchers have tried to use bubbles to help increase local heating [3–9]. Microbubbles can be introduced into tissue by injection of ultrasound contrast agents (UCA) containing bubbles, using high rarefactional pressure to trigger cavitation for generation of bubble nuclei or by using acoustic droplet vaporization (ADV) where liquid droplets are injected and then vaporized to bubbles in the HIFU focal region [10]. Among these techniques, use of UCA and ADV droplets requires lower acoustic pressure. While the bubbles are delivered to the whole body of the patient when using UCA, the ADV droplets technique ensures that the bubbles only exist in the focal region and therefore eliminates the possible shift of the HIFU focus and other effects due to the UCA bubbles being throughout the ultrasound propagation path [6].

Phantom and animal experiments have confirmed that ADV-assisted HIFU can achieve a 3–15 times larger lesion than HIFU alone [11, 12], so the number of HIFU exposures required to treat the whole tumor can be significantly reduced. The bubbles formed in the HIFU focal region by this technique help increase heat deposition based on three mechanisms: (1) bubble oscillations driven by ultrasound, and corresponding viscous heating; (2) the high frequency mechanical waves generated by the bubble oscillation; (3) the trapping of the ultrasound in the bubble network and the correspondingly lengthened propagation path and increased absorption of the tissue. Meanwhile, as bubbles are so efficient at converting energy, the ultrasound energy attenuates much faster along the propagation path in the presence of bubbles, leading to much lower ultrasound energy at distal locations [6, 13, 14]. From the mechanisms of bubbles increasing heat deposition, it could be easily speculated that the size and morphology of the thermal lesions are dependent on both the nature of the bubble region (morphology of the region, the bubble size and concentration, etc.) and the ultrasonic field.

In addition to triggering the droplets in the HIFU focus to help increase thermal absorption, ADV can create a bubble shield against ultrasound wave propagation to protect normal tissue [13], or to further confine the ultrasound wave in the targeted treatment region [15]. The high impedance difference of the bubble region leads to more reflection and scattering over the bubble cloud surface and the redistribution of the ultrasound energy. The acoustic impedance (density × sound speed) of the bubble region is dependent on bubble size distribution, bubble concentration and the ultrasound frequency [16].

From these previous studies, we can see that the bubble region dimension and the density and size of the bubbles inside are critical to the shape of the subsequent thermal lesions formed by HIFU. As the ADV droplets can provide a way to create desired shapes of bubble clouds, understanding how the local bubble region characteristics may influence the thermal lesions by HIFU would be essential for planning a much more efficient ADV enhanced HIFU treatment strategy. Meanwhile, it also provides a way to treat the tumor according to the actual tumor shape, while protecting the critical, normal tissue structures outside the tumor.

Thus, in this study, different shapes of bubble clouds were formed by the ADV technique with pulse wave (PW) ultrasound from the same transducer prior to the continuous wave (CW) HIFU thermal treatment. In this paper, we analyze and discuss the influences of droplet concentration and acoustic pressure amplitude on the shape and volume of the bubble clouds and associated thermal lesions.

Materials and methods

Experimental setup

The experimental setup is illustrated in Fig. 1a. The HIFU transducer was driven by the RF amplifier (AG1006, T&C Power Conversion, Rochester, NY). The input voltage to the transducer was monitored by an oscilloscope (MSO-X 2022A, Agilent Technologies, Inc., Santa Clara, CA). Two function generators (33210A, 33220A, Agilent Technologies, Inc., Santa Clara, CA) were used to send RF signals to the amplifier. One function generator gated the other to assure that signals of a certain number of pulses or a certain duration of continuous wave RF were input to the amplifier.

The HIFU transducer used in the study was a single element spherically focused transducer (H-108, Sonic Concepts, Inc., Bothell, USA), with a center frequency of 2.7 MHz. The aperture and focal length of the transducer were 60 mm and 50 mm, respectively. The transducer was calibrated for applied voltages of 1–21 V using a fiber optic hydrophone (FOH/1, Precision Acoustics Ltd., UK). For the higher input voltages, the forward electrical power was measured by an internal circuit in the power amplifier and the acoustic power from the transducer was calculated based on the transducer specification of its efficiency. The acoustic power was used to calculate the focal pressure using a modified KZK equation for low F-number transducers [17]. The output pressure of the transducer is a function of input voltage and its peak positive and negative values are shown in Fig. 1b. For the 150 V pulse wave, the electric power was extrapolated from the measured electric power at lower voltages (20–60 V). The corresponding focal pressure was then calculated from the modified KZK equation with this input power. As the transducer doesn't fully ring up at the three cycles pulses (the PW mode), the calculated pressure was scaled by the same parameter obtained from the calibration for lower input voltages (0.87 for peak positive and 0.85 for peak negative). Due to the nonlinear nature of the ultrasonic waveform, the input voltage is used in the following text. The corresponding acoustic pressures in gel phantom were also calculated, and listed in Table 1 with comparison to those in water.

The transducer and the phantom samples were immersed in 37 °C water. The focus of the transducer was positioned at a depth of 40 mm in the phantom. A B-mode scanner (MyLab 90, Esaote, Italy) was aligned with the focal axis of the HIFU transducer to

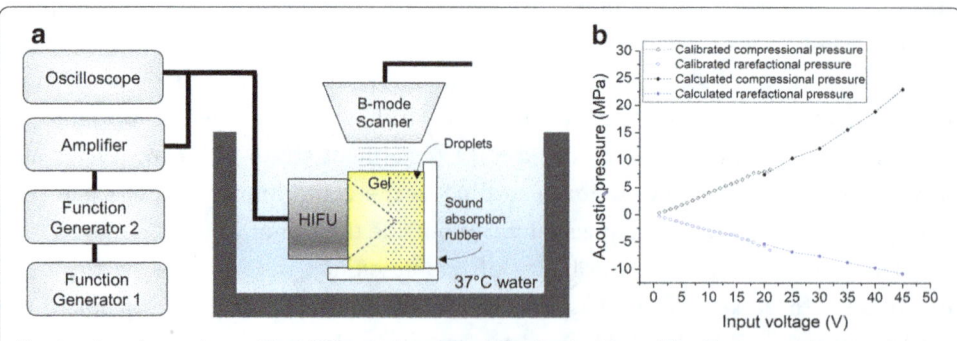

Fig. 1 **a** Experimental setup. The HIFU transducer was driven by the RF amplifier. The input voltage to the transducer was monitored by an oscilloscope. Two function generators were used to send a certain number of pulses or a certain duration of continuous wave RF signals to the amplifier. **b** The output pressure of the transducer as function of input voltage. For applied voltages of 1–21 V, the transducer was calibrated by a hydrophone. For higher input voltages, the pressures were calculated by a modified KZK equation [17]

Table 1 Estimated focal pressures of H-108 in water and in gel using a modified KZK equation [17]

Input voltage V	In water		In gel	
	Compressional pressure MPa	Rarefactional pressure MPa	Compressional pressure MPa	Rarefactional pressure MPa
20	7.3	−5.3	5.0	−3.7
25	10.4	−6.8	7.0	−4.7
30	12.2	−7.6	8.3	−5.3
35	15.6	−8.7	10.6	−6.1
40	18.9	−9.7	12.8	−6.8
45	23.0	−10.7	15.5	−7.5
150	136.9	−15.9	105.8	−13.8

The water density, the sound speed and sound attenuation coefficient in water used for calculations were 993 kg/m^3, 1523.7 m/s [18] and 0.04 Np/m [19], respectively. The phantom density, sound speed and sound attenuation coefficient in phantom used were 995 kg/m^3 [20], 1543 m/s [20] and 6.8 Np/m [20], respectively

monitor and record the bubble cloud created. The B-mode scanner was set to operate at the lowest output power to avoid any effects during the experiments. The images of the phantom were recorded both before and after PW and/or CW ultrasound exposure.

Preparation of perfluoropentane droplets and tissue-mimicking phantom

Lipid coated perfluoropentane droplets were prepared according to a previously published method [10]. A 750 μL volume of lipid blend was made by dissolving 1.9 mg DPPC (1,2-dipalmitoyl-*sn*-glycero-3-phosphocholine, 850355P, Avanti Polar Lipids, Inc., Alabaster, AL) and 0.08 mg DPPA (1,2-dipalmitoyl-*sn*-Glycero-3-Phosphate, 830855P, Avanti Polar Lipids, Inc., Alabaster, AL) in propylene glycol and 8:1 (v/v) saline-glycerol (Glycerol, Shanghai Chemical Reagent Co,. Ltd., China). Then 250 μL perfluoro-*n*-pentane (09-6182, Strem Chemicals, Inc., Newburyport, MA) was added to the lipid blend. The resulting mixture was sonicated for 30 s to produce the droplet emulsion. The droplet size distribution of the emulsion was measured by a Coulter counter (Multisizer 3, Beckman Coulter, Inc., Fullerton, CA) and can be found in Additional file 1. The droplet concentration in the emulsion was 3.41×10^{10}/mL, and the resulting mean diameter of the droplets was 0.89 μm.

Tissue-mimicking phantoms that can visualize thermal lesions were made by the same procedure as in previously published literature [20]. The phantom consists of 31.4% (v/v) water, 35% (v/v) egg white, 33% acrylamide solution (30% w/v, 19:1 acrylamide/bis-acrylamide, B161054, Decent Biotech (Shanghai) Co., Ltd., China), 0.5% of 10% (w/v) ammonium persulfate (Shanghai Maibio Co., Ltd, China) and 0.1% TEMED (*N,N,N',N'*-tetramethylethylenediamine, 15524010, Thermo Fisher Scientific Inc., Waltham, MA). The gel solution was degassed to remove potential gas nuclei. Different volumes of droplet emulsion were added to the gel solution to achieve different droplet concentrations inside the phantom. Each phantom had a size of 120 mm × 70 mm × 50 mm. The transducer was focused at a depth of 40 mm (along the 50 mm dimension) from the front surface of the phantom for each HIFU exposure. For all phantoms containing droplets, to eliminate any possible scattering of the droplets along the ultrasound wave propagation path, the phantom was composed of two halves, one proximal half was free of droplets, while the distal half of the phantom had the desired droplet concentration (as illustrated in Fig. 1a). The HIFU transducer was focused at the distal half, with each exposure placed at least 10 mm apart to avoid interference from each other.

Treatment strategies

Two treatment strategies were used in the experiments to investigate various characteristics of the bubble clouds and thermal lesions created in phantoms with different droplet concentrations. In the first strategy, the phantom was only exposed to 10 s of continuous wave ultrasound (CW alone). In the second strategy, the phantom was first exposed to a short pulse wave ultrasound to vaporize the ADV droplets in the phantom, then it was exposed to the 10 s of continuous wave to create thermal lesion (PW combined CW). Six different applied voltages were used for the CW ultrasound, 20 V, 25 V, 30 V, 35 V, 40 V and 45 V. The pulse wave used to vaporize droplets was 200 pulses with a pulse repetition frequency of 1 kHz, each pulse consisting of three cycles at 150 V. Five droplet concentrations were investigated in the experiment, 1.07×10^5/mL, 3.41×10^5/mL, 1.07×10^6/mL, 3.41×10^6/mL and 1.07×10^7/mL. Control experiments were conducted in phantoms free of droplets, which were exposed to the CW ultrasound at 30 V for 10 s and to 40 V for 60 s.

Measurement and data analysis

All B-mode images were converted to 8-bit images. Images of the same phantom before and after a given ultrasound exposure were subtracted. Then the threshold method of Isodata [21] was used to determine the outlines of the bubble clouds. The procedure divides the image into object and background by using an initial threshold, then calculates the averages of the pixels at or below the threshold and as well as those pixels above the threshold. The average of these two values are calculated, the threshold increased and the process repeated until the threshold is larger than the composite average. In the cases of high droplet concentration, the B-mode images following ADV attenuate significantly along the direction of the B-mode ultrasound propagation. Therefore, the lower part of the bubble cloud cannot be clearly seen and the upper outlines of the bubble clouds were quantified. This was acceptable as the bubble clouds are expected to be symmetric around the propagation axis of the transducer. The method is illustrated in Fig. 2. Then the outlines of the bubble clouds in the repeated experiments were aligned using the matching method of normalized correlation coefficient [22], which is widely used in pattern matching and object recognition. The bubble cloud outlines obtained with the same experimental conditions were averaged, and the standard deviations were obtained accordingly.

The volume of the bubble cloud was calculated by the equation

$$V = \int_0^L \pi r(z)^2 dz \tag{1}$$

where L is the total length of the bubble cloud, and $r(z)$ is the outline curve of the bubble cloud.

After the ultrasound exposure and B-mode imaging, the phantom was taken out of the water and dissected to a thickness such that the thermal lesions could be visually

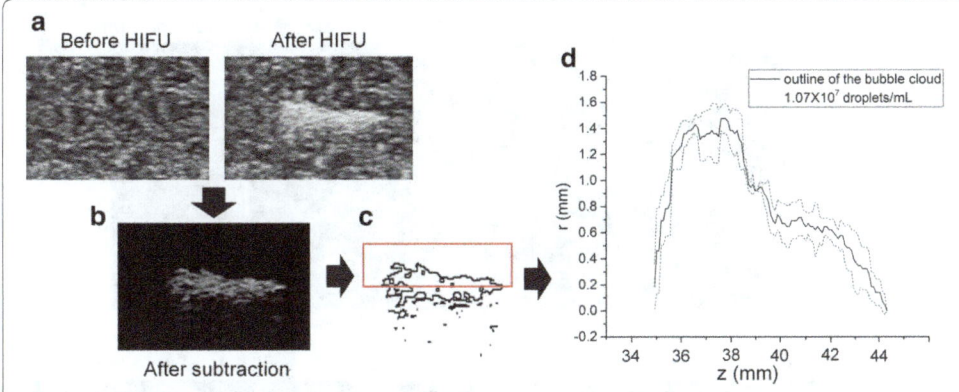

Fig. 2 **a** B-mode images of the phantom before and after HIFU treatment with the associated bubble cloud appearing in the latter. **b** The subtraction of the two images in (**a**). **c** The outline of the bubble clouds determined using methods described in the text. The red square marks the upper half of the bubble cloud. **d** The averaged outline over repeated experiments was obtained (solid line). The dashed lines represent the averaged outline ± one standard deviation of the outline (n ≥ 4)

observed. To clearly distinguish the thermal lesion from the bubble cloud, the phantom was placed in a container and pressurized to 3 MPa for 1 min. The bubbles in the phantom were thus compressed, leaving thermal lesions clearly seen. The thermal lesions were photographed and the same image processing method for quantifying the shape of the bubble clouds was used to obtain the outlines of the thermal lesions and calculate their volumes. Student's t-test was used to determine whether there were significant differences between experimental groups.

Results and discussion

Effect of pulse wave (PW) and continuous wave (CW) ultrasound on ADV cloud production

During clinical HIFU treatment, to generate enough heating effect, continuous wave (CW) is usually used. In these cases, the droplets could also be vaporized. Then the vaporized bubbles would oscillate and undergo some of the following processes: (1) some collapse violently; (2) some expand to larger bubbles due to rectified diffusion, then may break into several smaller bubbles; (3) some would merge and become larger bubbles. Thus, using continuous wave ultrasound may result in distribution of bubble sizes in the bubble cloud.

The bubble clouds created with CW ultrasound alone in phantoms with various droplet concentrations are shown in Fig. 3a (row 1). It can be seen that after the phantoms were exposed to 10 s of CW ultrasound with 30 V input voltage, the size of the bubble clouds initially increased slowly with increasing droplet concentration. The peak volume was reached at the concentration of 1.07×10^6/mL (Fig. 3e) then decreased with further increases in concentration.

In order to study the characteristics of ADV bubble clouds formed by PW ultrasound, the droplets were triggered by the HIFU transducer. The phantoms with different droplet concentrations were exposed to 200-pulse PW with an input voltage of 150 V. Example B-mode images obtained are shown in Fig. 3a (row 2). It can be seen that when the droplet concentration was as low as 1.07×10^5/mL, few droplets were vaporized by the exposure, shown as the several echogenic dots in the B-mode

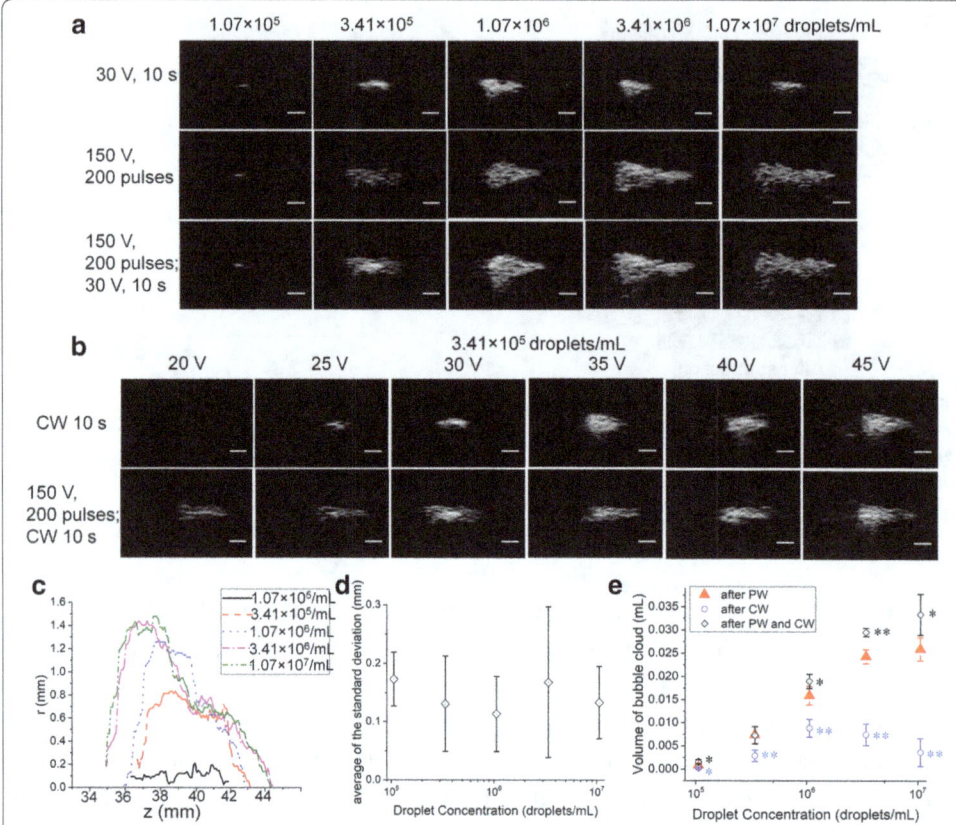

Fig. 3 **a** The bubble clouds created after 10 s exposure to continuous wave (CW) at 30 V (row 1), after exposure to 200 pulses (3 cycles each) of pulse wave (PW) at 150 V (row 2), exposure to 200 pulses of pulse wave at 150 V followed by 10 s of continuous wave at 30 V in the phantoms (row 3). The different droplet concentrations used were 1.07×10^5/mL, 3.41×10^5/mL, 1.07×10^6/mL, 3.41×10^6/mL, and 1.07×10^7/ mL, respectively. **b** The bubble clouds created after 10 s exposure to continuous wave (row 1), exposure to 200 pulses of pulse wave at 150 V followed by 10 s of continuous wave (row 2). The input voltage of the continuous wave were 20 V, 25 V, 30 V, 35 V, 40 V, and 45 V, respectively. The droplet concentration in the phantom was 3.41×10^5/mL. The scale bar represents 2 mm. **c** The averaged outline of the bubble clouds created after the exposure to 200 pulses of pulsed wave at 150 V in phantom with droplet concentration of 1.07×10^5/mL, 3.41×10^5/mL, 1.07×10^6/mL, 3.41×10^6/mL, and 1.07×10^7/mL, respectively. **d** The averaged standard deviation along bubble cloud outline in phantoms with different droplet concentrations after the exposure to 200 pulses of pulse wave at 150 V. **e** The volume of bubble cloud created in cases of different droplet concentrations after PW and after CW exposure. Colored * and ** represents $p \leq 0.05$ and $p < 0.01$ for significance of difference between the group of this color and the group of the PW treatment

image. In the case of 3.41×10^5/mL, more droplets were vaporized in the focal area resulting in a larger elliptically-shaped bubble cloud. In the case of 1.07×10^6/mL, the bubble clouds had a teardrop shape, with a wider proximal area moving toward to the transducer. In the cases of higher droplet concentration, the bubble clouds had a more triangular shape, the 'head' of the bubble cloud is wider and flatter along radial direction, and was closer to the transducer. (At the higher bubble concentrations, the appearance of this cloud in the ultrasound image is likely affected by the attenuation.) The averaged outlines of the bubble clouds taken from the proximal side with respect to the ultrasound imaging are shown in Fig. 3c where the progression of the proximal side of the cloud can be seen as increasingly wider and moving toward the transducer. The average of the standard deviation along the length of the bubble cloud outline was calculated (Fig. 3d) and there was no statistical difference across the droplet concentrations tested, i.e. the width of the transition boundary was similar in all cases. Since the bubble cloud has a larger size in higher droplet concentration cases, the corresponding relative error is lower and therefore better reproducibility is achieved in generating the bubble cloud shape. This could be because at low concentrations, with a limited number of droplets in the focal zone, the bubble clouds created depend much more on the random spatial distribution of the droplets, which varied from case to case. The evolution and variety of bubble cloud shapes for different droplet concentrations might be explained as follows. When the first ultrasound pulse arrives, the acoustic field established in the phantoms is almost the same since the droplets have little influence on the propagation of the ultrasound. But after the first pulse, different numbers of bubbles are created in the same focal area depending on the droplet concentration. A higher bubble density (void fraction) leads to a larger impedance difference, which means more acoustic energy is scattered by the surface of the bubble cloud, so the pressure in front of the bubble cloud is higher in the higher concentration case, leading to a larger volume of bubble cloud after subsequent ultrasound pulses. This implies that less PW power would be required to achieve a same volume of bubble cloud in the presence of a higher droplet concentration, a potentially safer means to the same end.

Compared with the bubble clouds created with the CW only (Fig. 3e), using the PW with an input voltage of 150 V for 200 pulses resulted in a much larger bubble cloud ($p < 0.01$ for droplet concentrations $\geq 3.41 \times 10^5$/mL, $p < 0.05$ for all cases). It can also be seen in Fig. 3a (rows 1 and 2), that the echogenicity created using PW are not as great as in the CW case for phantoms with the same droplet concentration. Significant differences were found in the mean gray value of the bubble cloud created after CW and after PW for droplet concentrations from 3.41×10^5/mL to 3.41×10^6/mL ($p < 0.05$). (Comparison of the results for the lowest and highest concentrations may have been affected by the relatively small size of the bubble clouds for the CW cases.) This difference in echogenicity indicates a likely higher impedance difference in those CW cases. Since the initial droplet concentration is the same, this could imply a larger void fraction has been generated. For the CW case, the temperature of phantom increases, so the thermal expansion of the bubbles may lead to a lower impedance. In addition, during the prolonged bubble oscillation, there may be a net inflow of gas into the bubbles from the surrounding medium due to rectified diffusion [23, 24]. And this process could also help bubbles grow larger. In both cases, the bubbles could break up into smaller ones but the overall void fraction would still increase.

Similar effects were observed when PW is combined with the CW as shown in third row of Fig. 3a. These are the same phantoms shown in Fig. 3a (row 2) but exposed to CW immediately after the PW exposure without moving the phantom. After exposure to PW and CW, the pixel value inside the bubble clouds were significantly higher than those exposed to PW alone, in all cases $p < 0.05$. The estimated volume of the bubble cloud is also significantly larger in these cases than in PW only cases when droplet concentrations were 1.07×10^5/mL, 1.07×10^6/mL, 3.41×10^6/mL and 1.07×10^7/mL ($p \leq 0.05$), which means more bubbles were formed outside of the previous bubble cloud during the CW exposure in these cases. When the droplet concentration was 3.41×10^5/mL, as can be seen from Fig. 3a, the high echogenic regions after CW were all located inside the bubble cloud created by PW only. This causes the insignificant difference between the quantified bubble cloud volume formed by PW only and the combined PW and CW treatment. While for higher droplet concentrations, the bubble density formed from PW only was so high that the acoustic waves from the subsequent CW exposure were scattered outside the preformed bubble cloud and vaporized local droplets and thus resulted in a greatly increased bubble cloud volume. When the concentration of the droplets is as low as 1.07×10^5/mL, the volume of the bubble clouds quantified are very small and there is significant difference among the groups. The quantified volume of bubble cloud formed by PW is larger than that formed by the CW only, but is smaller than the combined exposure. This is different from the case for the droplet concentration of 3.41×10^5/mL because the sparsely distributed bubbles formed by PW are not totally observed by the B-mode image, but are visible after being expanded by the subsequent CW exposure.

The influences of varying the CW voltage are shown in Fig. 3b. The droplet concentration is 3.41×10^5/mL and the phantoms are exposed to CW alone (row 1) and to PW combined with CW (row 2). The input voltages of CW exposures were 20 V, 25 V, 30 V, 35 V, 40 V, and 45 V, respectively. It can be seen that when the input voltage was 20 V, no echogenic area was observed in the B-mode image for the CW alone case, indicating that no droplets were apparently vaporized and the threshold was not achieved even in the focal area. In the cases of 25 V and 30 V, ellipsoidal bubble clouds were created. From 35 to 45 V, the bubble cloud became larger with increasing input voltage, with the 'head' of the bubble cloud moving toward the transducer, resulting in a teardrop shaped bubble cloud.

Comparing the bubble clouds created by PW combined with CW to CW alone, these two treatment strategies were quite different when the CW voltage was low. The volume of the bubble clouds created by PW combined with CW was larger than that from CW alone ($p < 0.05$) when the CW input voltage ≤ 30 V. With increasing CW voltage (≥ 35 V), no significant difference was found in the volume of the bubble cloud ($p > 0.05$). No significant difference was found in the mean gray value of the bubble cloud ($p > 0.05$), indicating that applying the PW before the CW has little influence in these cases.

For the cases studied here, it can be seen that bubble clouds of different shapes, sizes and densities can be formed depending on a range of ultrasound parameters and droplet concentrations. The vaporization of droplets is dependent on the pressure amplitude, but the higher pressure amplitude in the short pulse used for PW here should still generate little heating while producing a much larger bubble cloud. Different shapes and sizes over various droplet concentrations were also seen, likely due to the range of impedance differences produced. We have previously shown how the changing impedance in the medium, and the associated scattering, due to the evolving bubble cloud can produce uniquely shaped clouds after a number of pulses are applied [25]. The void fraction of bubbles achieved in CW is also pressure dependent where the additional effects of temperature increases and rectified diffusion may also play a role. It is the dynamic interaction of the acoustic pressure field and the ADV bubbles that lead to a great variety of bubble cloud characteristics. This will definitely affect the lesion shape formed during the HIFU treatment; thus the influence of bubble clouds on the final thermal lesion is examined next.

Observations on the thermal lesions formation

The thermal lesions created in the above experiments were also studied to investigate how the different bubble clouds changed the formation of these lesions. Due to the coexistence of the bubbles and the thermal lesion inside the phantom after the HIFU exposures, it is hard to distinguish the lesion from the bubble cloud from direct visual observation. Therefore, all phantoms were subsequently placed in a chamber and a pressure of 3 MPa was applied for 1 min. As shown in Fig. 4, by this technique, the bubbles in the phantom are compressed, leaving thermal lesions clearly visible. The thermal lesions formed in the phantoms were then obtained using this technique and quantified with results shown in Fig. 5.

For comparison purposes, experiments were also conducted in a phantom without any droplets (Fig. 5a). It was found that only after increasing the input voltage to 40 V with a duration of 60 s CW could a small ellipsoidal thermal lesion be observed. This shape is a typical thermal lesion created by HIFU in the absence of bubbles [1]. No echogenic regions were observed in B-mode images of the phantoms, indicating that no cavitation phenomena occurred under these conditions at least at the level of imageable bubble production.

For the thermal lesions formed after 10 s exposure to CW in phantoms with different droplet concentrations (Fig. 5b, row 1), the lesions created had a spherical shape with a small distal extension, and a range of the sizes. The lesion volume reached a peak at the concentration of 3.41×10^5/mL (see Fig. 5d), and then decreased with the increasing concentration. This trend is similar to the bubble cloud volume variation seen in Fig. 3e.

When the phantoms were exposed to the PW before the CW, the shape and position of the thermal lesions changed significantly depending on the droplet concentration, as seen in Fig. 5b (row 2). At the two lowest droplet concentrations, the thermal lesions formed are quite similar to those formed in the CW only group (row 1). And the shape is also similar. But with increasing droplet concentration, the thermal lesions became larger, and the lesions changed from a teardrop to the vertically-oriented ellipsoidal shape in the proximal portion of the focal area. As the droplet concentration reached 3.41×10^6/mL, the thermal lesion size started to decrease and moved closer to the transducer, and the shape got thinner in the axial dimension. With a droplet concentration of 1.07×10^7/mL, the thermal lesions were even thinner, only present in the layer close to

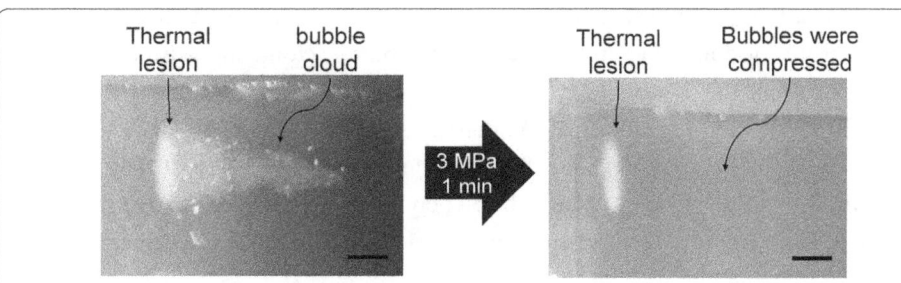

Fig. 4 After the HIFU treatment, each phantom was placed in a chamber and pressurized to 3 MPa for 1 min to compress the bubbles, leaving thermal lesion clearly seen

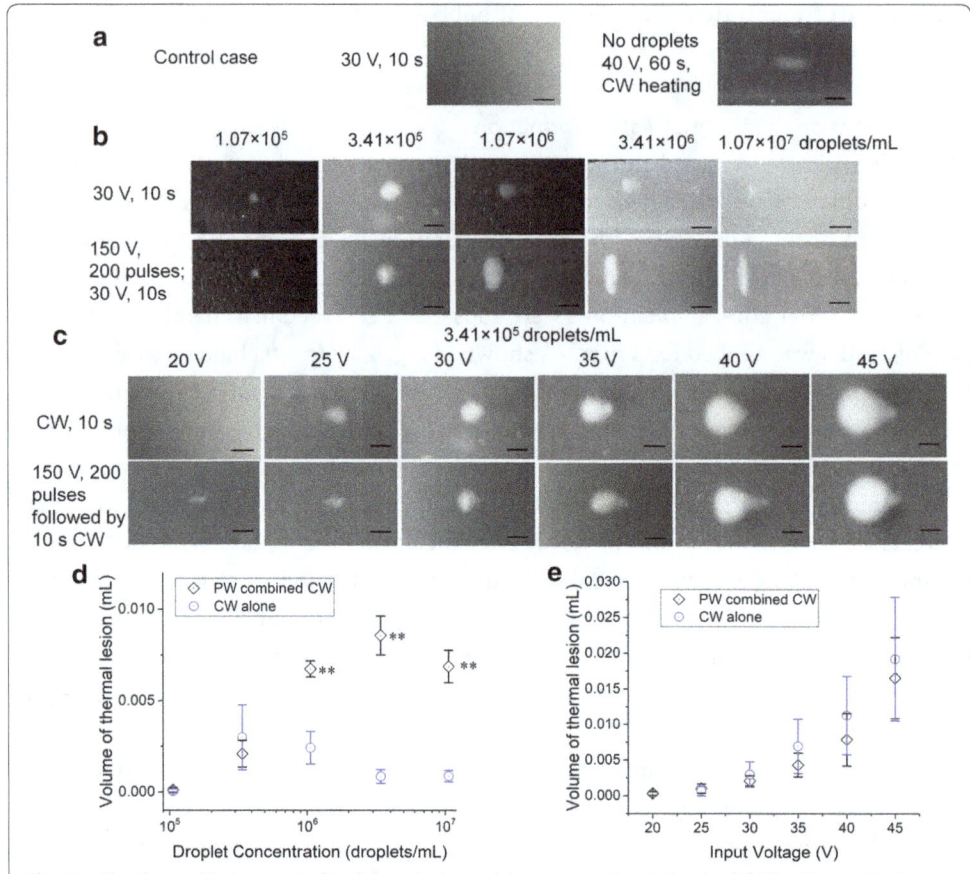

Fig. 5 a The thermal lesion created in the control case (phantoms without droplets). **b** The thermal lesion created after (row 1) 10 s exposure to continuous wave (CW) at 30 V, (row 2) exposure to 200 pulses of pulse wave at 150 V followed by 10 s of continuous wave at 30 V in the phantoms with different droplet concentrations, which were 1.07×10^5/mL, 3.41×10^5/mL, 1.07×10^6/mL, 3.41×10^6/mL, and 1.07×10^7/mL, respectively. **c** The thermal lesion created after (row 1) 10 s exposure to continuous wave, (row 2) exposure to 200 pulses of pulse wave at 150 V followed by 10 s of continuous wave. The input voltage of the continuous wave were 20 V, 25 V, 30 V, 35 V, 40 V, and 45 V, respectively. The droplet concentration in the phantom was 3.41×10^5/mL. **d** Volume of the thermal lesions as function of droplet concentration ($n \geq 4$). **e** Volume of the thermal lesions in phantom with droplet concentration of 3.41×10^5/mL as function of CW input voltage ($n \geq 4$). The scale bar represents 2 mm. ** Represents $p < 0.01$ between PW combined with CW and CW alone

the proximal surface of the bubble cloud. This is due to the high attenuation of the high bubble concentration. Thus the acoustic pressure decreases dramatically along the propagation path and only in the first layers of the bubble cloud region is the acoustic energy absorbed and transformed to thermal energy to create a thermal lesion. In addition, the high concentration of bubbles also leads to a lower thermal conductivity, which helps shape the thermal lesion at the proximal area of the bubble cloud and generate these distinct geometries for the thermal lesions.

According to Fig. 5d, the volume of the thermal lesion is significantly increased using the PW combined CW treatment for the high concentration cases ($p < 0.01$ for droplet concentration $\geq 1.07 \times 10^6$/mL). With a droplet concentration of 3.41×10^6/mL, the volume created by PW combined CW treatment is more than 10 times the volume created by CW alone.

The thermal lesions formed using different CW input voltages are shown in Fig. 5c. The droplet concentration in the phantom is 3.41×10^5/mL. The duration of continuous wave exposure is 10 s. And the CW input voltages were 20 V, 25 V, 30 V, 35 V, 40 V, and 45 V, respectively. It can be seen in Fig. 5c that PW triggering of the droplets did help form the lesions in the phantom with an input voltage of only 20 V, which failed when no PW was used. But the results also show that using PW to pre-trigger droplets does not always result in a larger lesion. For this droplet concentration and across the range of CW voltages used, the volumes of the thermal lesions produced by both treatments had no statistical difference ($p > 0.05$). The second image in Fig. 3a row 2 illustrates the bubble cloud shape created after PW alone for the 3.41×10^5/mL droplet concentration. Comparing this to all the bubble clouds formed after combined treatment with the same PW but different CW voltages (Fig. 3b row 2), significant difference was found in echogenicity ($p < 0.05$), indicating the higher void fraction inside the bubble clouds formed after CW, which may have actually limited the size of the thermal lesions.

In summary, comparing to the CW only cases, the PW combined CW strategy could lower the CW input voltage required to create thermal lesions of similar volume, which could improve the safety of HIFU treatment. This would reduce the thermal dose to the overlying tissue and thus shorten cooling time required, leading to faster treatments. The PW combined CW strategy also led to larger volume of thermal lesion in the cases of the high droplet concentration. At the droplet concentration of 3.41×10^6/mL, Fig. 5d shows that for the same CW exposure (10 s at 30 V), the volume of lesion produced by pre-triggering the droplets was 10 times larger, meanwhile no thermal lesion can be created in a phantom without droplets (Fig. 5a).

The results shown here have demonstrated that both the shape and the size of the thermal lesions can be changed by using the high amplitude short pulses of PW ultrasound before the significant thermal energy supplied by the CW ultrasound of HIFU. In real clinical treatments, the shape of the tumors are usually irregular. So to fully cover the tumor while sparing the important organs and normal tissues, a careful planning of the successive thermal lesions is critical since the lesion shapes are sensitive to the bubble clouds formed in the ADV technique. The PW strategy could help enlarge the thermal lesions or change the lesion shapes, which depends on the local droplet concentrations.

In this study, we used 200 pulses of PW ultrasound at an input voltage of 150 V to create a bubble area as large as possible. However, in real situations, both the number of pulses and the input voltage of the pulse could be adjusted to further manipulate the shape of the thermal lesions. Numerical models that couple the propagation of the acoustic waves, the dynamic formation of the bubble clouds and the thermal deposition and heat transfer would be a powerful treatment planning tool and aid in new treatment modality design. A numerical model has been proposed to calculate the ADV bubble cloud evolution during the PW exposure [25]. Simulation of the thermal lesion formation during the CW exposure in ADV assisted HIFU is now under investigation. This simulation work could be used for treatment planning in the future.

In this experiment, tissue-mimicking phantoms that can visualize thermal lesions were used to denote the thermal dose accumulated during the treatment, which is commonly used in experimental research on HIFU therapy [6, 11, 26]. The limitation of this approach is that it provides little detail about temperature distribution and variation during the treatment. In comparison, MRI can monitor the temperature distribution and record the temperature variation over time during the treatment, which is really helpful in understanding the complicated process of the highly dynamic interaction of the acoustic field, bubble cloud evolution and temperature field in ADV assisted HIFU treatment. In the future, a wider range of acoustic parameters would be needed to determine how robust the modeling performs under differing conditions. Therefore, more work needs to be done to reveal the underlying mechanism of this promising treatment technique and how to optimize this approach using ADV, which could in return help optimizing the treatment and benefit more patients.

Conclusion

The influence of ADV bubbles on the formation of HIFU thermal lesions was investigated in this study. Two treatment strategies were examined, CW ultrasound heating alone and PW ultrasound combined with CW. It was found that with the same PW exposure, the bubble cloud created in phantom with high droplet concentration is much larger likely due to the strong reflection at the interface of high impedance difference. The shape of the bubble cloud progresses from small dots to teardrop to triangular with increasing concentrations of droplets. The cloud reproducibility was better at higher droplet concentrations.

The thermal lesions created in the phantoms with high droplet concentration were much larger in the case of PW combined with CW ultrasound treatment than those created by CW alone. By using the PW ultrasound to trigger the bubbles first, the thermal lesion created in the experiments varies from small dot to teardrop to vertically-oriented ellipsoid. It has been demonstrated that the size and morphology of the thermal lesion could be regulated by changes in both the acoustic pressure and the droplet concentration. The studies confirmed the feasibility of thermal lesion shape manipulation and offered new possibilities for optimization of HIFU treatment according to different requirements.

Abbreviations
ADV: acoustic droplet vaporization; HIFU: high intensity focused ultrasound; PW: pulse wave; CW: continuous wave; UCA : ultrasound contrast agent.

Authors' contributions
YX did all the experiments, analyzed experimental data and drafted the manuscript. AZ helped with the design of the experiment, data analysis and manuscript revision. LXX helped with the experimental data analysis and manuscript revision. JBF helped with the design of the experiment, data analysis and manuscript revision. All authors read and approved the final manuscript.

Acknowledgements
This work is supported by the Chinese government "111 Project" (B08020), National Natural Science Foundation of China (NSFC 51476101, 51521005, 51876129), Science and Technology Commission of Shanghai Municipality (16441903100), and Ministry of Science and Technology of China (2016YFC0106200).

Competing interests
The authors declare that they have no competing interests.

Consent for publication
Not applicable.

References

1. Kennedy J, ter Haar G, Cranston D. High intensity focused ultrasound: surgery of the future? Br J Radiol. 2003;76:590–9.
2. Napoli A, Anzidei M, Ciolina F, et al. Mr-guided high-intensity focused ultrasound: current status of an emerging technology. Cardiovasc Interv Radiol. 2013;36:1190–203.
3. Lafon C, Zderic V, Noble ML, et al. Gel phantom for use in high-intensity focused ultrasound dosimetry. Ultrasound Med Biol. 2005;31:1383–9.
4. Takegami K, Kaneko Y, Watanabe T, et al. Heating and coagulation volume obtained with high-intensity focused ultrasound therapy: comparison of perflutren protein-type a microspheres and mrx-133 in rabbits 1. Radiology. 2005;237:132–6.
5. Umemura S-I, Kawabata K-I, Sasaki K. In vivo acceleration of ultrasonic tissue heating by microbubble agent. IEEE Trans Ultrason Ferroelectr Freq Control. 2005;52:1690–8.
6. Tung Y-S, Liu H-L, Wu C-C, et al. Contrast-agent-enhanced ultrasound thermal ablation. Ultrasound Med Biol. 2006;32:1103–10.
7. Luo W, Zhou X, Ren X, et al. Enhancing effects of sonovue, a microbubble sonographic contrast agent, on high-intensity focused ultrasound ablation in rabbit livers in vivo. J Ultrasound Med. 2007;26:469–76.
8. Luo W, Zhou X, He G, et al. Ablation of high intensity focused ultrasound combined with sonovue on rabbit vx2 liver tumors: assessment with conventional gray-scale us, conventional color/power doppler us, contrast-enhanced color doppler us, and contrast-enhanced pulse-inversion harmonic us. Ann Surg Oncol. 2008;15:2943–53.
9. Coussios C, Farny C, Ter Haar G, et al. Role of acoustic cavitation in the delivery and monitoring of cancer treatment by high-intensity focused ultrasound (hifu). Int J Hyperthermia. 2007;23:105–20.
10. Kripfgans OD, Fowlkes JB, Miller DL, et al. Acoustic droplet vaporization for therapeutic and diagnostic applications. Ultrasound Med Biol. 2000;26:1177–89.
11. Zhang M, Fabiilli ML, Haworth KJ, et al. Acoustic droplet vaporization for enhancement of thermal ablation by high intensity focused ultrasound. Acad Radiol. 2011;18:1123–32.
12. Zhu M, Jiang L, Fabiilli ML, et al. Treatment of murine tumors using acoustic droplet vaporization-enhanced high intensity focused ultrasound. Phys Med Biol. 2013;58:6179.
13. Lo AH, Kripfgans OD, Carson PL, et al. Spatial control of gas bubbles and their effects on acoustic fields. Ultrasound Med Biol. 2006;32:95–106.
14. Moyer LC, Timbie KF, Sheeran PS, et al. High-intensity focused ultrasound ablation enhancement in vivo via phase-shift nanodroplets compared to microbubbles. J Ther Ultrasound. 2015;3:7.
15. Kripfgans OD, Zhang M, Fabiilli ML, et al. Acceleration of ultrasound thermal therapy by patterned acoustic droplet vaporization. J Acoust Soc Am. 2014;135:537–44.
16. Commander KW, Prosperetti A. Linear pressure waves in bubbly liquids: comparison between theory and experiments. J Acoust Soc Am. 1989;85:732–46.
17. Rosnitskiy P, Yuldashev P, Vysokanov B, et al. Setting boundary conditions on the khokhlov–zabolotskaya equation for modeling ultrasound fields generated by strongly focused transducers. Acoust Phys. 2016;62:151–9.
18. Del Grosso V, Mader C. Speed of sound in pure water. J Acoust Soc Am. 1972;52:1442–6.
19. Pinkerton J. A pulse method for the measurement of ultrasonic absorption in liquids: results for water. Nature. 1947;160:128–9.
20. Takegami K, Kaneko Y, Watanabe T, et al. Polyacrylamide gel containing egg white as new model for irradiation experiments using focused ultrasound. Ultrasound Med Biol. 2004;30:1419–22.
21. Ridler T, Calvard S. Picture thresholding using an iterative selection method. IEEE Trans Syst Man Cybern. 1978;8:630–2.
22. Silver WM. Normalized correlation search in alignment, gauging, and inspection. In: Proceedings of the image pattern recognition: algorithm implementations, techniques, and technology, 1987. International Society for Optics and Photonics.
23. Hsieh DY, Plesset MS. Theory of rectified diffusion of mass into gas bubbles. J Acoust Soc Am. 1961;33:206–15.
24. Crum L, Hansen G. Growth of air bubbles in tissue by rectified diffusion. Phys Med Biol. 1982;27:413.
25. Xin Y, Zhang A, Xu LX, et al. Numerical study of bubble area evolution during acoustic droplet vaporization-enhanced hifu treatment. J Biomech Eng. 2017;139:091004.
26. Zhang P, Kopechek JA, Porter TM. The impact of vaporized nanoemulsions on ultrasound-mediated ablation. J Ther Ultrasound. 2013;1:2.

10

An efficient optic cup segmentation method decreasing the influences of blood vessels

Chunlan Yang[1*], Min Lu[1], Yanhua Duan[1] and Bing Liu[2]

*Correspondence:
clyang@bjut.edu.cn
[1] College of Life Science
and Bioengineering, Beijing
University of Technology,
Beijing 100124, China
Full list of author information
is available at the end of the
article

Abstract

Background: Optic cup is an important structure in ophthalmologic diagnosis such as glaucoma. Automatic optic cup segmentation is also a key issue in computer aided diagnosis based on digital fundus image. However, current methods didn't effectively solve the problem of edge blurring caused by blood vessels around the optic cup.

Methods: In this study, an improved Bertalmio–Sapiro–Caselles–Ballester (BSCB) model was proposed to eliminate the noising induced by blood vessel. First, morphological operations were performed to get the enhanced green channel image. Then blood vessels were extracted and filled by improved BSCB model. Finally, Local Chart-Vest model was used to segment the optic cup. A total of 94 samples which included 32 glaucoma fundus images and 62 normal fundus images were experimented.

Results: The evaluation parameters of F-score and the boundary distance achieved by the proposed method against the results from experts were 0.7955 ± 0.0724 and 11.42 ± 3.61, respectively. Average vertical optic cup-to-disc ratio values of the normal and glaucoma samples achieved by the proposed method were 0.4369 ± 0.1193 and 0.7156 ± 0.0698, which were also close to those by experts. In addition, 39 glaucoma images from the public dataset RIM-ONE were also used for methodology evaluation.

Conclusions: The results showed that our proposed method could overcome the influence of blood vessels in some degree and was competitive to other current optic cup segmentation algorithms. This novel methodology will be expected to use in clinic in the field of glaucoma early detection.

Keywords: Optic cup, Digital fundus image, Segmentation, Blood vessel, BSCB model

Background

Optic cup is an important retinal structure in ophthalmologic diagnosis such as glaucoma [1, 2]. Glaucoma is the second leading disease to blindness worldwide [3]. Given the lack of visual symptoms in the early stages, several studies showed that more than 90% of the patients were unaware of this disease until it has developed into the severe stages [4–6]. Since it is time cost and the precision of diagnosis by manual is limited, then the automatic technique for disease detection such as computer aided diagnosis (CAD) system is strongly needed [7].

With the development of computer science, medical image processing technique has been successfully applied to clinical diagnosis and treatment. Currently, digital fundus image has been widely used in many hospitals. Thus, it is possible to develop the CAD system used for ophthalmologic diagnosis based on fundus image processing technique [8–11].

The OD includes two distinct parts, namely, a circumjacent zone called the rim and a central bright region called the optic cup [12]. The optic cup can be divided into nasal and temporal regions. The former is generally occluded by the main blood vessels. Physiologically, the loss in optic nerve fibers (ONF) leads to the enlargement of optic cup which is called large cupping, and atrophy of neuroretinal rim which is called rim loss. Large cupping and rim loss are two important indicators of glaucoma. They could be detected by measuring the vertical optic cup-to disc ratio (CDR) [13]. Once more ONF disappear, the optic cup will become larger with respect to the optic disc, which corresponds to CDR increasing. In generally, an abnormal vertical CDR indicate a high risk of glaucoma. Therefore, automatic segmentation of the optic cup is crucial for CAD system for glaucoma.

To date, there are few studies on optic cup segmentation algorithms, which include thresholding-based methods [14–16], region growing [17, 18], model-based methods (active contour models or snakes [19, 20], level sets [21, 22], and elliptical shape model [23]), anatomical evidence-based methods [24], and superpixel classification [25–27]. Optic cup segmentation is challenged because the intensity has a sudden change in areas which blood vessels pass across the cup-disc boundary. However, most studies haven't solved this problem effectively. Thus, inpainting of the blood vessels is an important step in optic cup segmentation.

This study aims to develop an efficient method to overcome the aforementioned problem in optic cup segmentation. Morphological operations were first performed to get the enhanced green channel image. Then, blood vessels were extracted and filled by using an improved Bertalmio–Sapiro–Caselles–Ballester (BSCB) model. Finally, local chart-vest (LCV) model was used to segment the optic cup.

The remainder of this paper is organized as follows: first, we introduced the image data used in this study and presented the proposed methodology, an efficient optic cup segmentation method decreasing the influences of blood vessels. The experimental results are then provided, followed by discussions and conclusions.

Methods

Data acquisition

Digital fundus images were acquired from local ophthalmologic hospital. The 24-bit color images were captured using digital fundus camera (Canon CR-DGi) with the array size of 1440 × 960 pixels. The photographic angle of the fundus camera was set to 60°, and the optic disc was adjusted at image center. The average optic cup boundary identified by two experts was taken as the ground truth for the following evaluation. All the experimented 94 images were acquired by the same instrument and the subjects were collected randomly to take the ocular disease screening. A total of 94 images including 32 patients with glaucoma (18 male, 14 female) and 62 healthy subjects (34 male, 28 female) were included in our experiment. For the 32 patients with glaucoma, the ages were ranged from 35 to 58 years old (45.23 ± 3.31 years). None of the left 62 normal subjects had history of hypertension, nor cardiovascular disease and diabetes. Their ages were ranged from 31 to 63 years old (43.51 ± 5.27 years). In addition, 39 glaucoma images from the public dataset RIM-ONE (An Open Retinal Image Database for Optic Nerve Evaluation) were also used for evaluation [28].

Optic cup segmentation

The edge of optic cup is more difficult to identify compared with that of optic disc, primarily because the image is blurred where the blood vessels pass across the optic cup. After preprocessing such as image enhancement, blood vessels were extracted and inpainting. Then the optic cup was segmented by using LCV model.

Preprocessing

Among the image components of color which are red (R), green (G) and blue (B), the G channel shows the optimal image contrast for the optic cup (see Fig. 1). Therefore, the G channel image U_G was selected for the subsequent processing.

In the proposed algorithm, both top-hat and bottom-hat transformations were applied to enhance the image contrast [27]. Top-hat transformation refers to the subtraction of the opening operation result from the image itself. By contrast, bottom-hat transformation refers to the subtraction of the image from the result of closing operation. Both top-hat and bottom-hat transformations are based on a predefined neighborhood or structuring element (*SE*). The above two transformations are illustrated as Eqs. (1) and (2), respectively.

$$T_{hat}(U_G) = U_G - (U_G \circ SE) \tag{1}$$

$$B_{hat}(U_G) = (U_G \cdot SE) - U_G \tag{2}$$

Here, a structuring element with a size of 5×5 pixels was used (more details of size selection were provided in "Results" section). The above two transformations could be represented as Eq. (3), the boundary of optic cup become clearer after the operations (see Fig. 2).

$$U = (U_G + T_{hat}(U_G)) - B_{hat}(U_G) \tag{3}$$

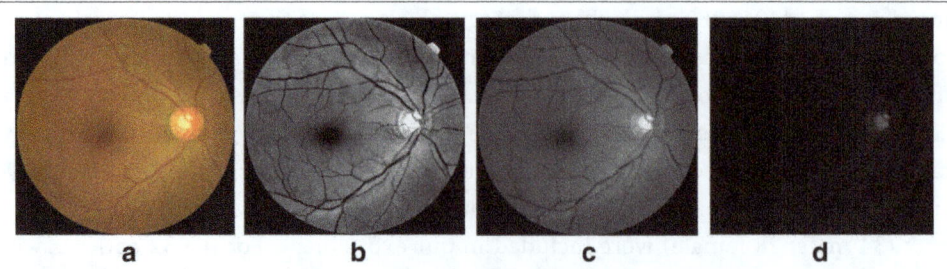

| a | b | c | d |

Fig. 1 Digital fundus images with different color channels. **a** Original image, **b** red channel image, **c** green channel image, **d** blue channel image

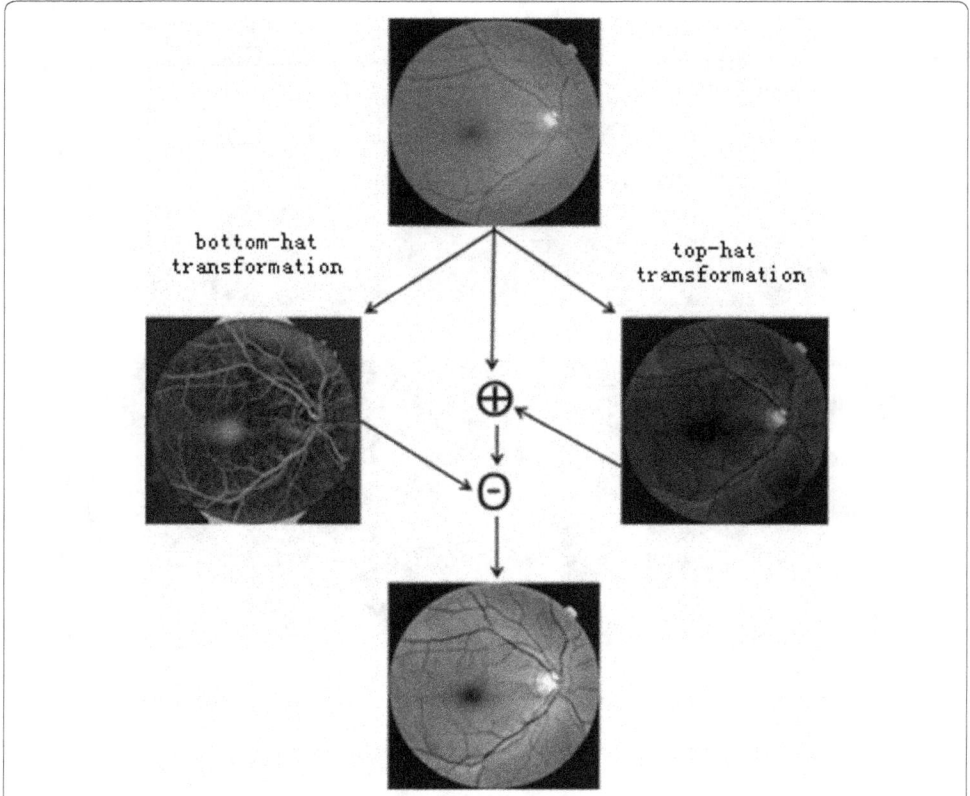

Fig. 2 Flowchart of image contrast enhancement by top-hat and bottom-hat transforms. The top is the original image and the bottom is the result

Blood vessel extraction

Since the intensities of blood vessels were lower than those of background and other structures, the intensities of blood vessels will become higher after median filtering. Therefore, the blood vessels could be possibly identified based on the intensity difference. In order to capture the change of image intensity, the contrast enhanced image U was subtracted from the median filtered image U_{med}. Then the intensity differential image could be acquired which was represented by Eq. (4).

$$U_{sub} = U_{med}(i, j) - U(i, j) \qquad (4)$$

The binary image U_D was generated according to the value of U_{sub} [29]. Only the intensities of pixels with corresponding $U_{sub} > 0$ were set to 1 according to Eq. (5). These pixels were considered belonging to blood vessels (see Fig. 3d).

$$U_D(i, j) = \begin{cases} 1, & \text{if } U_{sub} > 0 \\ 0, & \text{otherwise} \end{cases} . \qquad (5)$$

In this study, the size of median filtering was set to 9×9 pixels (more details of size selection were provided in "Results" section).

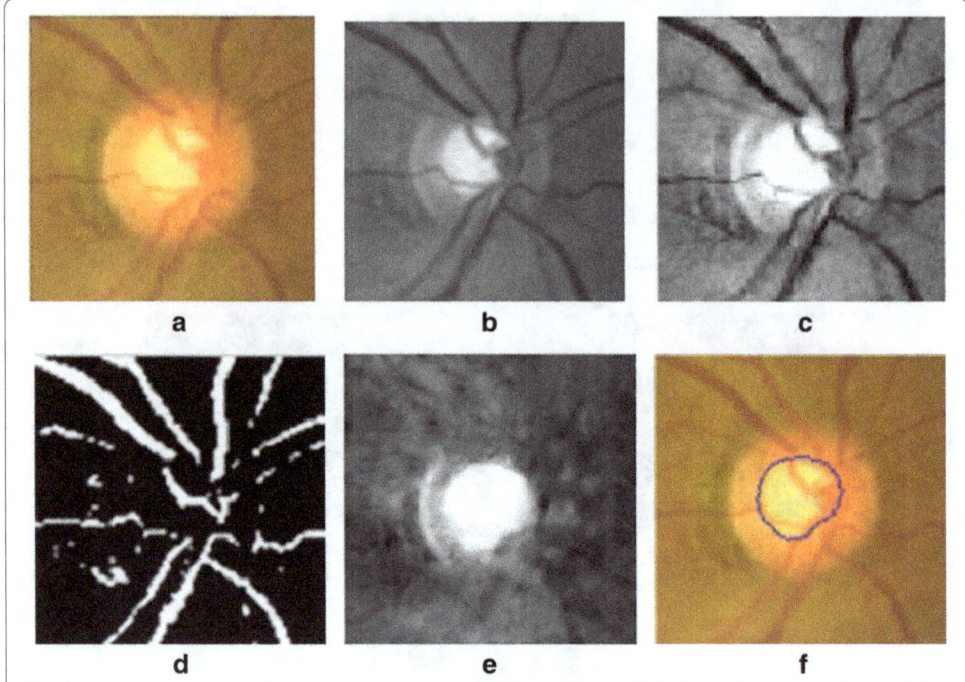

Fig. 3 Key steps in proposed optic cup segmentation. **a** Original image, **b** G channel image, **c** enhanced G channel image, **d** blood vessel identification, **e** result of blood vessel inpainting, **f** optic cup identification

Blood vessels inpainting

Bertalmio et al. [30] proposed the novel BSCB model for image inpainting in 2000 based on theory of partial differential equation (PDE). The BSCB model uses Laplace operator to measure the neighborhood information for image inpainting. It smoothly propagates the information to the same region along the direction of isophote. At the same time, an anisotropic diffusion function is adopted to prevent the prolongation lines from crossing one to another. Generally, the model comprises two steps: inpainting and diffusion. In this study, the improved BSCB model was developed, that is, the neighborhood intensities were used as the propagation information, instead of using only one single pixel.

After the above processing, the intensity will become more uniform within the area of optic cup. The influence caused by the blood vessels will be decreased. The following were the detailed descriptions about the processing procedure.

Assuming Ω is the region to be inpainted and ∂D is the boundary of Ω. The BSCB model can be described by Eqs. of (6) and (7) [31, 32].

$$\frac{\partial U}{\partial t} = \nabla L \cdot \vec{T} \tag{6}$$

$$\frac{\partial U}{\partial t} = g_\varepsilon k |\nabla U|. \tag{7}$$

Equation (6) represents inpainting, where ∇L is the propagation information and $\vec{T}$ is the isophote direction. Equation (7) is used for diffusion, where k is the Euclidean curvature of the isophote and Ω^ε is the dilation of Ω with a balling radius of ε, and g_ε is a smoothing function of Ω^ε.

To be easier understanding, Eqs. of (6) and (7) can be comprehensively described as follows.

$$U^{n+1}(i,\,j) = U^n(i,\,j) + \Delta t U^n_\tau(i,\,j), \quad \forall(i,\,j) \in \Omega, \tag{8}$$

$$U^{n+1}(i,\,j) = U^n(i,\,j) + \Delta t g_\varepsilon(i,\,j) k(i,\,j,\,n) |\nabla U(i,\,j,\,n)|, \quad \forall(i,\,j) \in \Omega^\varepsilon, \tag{9}$$

Here $U^{n+1}(i,\,j)$ is the value of pixel intensity located at $(i,\,j)$ in the n-th iteration image which $U^n_\tau(i,\,j) = \nabla L \cdot \vec{T}$. Notably, $U^0(i,\,j) = U(i,\,j)$ and $\lim\limits_{n\to\infty} U^n(i,\,j) = U_r(i,\,j)$, where $U^0(i,\,j)$ is the input image, $U_r(i,\,j)$ is the output of the algorithm, and Δt is the improvement rate.

In the traditional BSCB model, propagation information $L^n(i,\,j)$ is substituted by the discrete Laplace operator, which is shown in Eq. (10).

$$L^n(i,\,j) = u^n_{xx}(i,\,j) u^n_{yy}(i,\,j). \tag{10}$$

Although texture could be the transfer information for image inpainting, it is not necessary in optic cup segmentation. In contrast, local image information is more effective. Therefore, using neighborhood mean value as transfer information may eliminate the influence of noise. According to the above deduction, $u^n_{xx}(i,\,j)$ and $u^n_{yy}(i,\,j)$ were replaced by $\overline{u^n_{xx}}(i,\,j)$ and $\overline{u^n_{yy}}(i,\,j)$ in the following vascular inpainting, represented by Eqs. (11) and (12) respectively. In this study, the size of neighborhood was set to 3×3 pixels.

$$\overline{u^n_{xx}}(i,\,j) = \frac{1}{9} \sum_{m=i-1}^{i+1} \sum_{n=j-1}^{j+1} u^n_{xx}(m,\,n) \tag{11}$$

$$\overline{u^n_{yy}}(i,\,j) = \frac{1}{9} \sum_{m=i-1}^{i+1} \sum_{n=j-1}^{j+1} u^n_{yy}(m,\,n) \tag{12}$$

Optic cup boundary identification

Wang et al. [33] proposed a LCV model which included local statistical information in level set based segmentation framework. In the algorithm, extended structure tensor (EST) was combined which intensity inhomogeneity could be decreased effectively. In this study, the above-mentioned LCV model was used for the following optic cup segmentation.

First, the centroid $(x_c,\,y_c)$ was selected from the region with relative high intensity, which was acquired according to the following criteria represented as Eq. (13)

$$x_c = \frac{1}{N} \sum_{i=1}^{N} x_i, \quad y_c = \frac{1}{N} \sum_{i=1}^{N} y_i \tag{13}$$

Here N is the total number of pixels within the target region.

In this step, two parameters need dynamic adjusted in the LCV model, i.e., α and μ. Commonly, α was set to 0.1 or 1, depending on whether the image has intensity inhomogeneity. For μ, two corresponding values are adopted: 0.01×255^2 and 0.1×255^2. If several targets need to be detected, μ should be small, and vice versa. Because the optic cup area has intensity inhomogeneity and only the optic cup be the target, the values of α and μ were then set to 0.1 and 0.1×255^2 respectively.

Evaluation

The experimental results were evaluated by three statistical criteria, namely, F-score (area-based), distance (curve-based) and vertical CDR, which were explained in "F-score (F)", "Distance (D)", and "Vertical CDR" sections.

Besides the proposed algorithm, the experiments were also performed by two other known methods, proposed by Joshi et al. [17] and Liu et al. [21]. There are two reasons why the above methods were selected to be compared. First, since the optic cup segmentation could be classified into framework of region based, edge based and hybrid, the proposed algorithm and another two selected methods all belong to region based method. Second, the proposed method aimed at decreasing the influence of blood vessels, which was similar to the other selected methods putting forward to the solutions.

In details, Joshi et al. [17] imposed the expected symmetry of optic cup region by setting threshold to recover the under-segmentation areas located in the blood vessels. Liu et al. [21] used ellipse to redraw the optic cup boundary after employing a combinative algorithm with level set and threshold setting, which was to weaken the noising by blood vessels.

For each method, the evaluation of F-score and distance were compared. The vertical CDR values were presented against manual segmentation results achieved by experts.

F-score (F)

The pixel-wise precision and recall values were computed to assess the overlap area between the computed region and the ground truth. These values were defined as Eqs. (14) and (15),

$$precision = \frac{TP}{TP + FP} \tag{14}$$

$$recall = \frac{TP}{TP + FN} \tag{15}$$

where TP, FP, and FN represented the number of true positive, false positive, and false negative pixels respectively. The harmonic mean of the precision and recall values, called F-score (F), was computed to better appreciate the results. The F-score was expressed in Eq. (16),

$$F = 2 \frac{precision \cdot recall}{precision + recall} \qquad (16)$$

Given that both recall and precision are evenly weighted, the F-score value lies between 0 and 1, and the recall, precision, F-score should be all ideally close to 1.

Distance (D)

To assess the accuracy of the boundary, a curve-based evaluation is performed. Let C_e be the boundary identified by the ophthalmologist and C_m be the boundary achieved by the proposed algorithm. The distance (D) which was computed in pixels between two curves was expressed as Eq. (17),

$$D = \frac{1}{n} \sum_{\theta=1}^{\theta_n} \left| d_e^\theta - d_m^\theta \right|, \qquad (17)$$

Here, n is the number of angles, the distance from centroid of C_e to the points on C_e in direction of θ was defined as d_e^θ, similar with the definition of d_m^θ. D should be close to 0 for an accurate algorithm.

Vertical CDR

To estimate the vertical CDR, the optic disc must be segmented ahead. Compared to optic cup, optic disc segmentation is relatively easier. In this study, to ensure the accuracy of the CDR and evaluate the effectiveness of the proposed algorithm, the optic disc was manually delineated by experts. The vertical CDR was calculated by Eq. (18) proposed by Gloster et al. [34],

$$CDR = C_V / D_V \qquad (18)$$

where C_V represented the vertical diameter of optic cup, and D_V represented the vertical diameter of optic disc. C_V and D_V were determined by the distances from the top to the bottom of these diameters, as shown in Fig. 4. In clinic, the risk of suffering glaucoma increases with the value of CDR.

Results

A total of 94 digital fundus images, including 32 glaucoma images, were experimented using the proposed algorithm. The algorithm was individually evaluated against manual segmentation results by expert-1 and expert-2, and also against the average expert marking called expert-X. For comparison, the results obtained by two other known methods of Joshi et al. [17] and Liu et al. [21], were also presented in evaluation of F-score and distance.

The optic cup segmentation results were shown in Fig. 5. In order to compare with the other above mentioned methods, more examples were shown in Fig. 6. The evaluation results of F-score and distance were shown in Tables 1 and 2. The values of CDR achieved by experts and the proposed method were also compared shown in Fig. 7. The statistical results of CDR for normal and glaucoma data were listed in Table 3.

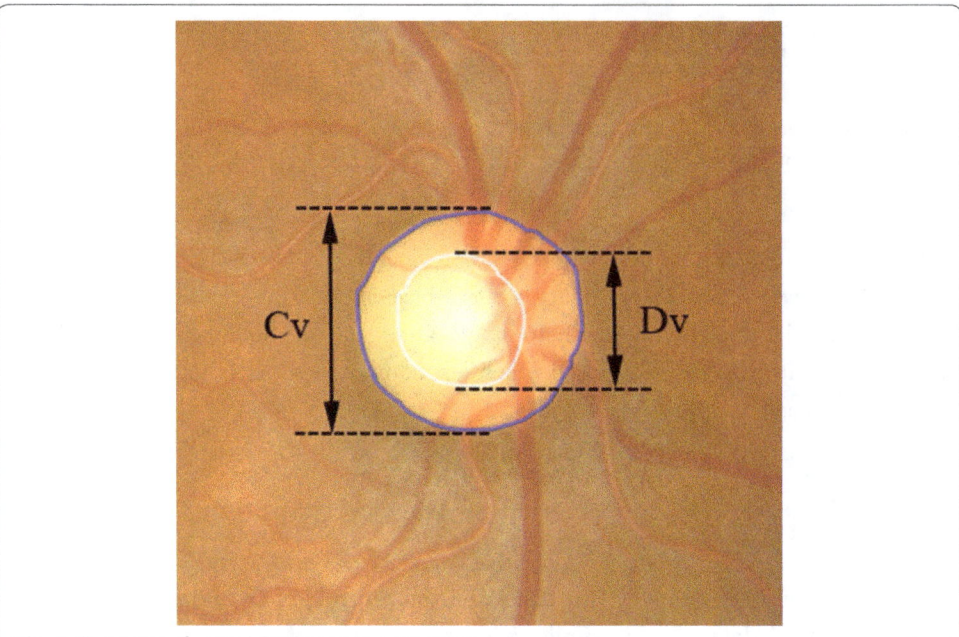

Fig. 4 Illustration of vertical CDR measurement criteria

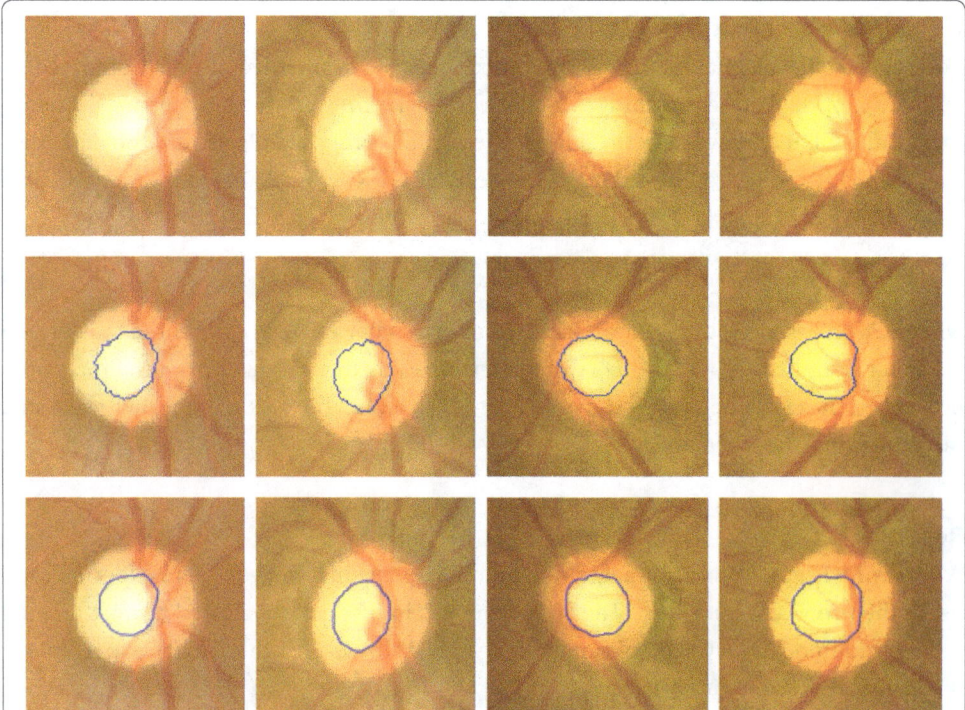

Fig. 5 The optic cup segmentation results. From top to bottom are original images, results achieved by our proposed method and by experts respectively

The selections of parameters used in the algorithm were also evaluated. For the SE used in morphometric operations of preprocessing, the size from 2×2 pixels to 15×15 pixels have been experimented. Representative results were shown in Fig. 8 and

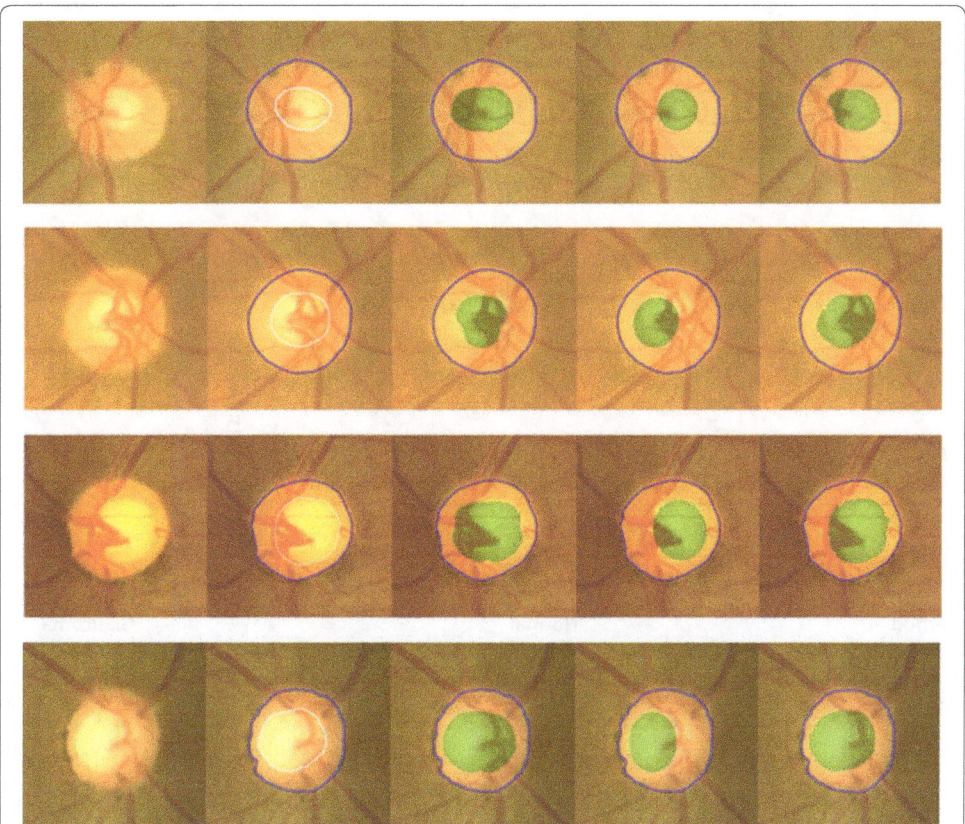

Fig. 6 Illustration of comparisons among the segmentation results. From left to right are original images, results achieved by experts, by Joshi et al., by Liu et al. and our proposed method

Table 1 F-score acquired by different methods

Expert	Joshi et al. [17]	Liu et al. [21]	Proposed
Expert-1	0.6125 ± 0.1139	0.7010 ± 0.0901	0.7955 ± 0.0724
Expert-2	0.6235 ± 0.0999	0.7245 ± 0.1032	0.7780 ± 0.0794

Table 2 Boundary distance in radial direction acquired by different methods (in pixels)

Expert	Joshi et al. [17]	Liu et al. [21]	Proposed
Expert-1	22.82 ± 5.00	16.78 ± 3.95	11.42 ± 3.61
Expert-2	20.82 ± 4.08	15.78 ± 4.40	12.32 ± 3.71

Table 4. Representative results of median filtering (window size is from 3×3 pixels to 15×15 pixels, interval of 2×2 pixels) were shown in Fig. 9 and Table 5.

From the results of Figs. 5 and 6, the proposed algorithm achieved the satisfied segmentation results which were better than the methods proposed by Joshi et al. [17] and Liu et al. [21]. Although both of them also considered the influence caused by blood vessels and proposed the corresponding ways trying to decrease the influence, our proposed method showed competitive to overcome the difficulty in areas with a lot of blood

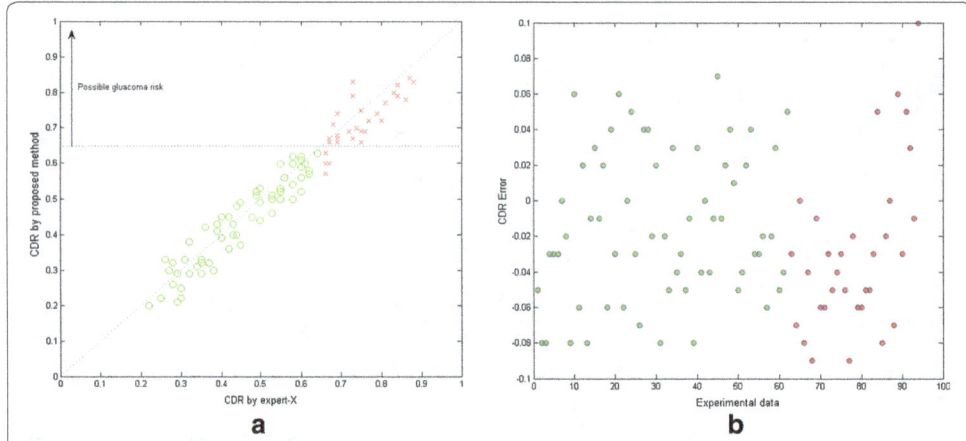

Fig. 7 Comparison results of CDR achieved by experts and the proposed method. **a** CDR values achieved by experts vs. by proposed method, **b** errors between the results by experts and the proposed algorithm

Table 3 Average CDR values for normal and glaucoma

Method	Normal	Glaucoma
Proposed	0.4369 ± 0.1193	0.7156 ± 0.0698
Expert	0.4516 ± 0.1176	0.7444 ± 0.0666

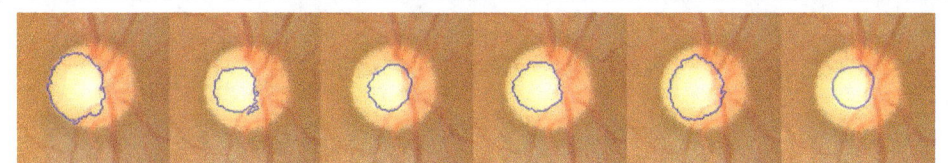

Fig. 8 Comparison results with different SE size. From left to right are optic cup segmentation results acquired without morphometric operation, with size of 2×2 pixels, 5×5 pixels, 10×10 pixels, 15×15 pixels and by the experts

Table 4 The precisions of optic cup segmentation with different size of structuring element used in the morphometric operations

Image type	2×2 pixels	5×5 pixels	10×10 pixels	15×15 pixels
Normal	0.7950 ± 0.0600	0.8914 ± 0.0300	0.6964 ± 0.0600	0.5900 ± 0.1600
Glaucoma	0.5753 ± 0.2000	0.7218 ± 0.0400	0.4415 ± 0.2100	0.2519 ± 0.1100

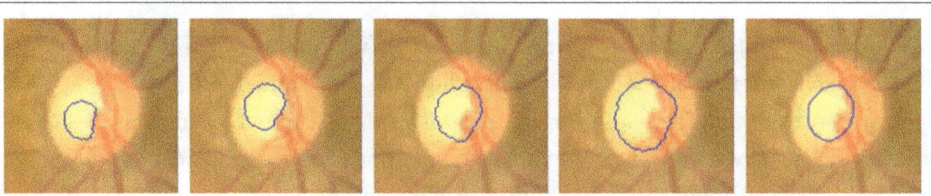

Fig. 9 Comparison results with different window size of median filtering. From left to right are optic cup segmentation results acquired with size of 5×5 pixels, 7×7 pixels, 9×9 pixels, 11×11 pixels and by the experts

Table 5 The precisions of optic cup segmentation with different size of median filtering

Image type	5 × 5 pixels	7 × 7 pixels	9 × 9 pixels	11 × 11 pixels
Normal	0.5250 ± 0.0600	0.7114 ± 0.0300	0.8914 ± 0.0300	0.6900 ± 0.1600
Glaucoma	0.3951 ± 0.2000	0.5618 ± 0.0400	0.7218 ± 0.0400	0.5219 ± 0.1100

Table 6 Comparison with the recent algorithms based on public database RIM-ONE

Criteria	Haleem et al. [35]	Al-Bander et al. [36]	Bechar et al. [37]	Proposed	Expert
F-score	–	0.6903	0.8643	0.8127	–
CDR	0.60 ± 0.17	–	–	062 ± 0.17	0.66 ± 0.18

vessels. In addition, both the maximum F-score and minimum distance of our proposed method were acquired. That is, the accuracy of the proposed method is better than the other two. Furthermore, the standard deviations of F-score and distance were also lower which showed the robustness of the proposed method.

From results of Fig. 8 and Table 4, the size of SE was suggested to be selected as 5 × 5. From results of Fig. 9 and Table 5, the window size of median filtering was suggested to be selected as 9 × 9.

In order to evaluate whether our proposed method is feasible and comparable to the most recent published optic cup segmentation algorithms [35–37], also 39 glaucoma images from the public database RIM-ONE were experimented. F-score and CDR acquired by our proposed method and these referenced algorithms were listed in Table 6. From the results, F-score acquired by our proposed method is acceptable and the corresponding CDR is more close to the manual results by experts.

Discussion

The proposed algorithm and the other two above mentioned methods used for comparison all put forward the solutions aimed at reducing optic cup segmentation errors caused by blood vessels. Given the figures listed in results, it can be concluded that the proposed method will give rise to higher F and lower D, which was hence better in optic cup segmentation. Joshi et al. [17] imposed the expected symmetry of cup region in nasal and temporal side after threshold processing to recover the vessel errors. A vertical axis of symmetry passing through optic disc center was considered and nasal region was obtained by mirroring the temporal region. Therefore, the segmentation results were decided completely by the segmentation quality of temporal region. However, most of the cups are not symmetrical about the vertical axis passing through optic disc center. What is more, as the acquisition conditions like photographic angle changes, the shapes of both sides are various which also causes the uncertainty of the relative size on both sides, leading to under-segmentation or over-segmentation of nasal side after the operation of symmetry. Liu et al. [21] applied ellipse fitting to redraw the cup boundary for weakening vessel influence in the segmentation process. Nevertheless, the ellipse fitting was implemented after employing combinative algorithm of level set and color intensity thresholds which didn't take the vessel into account because of its low intensity. Thus only the minor remission of the errors caused by vessels could be reached after fitting, serious under-segmentation still cannot be avoided. By contrast, the proposed scheme firstly fills the blood vessels by an improved BSCB model, making the whole cup area more uniform for the following segmentation,then the optic cup is determined through the LCV model approach which can capture the unsharp boundary of cup, reducing under-segmentation or over-segmentation to a great extent compared with other methods.

For the CDR evaluation in Table 3, the CDR values by the proposed algorithm are little smaller than those by expert, which indicates that the cup region determined by the proposed method was little smaller toward upper cup edge and lower cup edge than that determined by expert. The CDR values are little too small caused the misdiagnosis of four patients with glaucoma as normal in the experiment. The reason is that, for the dense vascular area near the cup edge, the grayscale is close to that of the neuroretinal rim after inpainting, leading to little local insufficient segmentation. But the difference of CDR between the proposed method and expert is small, thus the CDR was in the range of normal.

To sum up, we have proposed a novel automatic optic cup segmentation algorithm based on inpainting of blood vessels in this study. Compared with other techniques, the proposed methodology has several advantages.

1. The proposed algorithm repaired the vascular regions by using an improved BSCB model, reducing the errors caused by blood vessels to a large degree and improving the segmentation accuracy;
2. In LCV model, the centroid point located in area with higher intensity was selected as the starting point, which realized the automatic selection of the seed point. In addition, the loose selection of circular radius make the initial contour easily be automatically selected.
3. The proposed algorithm realized fully automatic segmentation with high accuracy, which reduced the human intervention of the traditional semi-automatic method.

The proposed method also has some limitations. First, the region of optic cup segmented by the proposed method was slightly smaller than that identified by experts, especially in the dense vascular area. It may be because parts of the image intensities near the edge of optic cup are close to these of the neuroretinal rim after inpainting. Accordingly, local insufficient segmentation will be happened. Second, the robustness of the algorithm is expected to improve. Since the used experimental data were collected from the same type of camera, more data from different type of camera should be tested and the robustness of the algorithm will be improved.

Conclusion

This study proposed a novel optic cup segmentation method based on inpainting of blood vessels using an improved BSCB model. The proposed algorithm realized fully automatic segmentation, which captured the optic cup boundary more accurately compared with other previous methods also dedicating to reduce the influence of blood vessels. Future work is needed to experimented more samples acquired from different type of camera and also developed more robust algorithms.

Authors' contributions

CY, ML and YD made contributions to the conception, design and analysis of data. ML, YD and BL made contributions to the data processing. CY made contributions to the revision of the final manuscript. All authors read and approved the final manuscript.

Author details

[1] College of Life Science and Bioengineering, Beijing University of Technology, Beijing 100124, China. [2] Department of Ophthalmology, Hospital of Beijing University of Technology, Beijing 100124, China.

Acknowledgements

This work was supported by Beijing Excellent Talents Project (No. 2011D005015000008) and project grants from Beijing Nova Program (xx2016120), National Natural Science Foundation of China (81101107), Natural Science Foundation of Beijing (4162008) and Beijing Municipal Education Commission (PXM2017_014204_500012), program for top young innovative talents of the Beijing Educational Committee (CIT&TCD201404053).

Competing interests

The authors declare that they have no competing interests.

References

1. Weinreb RN, Aung T, Medeiros FA. The pathophysiology and treatment of glaucoma: a review. JAMA. 2014;311(18):1901–11.
2. Lee S, Young M, Sarunic MV, et al. End-to-end pipeline for spectral domain optical coherence tomography and morphometric analysis of human optic nerve head. J Med Biol Eng. 2011;31(2):111–9.
3. Chen SLJ. Detection of the optic disc on retinal fluorescein angiograms. J Med Biol Eng. 2011;31(6):405–12.
4. Shen SY, Wong TY, Foster PJ, et al. The prevalence and types of glaucoma in Malay people: the Singapore Malay Eye Study. Invest Ophthalmol Vis Sci. 2008;49(9):3846–51.
5. Cheng JW, Cheng SW, Ma XY, et al. The prevalence of primary glaucoma in mainland China: a systematic review and meta-analysis. J Glaucoma. 2013;22(4):301–6.
6. Dirani M, Crowston JG, Taylor PS, et al. Economic impact of primary open-angle glaucoma in Australia. Clin Exp Ophthalmol. 2011;39(7):623–32.
7. Michelson G, Wärntges S, Hornegger J, et al. The papilla as screening parameter for early diagnosis of glaucoma. Dtsch Arztebl Int. 2008;105(34–35):583–9.
8. Hatanaka Y, Noudo A, Muramatsu C, et al. Automatic measurement of vertical cup-to-disc ratio on retinal fundus images. In: International conference on medical biometrics. Berlin: Springer; 2010. p. 64–72.
9. Xu Y, Hu M, Jia X, et al. Computer-aided diagnosis of glaucoma using fundus images. In: Proceedings of the 2014 international conference on mechatronics, electronic, industrial and control engineering; 2014.
10. Fujita H, Uchiyama Y, Nakagawa T, et al. CAD on brain, fundus, and breast images. In: International conference on medical imaging and informatics. Berlin: Springer; 2008. p. 358–66.
11. Yin F, Liu J, Wong DWK, et al. Automated segmentation of optic disc and optic cup in fundus images for glaucoma diagnosis. In: IEEE CBMS 2012: 25th international symposium on computer-based medical systems (CBMS). New York: IEEE; 2012. p. 1–6.
12. Zhang Z, Srivastava R, Liu H, et al. A survey on computer aided diagnosis for ocular diseases. BMC Med Inform Decis Mak. 2014;14(1):1.
13. Bock R, Meier J, Nyúl LG, et al. Glaucoma risk index: automated glaucoma detection from color fundus images. Med Image Anal. 2010;14(3):471–81.
14. Tangelder GJM, Reus NJ, Lemij HG. Estimating the clinical usefulness of optic disc biometry for detecting glaucomatous change over time. Eye. 2006;20(7):755–63.
15. Nayak J, Acharya R, Bhat PS, et al. Automated diagnosis of glaucoma using digital fundus images. J Med Syst. 2009;33(5):337–46.
16. Babu TRG, Shenbagadevi S. Automatic detection of glaucoma using fundus image. Eur J Sci Res. 2011;59(1):22–32.
17. Joshi GD, Sivaswamy J, Karan K, et al. Optic disk and cup boundary detection using regional information. In: 2010 IEEE international symposium on biomedical imaging: from nano to macro. New York: IEEE; 2010. p. 948–51.
18. Xiao D, Lock J, Manresa JM, et al. Region-based multi-step optic disk and cup segmentation from color fundus image. In: SPIE medical imaging. International Society for Optics and Photonics; 2013. p. 86702H–8.
19. Madhusudhan M, Malay N, Nirmala SR, et al. Image processing techniques for glaucoma detection. In: International conference on advances in computing and communications. Berlin: Springer; 2011. p. 365–73.
20. Joshi GD, Sivaswamy J, Krishnadas SR. Optic disk and cup segmentation from monocular color retinal images for glaucoma assessment. IEEE Trans Med Imaging. 2011;30(6):1192–205.
21. Liu J, Wong DWK, Lim JH, et al. ARGALI: an automatic cup-to-disc ratio measurement system for glaucoma detection and analysis framework. In: SPIE medical imaging. Bellingham: International Society for Optics and Photonics; 2009: 72603K–8.
22. Liu J, Wong DWK, Lim JH, et al. ARGALI: an automatic cup-to-disc ratio measurement system for glaucoma analysis using level-set image processing. In: 13th International conference on biomedical engineering. Berlin: Springer; 2009. p. 559–62.
23. Zhang Z, Liu J, Cherian NS, et al. Convex hull based neuro-retinal optic cup ellipse optimization in glaucoma diagnosis. In: 2009 annual international conference of the IEEE engineering in medicine and biology society. New York: IEEE; 2009. p. 1441–4.

24. Wong DWK, Liu J, Lim JH, et al. Automated detection of kinks from blood vessels for optic cup segmentation in retinal images. In: SPIE medical imaging. Bellingham: International Society for Optics and Photonics; 2009. p. 72601J–8.
25. Cheng J, Liu J, Xu Y, et al. Superpixel classification based optic disc and optic cup segmentation for glaucoma screening. IEEE Trans Med Imaging. 2013;32(6):1019–32.
26. Cheng J, Liu J, Tao D, et al. Superpixel classification based optic cup segmentation. In: International conference on medical image computing and computer-assisted intervention. Berlin: Springer; 2013. p. 421–8.
27. Xu Y, Liu J, Cheng J, et al. Efficient optic cup localization based on superpixel classification for glaucoma diagnosis in digital fundus images. In: 2012 21st international conference on pattern recognition (ICPR). IEEE; 2012. p. 49–52.
28. Fumero F, Alayón S, Sanchez JL, et al. RIM-ONE: an open retinal image database for optic nerve evaluation. In: 2011 24th international symposium on computer-based medical systems (CBMS). IEEE; 2011. p. 1–6.
29. Dougherty ER, Lotufo RA, The International Society for Optical Engineering SPIE. Hands-on morphological image processing. Washington: SPIE Optical Engineering Press; 2003.
30. Bertalmio M, Sapiro G, Caselles V, et al. Image inpainting. In: Proceedings of the 27th annual conference on computer graphics and interactive techniques. New York: ACM Press/Addison-Wesley Publishing Co.; 2000. p. 417–24.
31. Perona P, Malik J. Scale-space and edge detection using anisotropic diffusion. IEEE Trans Pattern Anal Mach Intell. 1990;12(7):629–39.
32. Catté F, Lions PL, Morel JM, et al. Image selective smoothing and edge detection by nonlinear diffusion. SIAM J Numer Anal. 1992;29(1):182–93.
33. Wang XF, Huang DS, Xu H. An efficient local Chan–Vese model for image segmentation. Pattern Recogn. 2010;43(3):603–18.
34. Gloster J, Parry DG. Use of photographs for measuring cupping in the optic disc. Br J Ophthalmol. 1974;58(10):850.
35. Haleem MS, Han L, van Hemert J, et al. A novel adaptive deformable model for automated optic disc and cup segmentation to aid glaucoma diagnosis. J Med Syst. 2018;42(1):20.
36. Al-Bander B, Williams BM, Al-Nuaimy W, et al. Dense fully convolutional segmentation of the optic disc and cup in colour fundus for glaucoma diagnosis. Symmetry. 2018;10(4):87.
37. Bechar MEA, Settouti N, Barra V, et al. Semi-supervised superpixel classification for medical images segmentation: application to detection of glaucoma disease. Multidimens Syst Signal Process. 2018;29(3):979–98.

11

Semi-automatic measurements and description of the geometry of vascular tree based on Bézier spline curves: application to cerebral arteries

Jarosław Żyłkowski[1], Grzegorz Rosiak[1] and Dominik Spinczyk[2]*

*Correspondence:
dspinczyk@polsl.pl
[2] Faculty of Biomedical Engineering, Silesian University of Technology, Roosevelta 40, Zabrze, Poland
Full list of author information is available at the end of the article

Abstract

Background: The geometry of the vessels is easy to assess in novel 3D studies. It has significant influence on flow patterns and this way the evolution of vascular pathologies such as aneurysms and atherosclerosis. It is essential to develop robust system for vascular anatomy measurement and digital description allowing for assessment of big numbers of vessels.

Methods: A semiautomatic, robust, integrated method for vascular anatomy measurements and mathematical description are presented. Bezier splines of 6th degree and continuity of C3 was proposed and distribution of control points was dependent on local radius. Due to main interest of our institution, the system was primarily used for the assessment of the geometry of the intracranial arteries, especially the first Medial Cerebral Artery division.

Results: 1359 synthetic figures were generated: 381 torus and 978 spirals. Experimental verification of the proposed methodology was conducted on 400 Middle Cerebral Artery divisions.

Conclusions: In difference to other described solution all proposed methodology steps were integrated allows analysis of variability of geometrical parameters among big number of Medial Cerebral Artery bifurcations using single application. This allows for determination of significant trends in the parameters variability with age and in contrary almost no differences between men and women.

Keywords: Geometry of cerebral vessels, Computer-aided assessment of geometry of cerebral vessels

Background

The vessel geometry is a main feature assessed by noninvasive, three-dimensional angiographic studies such as computed tomography angiography (CTA) and MRA (magnetic resonance angiography). Its influence on the development of vascular diseases such as aneurysms and atherosclerosis has been widely studied through last decades [1–7]. New possibilities for studying vascular anatomy arise due to rising number of performed angiographies. They are performed in individuals of all ages, with normal or altered anatomy or with vascular diseases such as aneurysms and atherosclerosis [8–10]. Large numbers of cases allow analyses of geometrical differences between age and sex groups and between patients with and without given diseases. Assuming that differences in average values of various groups (sex and age dependent) in the whole population represent general trends in geometry changes in individuals we could better understand this processes and their influence on the vascular pathologies evolution. This kind of strategy is being used in the cosmology for studying galaxy evolution where it is not possible to analyze evolution of particular galaxy. In medicine we still do not have CTA or MRA series covering whole lifetime of any individual. The main requirement for this type of studies if fast and robust system for vascular anatomy measurement and digital description allowing for assessment of great numbers of vessels.

Computer aided, three-dimensional studies of vascular anatomy have been done by many authors [4–6]. When analyzing their methods and results we found that all presented systems were combined with separated modules, each for individual step of preprocessing, measurement and calculations. The vessel analyses were robust but also labor-intensive. This feature makes the application of these system in the large-number studies not practical. In our system we reduced the number of modules to two: one for measurements and data storage and one for the data analysis and collating.

The system of the description of the vessel anatomy applied by our team is not new and is based on works of Italian and Irish teams [4–10]. It is based on centerlines of vessels and allows for proper and robust description of different types of the vascular structures (both straight and diverging vessels).

The aim of this study is to propose semiautomatic, robust, integrated and fast system for vascular anatomy measurements and mathematical description studies with large number of subjects. The application of a spline curve, consisting of Bezier segment for centerlines approximation is a new concept. Smoothing and determination of curvature and torsion was a basic concept of central lines (CLs) transformation into mathematical functions. This step was achieved by transformation into Bezier splines of degree 6. The continuity of C3 degree was proposed and distribution of control points was dependent on local radius. The system was primarily used for the assessment of the geometry of the intracranial arteries, especially the first MCA division being an area of interest of our institutions.

The paper is organized as follows: in "Methods" section, useful convention and information are introduced. Then, the proposed methodology steps are presented in details ("Preprocessing the contrast enhanced CT examination", "Analysis of the vessel cross section", "Determination of the centerlines", "Description of the centerlines", "Calculation of the division zones of the vessel", "Mathematical analysis of the centerlines"). "Materials and validation method" section describes the data set that was used and validation approach. "Results" section shows the outcomes in a different manner for both synthetic and clinical data. The obtained results are analyzed in reference to other works addressing the subject of vessel geometry in "Discussion" section. The last chapter is "Conclusion" section which summarizes results of the study.

Methods

The methodology steps are presented in the flow chart (Fig. 1) and are described in the next paragraphs of this section: preprocessing of the CTA study data, analysis of the vessel cross section, determination of the centerlines, calculation of the division zones of the vessel and finally numerical description of the vessel geometry.

Vessels topology

System of vessels topology is shown in Fig. 2. The location in the vascular tree was described using two integers. The first one was Level zero-based index of vessel, rising distally. The vessel designation on particular Level was started from number 1.

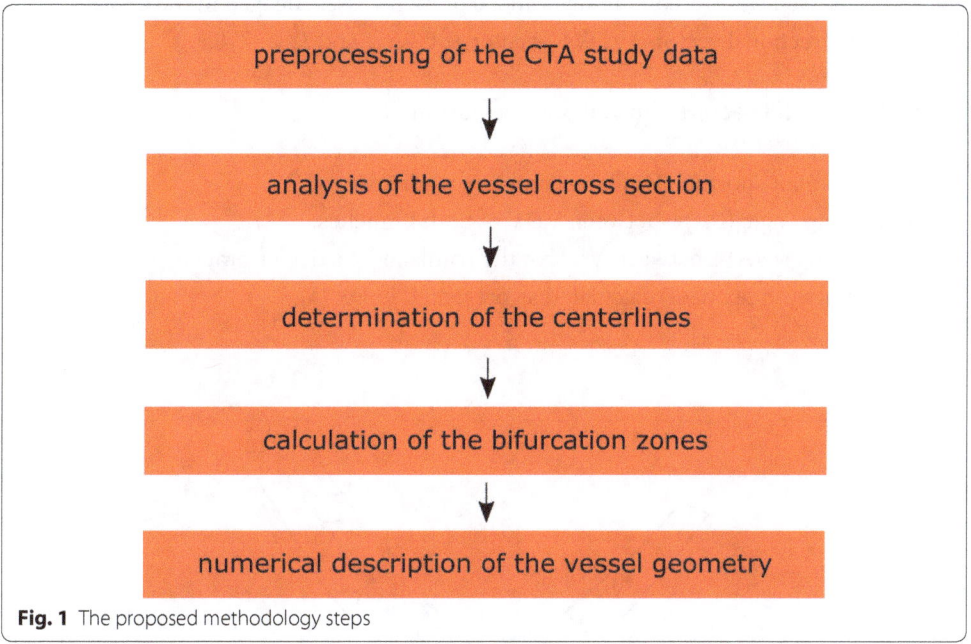

Fig. 1 The proposed methodology steps

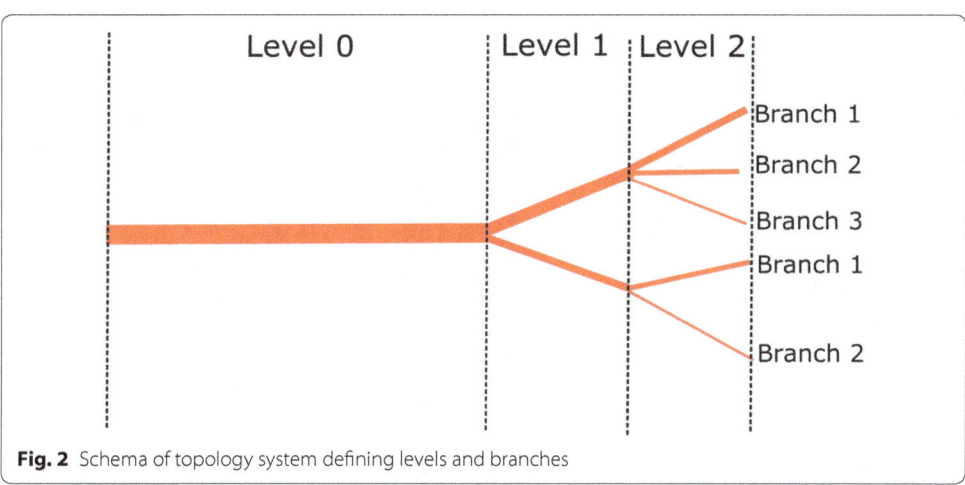

Fig. 2 Schema of topology system defining levels and branches

The following branches were sequentially numbered according to their declining cross section area.

Angles and coplanarity index calculation

Numerical description of the vessels division was based on concept of the bifurcation zone. Accordingly to recent works [11–17] it was defined as the set of points on the course of the centerlines of vessels forming the division. It includes trunk points (T_r) and branch points (B_r). The T_r and B_r points of the stem and branches are numbered from 0 to n (n ∈ N and n > 0). The numbering of the B_r correspond to the direction of the blood flow, while T_r points is in the opposite direction (Fig. 3).

Determined points were used to construct planes and vectors (Fig. 4):

- Division plane (DP)—containing points 0 of each vessel engaged in the division,
- Vessel directional vectors (VDV)—vector $B_r0 - B_r1$ and $T_r1 - T_r0$.

Defined plane and vectors allowed for calculation of:

- Division plane normal (DPN)—vector perpendicular to DP,
- Branching angle (BA)—between VDV of both branches,
- Vessel angle (VA)—between VDV of the trunk and particular branch,
- Coplanarity index (CoI) calculated as presented in Eq. (1)

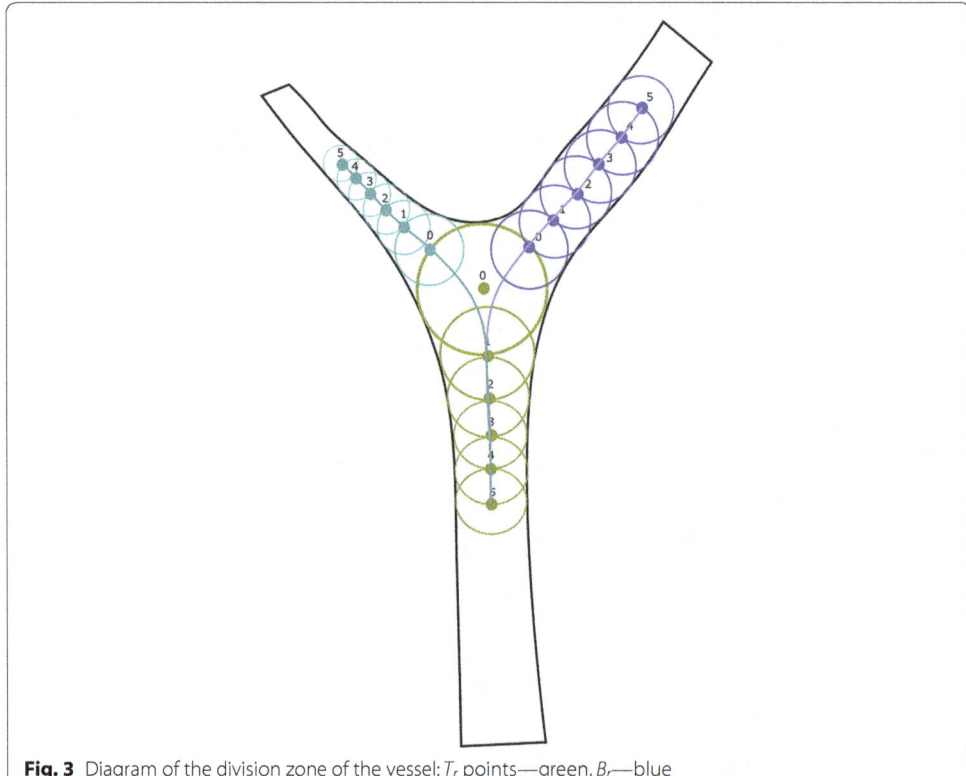

Fig. 3 Diagram of the division zone of the vessel: T_r points—green, B_r—blue

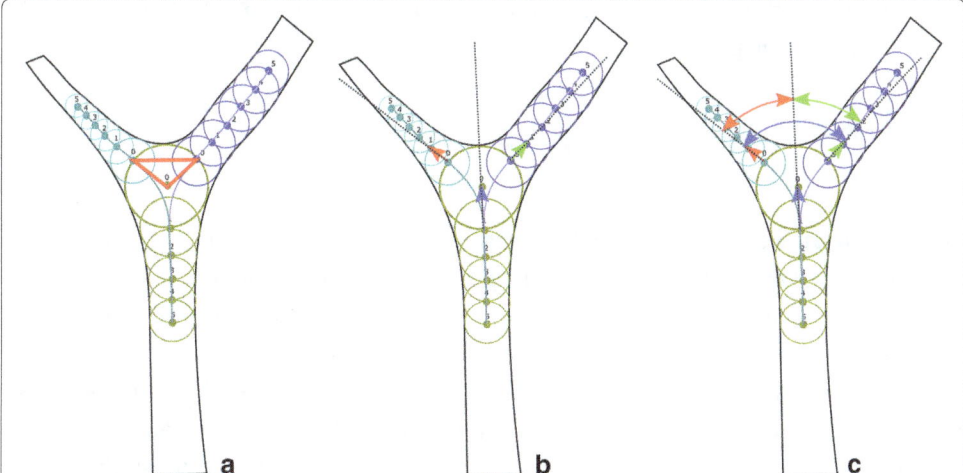

Fig. 4 Planes, vectors and geometrical parameters of bifurcation zone. **a** Bifurcation plane (red), **b** directional vectors: trunk (blue) and vessels (red and green), **c** angles: (blue arrow), vessels angles (red and green arrows)

$$CoI_N = 1 - \left| \frac{\angle(VDV_N, DPN)}{\frac{\pi}{2}} \right| \tag{1}$$

Curvature and torsion of the curve

The course of the curve in space is related to the concept of Frénet–Serret frame. Figure 5 shows the course of the spatial curve K. At point P, it passes in plane A after a momentary arc with center at point O. The torsion of the curve (τ) at point P is defined as:

$$\tau_P = \lim_{P' \to P} \frac{\alpha}{|P'P|}. \tag{2}$$

where α is the angle between the binomial at points P' and P. The analysis of the curvature and the approximation of the centered curve approximation in the division zone was made within the range of the arc involving the division of vessels (Arc P).

In order to determine the position on the central curve, the distance in mm was used. The calculation of the distance of the point and (l_i) from the beginning of the curve in mm was made discretely on the basis of the sum of the lengths of all the

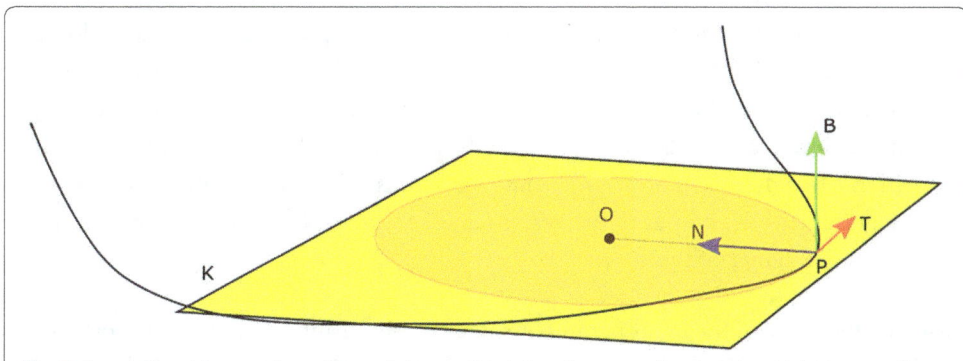

Fig. 5 Frenet–Serret frame; where: K—spatial curve; P point on the curve; A—plane in which the curve K moves at point P; T—tangent of curve K at point T; N—normal curve K at point P; B—binormal curve K at P

sections connecting the following points defining the curve and located before the point for which the position was calculated:

$$\text{Ł}_i = \sum_{n=1}^{i} |p_{n-1}p_n| \tag{3}$$

Preprocessing the contrast enhanced CT examination

The user sets volume of interests (VOI) in which the remaining measurements were performed. Within the VOI the resolution was raised to 0.2 mm ($\sim$ threefold typical resolution of CTA study) in each dimension and this step utilizes tricubic value approximation method. The vessels was segmented utilizing thresholding with two low levels of 100 Hounsfield Units (HU) up to 150 HU. Values of both levels were established during initial phase of the project. The lower values were chosen for strictly arterial phase studies, which was defined as enhancement of no more than 50 HU of deep Sylvian veins occupying space around MCA trunk [18, 19].

After segmentation step, calculation of maximum value of minimal distance from model edge (MOBBM) was performed in the VOI. The results of this step values were used by centerline detection algorithms. In the last step of this part of the analysis the user sets one start point on vessel level-0 and one endpoint on each vessel level-1.

Analysis of the vessel cross section

The purpose of this stage is to automatically orient the measuring plane perpendicular to the long axis of the vessel and measure diameter (minimal, maximal and average) and vessel section area.

The input parameters for the method, expressed in HU, are: the range considered as the core of the vessel $[C_{min};C_{max}]$, lower threshold value of border area of adjacent vessels—M_{min}, lower threshold value of the end of the region growing process—O_{dv}. To accomplish this goal the following specific steps are required:

- Initialization—consists of manually entering the vessel cross-section into a rectangular area.
- Finding the geometric center of the vessel P_{sc}.
- Finding the border of a vessel—the algorithm uses a specific way of distributing values across the vessel. The central part of the vessel exhibits clearly higher values, which, when away from the center, initially drop slightly, and then, closer to the edge, go down rapidly. In case of neighboring vessels when moving between them, a rise of values is observed after an initial drop subsequent to crossing of the vessels edge. Images in Fig. 6c, d show differences in the result of the algorithm when detection of neighboring vessels was set to on and off respectively.
- Smoothing of the border of the vessel P_0—uses the local average distance from the geometric center of the vessel.

Based on the value of the input parameters, the threshold and labeling of the core region of the vessel and the center of gravity are determined. After transformation of the P_{SC} to the global coordinate system, the algorithm of finding the edge of the vessel in n

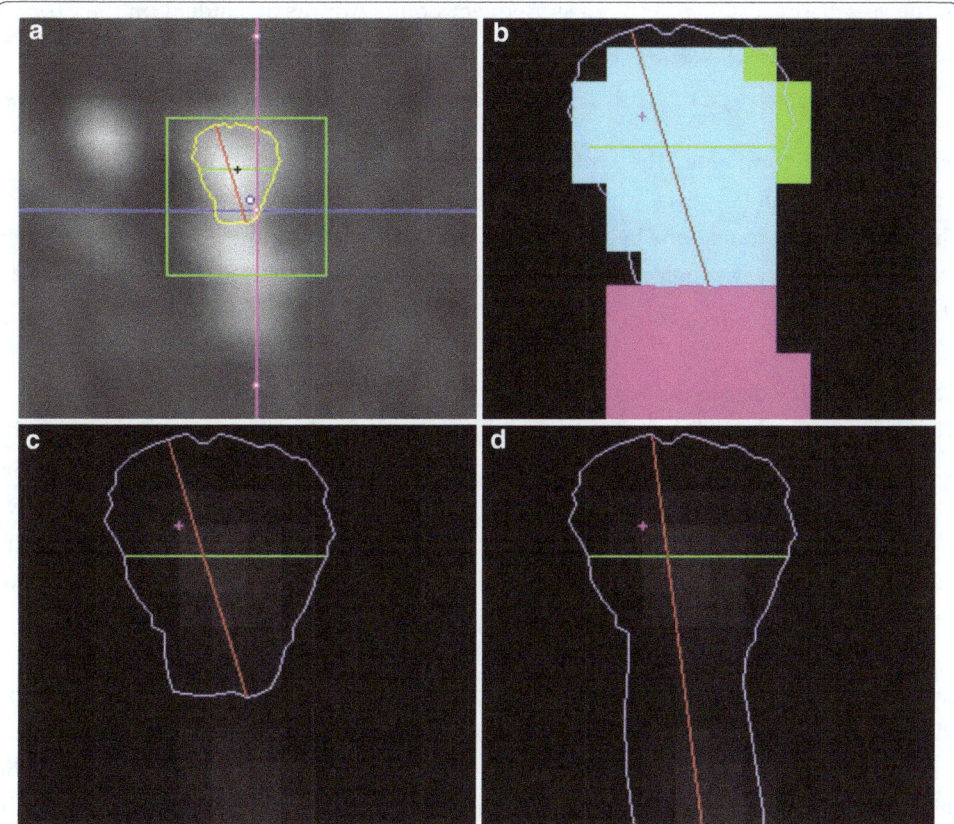

Fig. 6 Vessel cross-section analysis: a method of tracing the vessel (**a**), the effects of subsequent algorithm steps, dividing the section into groups of points and finding the border (**b**). Effect of switching on (**c**) and off (**d**) function preserving crossing close related vessels borders

directions with step $2\pi/n$ is n-folded. The result of this step is obtaining a set of n 3D points describing the shape of the border. The last step was to find the extremes and the average diameter and cross-sectional area. Extremities were found by analyzing the length of the sections between the border points with indices i and n + 2, where n is the number of directions in which edge points were searched, $i \in [0, n/2]$. The cross-sectional area (PPP) was calculated as the sum of the surface area of the triangles: (P_{sc}, P_{0i}, P_{0i+1}) where $i \in [0, n-1]$ and (P_{SC}, P_{0n-1}, P_{00}). The average diameter Δd was calculated from PPP by the formula:

$$\Delta d = \sqrt{\frac{4P_{pp}}{\pi}}. \tag{4}$$

Determination of the centerlines

The center curve calculation algorithm was used to determine a single central curve of the vessel between the indicated vessel cross-sections defined in the previous step. The curve was driven by points representing the local maxima of the minimum distance from the edge of the M_{OOBM} model, in the cross-sectional area perpendicular to the long axis of the vessel. The distance between subsequent points Δd was dependent on the M_{OOBM} value at P, from where the next point was sought:

$$\Delta d = M_{OOBM}(P) * f \tag{5}$$

where f is the crawl rate. The typical f values for this study were in the range [0.5;1]. Values above 1, and especially above 2 in case of neighboring vessels could cause undesirable algorithm transitions between vessels. Values below 0.5 were impractical as they were generating very rough centerlines, especially for smaller vessels with radii close to study resolution.

Calculation of the division zones of the vessel

The process of calculating division zones began with finding the point of decay of the central curves forming the division. The algorithm for each central curve forming the division found the last point for which the minimum distance from the second curve was less than the set value of d (assumed to be 0.2 mm—the spatial resolution of the auxiliary volumes). The split point was calculated as the geometric mean of the two points found. Its value was entered into the partition structure as a T_0.

Based on the location of the points found, the parameters of the partition zone were calculated. Split zone points were also used in further analysis of curvature and torsion (described in the following paragraph). The last steps in the division zone calculation were to determine whether it describes the actual division of the MCA or whether the aneurysm was removed—one of the component curves is the curve to the aneurysm. Of the identified zones, the one for which the split point was the furthest from the origin of the MCA trunk was chosen. This meant that it was a division zone describing the departure of the aneurysm dome from the parent vessel. The second curve of this zone was the curve of the parental aneurysm vessel. In this way, the aneurysm stem was identified. After this step, due to the scope of the analysis, zones other than those describing the actual division of the MCA were rejected.

Description of the centerlines

As in the case of vessel cross-section sections, a number (type of curve) identifying the type of vessel described by it was assigned to further identify each curve. For example, during measurements of MCA divisions, the curve of type 1 was a curve to the aneurysm sac, the curves of successive branches forming the MCA obtain consecutive umbers 2, 3 and so on. Additionally it was possible to mark up to 25 control points along the curve. These point were used during algorithms assessment phase in true CTA studies. This functionality was implemented by introducing editable control points in the view (both MPR and 3D) of the central curve.

Mathematical analysis of the centerlines

The assumption of mathematical analysis of the central curves was to determine correlations between distribution of the curvature and the torsion and evolution of vascular pathologies such as aneurysms and atherosclerosis. In order to calculate the curvature and the torsion of the center lines all curves were approximated with use of the Bezier curves. In order to partially decompose the crankshaft curve, it was decided to use the spline curves consisting of Bezier curves, connected at the site of the division of the vessel. The 6th grade curves and C3 continuity class were selected. The continuity class C3 was needed to ensure the continuity of the torsion function [20]. The 6th grade was a minimum grade allowing the creation of B-spline curves of any number of segments while maintaining this continuity class. The additional effect of the approximation with the parametric curves was the smoothing of the curves.

Materials and validation method

The purpose of this stage of the study was to assess the alignment and accuracy of the algorithms presented and to identify factors and conditions that could significantly impair the measurement values. Verification of the operation and accuracy of the algorithms presented above was made using synthetic models. Models were generated within a synthetic image data volume (SWD) dimension of $100 \times 100 \times 100\,\text{mm}$ and a 0.6 mm spatial resolution in all axes. The HU distribution on the vessel cross section was simulated by the formula:

$$HU(r) = C + C\left(\frac{1}{2} - \frac{1}{1 + e^{-\alpha(r-R)}}\right) \tag{6}$$

where C—HU at the boundary of the vessel, r—distance from the center of the vessel for which we make the calculation, R—radius of the vessel, a—slope coefficient of the HU in the perimeter of the vessel. In Fig. 7 surface charts of HU(r) function values across the vessel are presented for different values of the parameters.

Models were drawn within a measuring volume using a spatial brush. The brush worked within a cube area of 1.5R sides and entered a WSP value in the SWD according to the above pattern if the calculated value was higher than the current one in the voxel. For all synthetic models the following parameters were used: C = 150, a = 10.

Experimental verification of the proposed methodology was conducted on 400 MCA trunk divisions examined with the contrast enhanced CTA. Patients were randomly assigned to this group from set of over 4500 individuals examined in the Computer

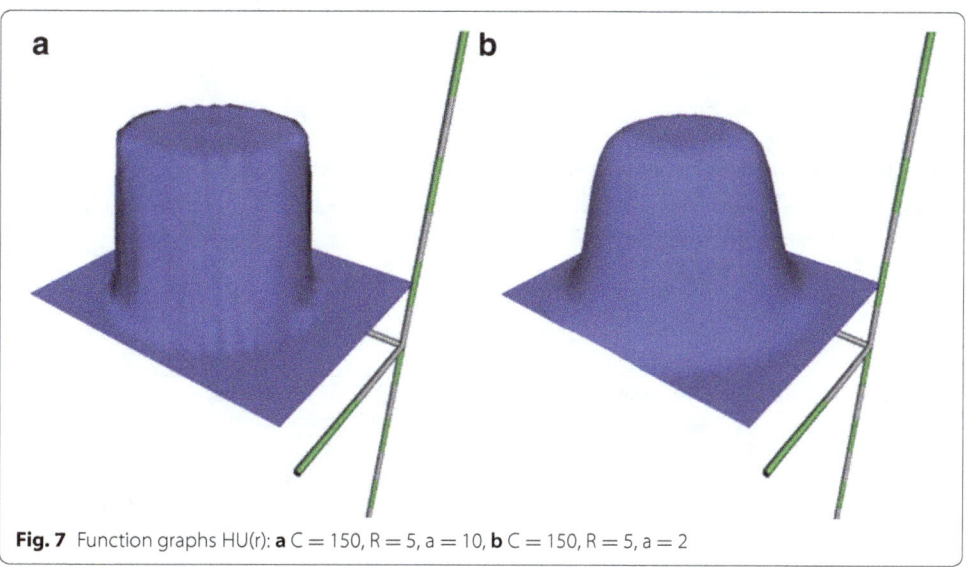

Fig. 7 Function graphs HU(r): **a** C = 150, R = 5, a = 10, **b** C = 150, R = 5, a = 2

Tomography Lab of the Second Department of Clinical Radiology Medical University of Warsaw between June 2006 and December 2016. For the measurements step 423 studies were assigned, but 23 (5.4%) of them were rejected due to issues concerning study: mixed arterio-venous phase or insufficient contrast enhancement of the vessels.

Results

Validation of the algorithms

Within the volume, 1359 synthetic figures were generated: 381 torus and 978 spirals. Each of the examined figures had different shape parameters expressing the simulated vessel diameter (D), the basal radius of the vessel (R1), and spiral addition (R2). The parameter ranges used for generation of models are summarized in Table 1.

Patterns that overlap synthetic vessels were rejected, those in which R1 < 2D or R2 < 2D.

Vessel diameter measurements validation

On each model, at 10 random points ($M_{min} = 150$, $O_{dv} = 160$, $C_{min} = 170$, and $C_{max} = 1500$) vessel diameter measurements were performed and recorded in the file along with model parameters. For each measurement mean D, standard mean deviation and root mean square (RMS) were calculated. The average calculated parameters for all measurements separately for toroidal and spiral models are summarized in Table 2 and in Fig. 8. The presented data show accuracy of diameter and cross-section area of the vessel measurements.

The results showed in Table 2 present very good mean RMS_D and RMS_P. In case of RMSD almost one order of magnitude lower than study resolution. For the same set of data, the maximal RMS_D was more than twice bigger than this resolution. The cause of this discrepancy was found in Fig. 8.

This analysis shows very good measurement accuracy for vessels with D > 1.5 mm. Below this value the accuracy of measurement is clearly decreasing, breaking down for

Table 1 Range of parameters used for generation of synthetic figures for validation of the division zones of the vessel algorithm

Model	D_{min}	D_{max}	D_{step}	$R1_{min}$	$R1_{max}$	$R1_{step}$	$R2_{min}$	$R2_{max}$	$R2_{step}$
Torus	1	5	0.25	5	30	1	Nd	Nd	Nd
Spiral	1.5	4	0.25	5	30	1	5	9	1

Table 2 Summarized results of vessel diameter measurements validation

Parameter	Torus N	Torus mean	Torus min	Torus max	Torus std dev
SD_D	381	0.03	0.00	0.33	0.051
SD_P	381	0.10	0.00	0.74	0.12
RMS_D	381	0.12	0.017	1.00	0.21
RMS_P	381	0.35	0.039	1.10	0.26
Parameter	**Spiral N**	**Spiral mean**	**Spiral min**	**Spiral max**	**Spiral std dev**
SD_D	978	0.058	0.0070	1.29	0.12
SD_P	978	0.19	0.032	0.90	0.24
RMS_D	978	0.082	0.013	1.43	0.13
RMS_P	978	0.31	0.058	1.48	0.29

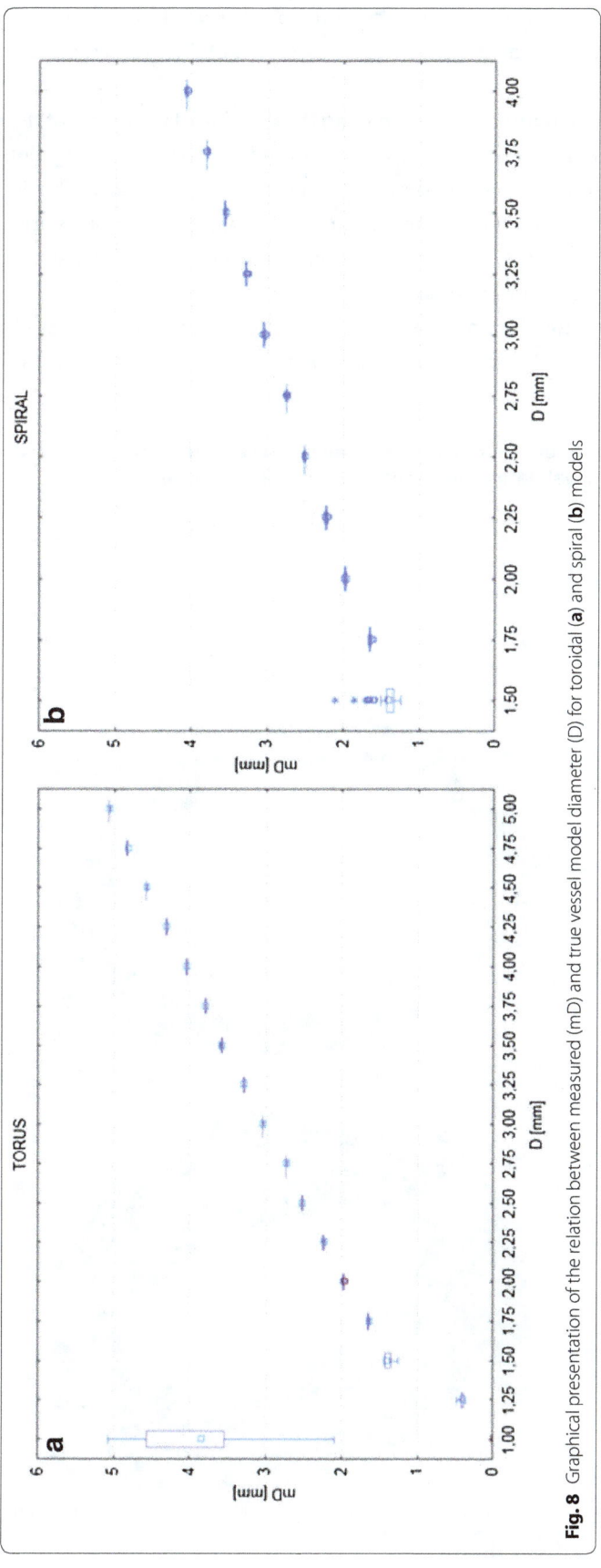

Fig. 8 Graphical presentation of the relation between measured (mD) and true vessel model diameter (D) for toroidal (**a**) and spiral (**b**) models

$D = 1$ mm. The result is consistent with the predictions based on the way in which the measurement algorithms and the spatial resolution of the synthetic data volume operate.

Validation of the algorithm for determining the position of the points of the centerlines
The evaluation of the precision of the algorithm in determining the points of the central curves was based on the evaluation of the mean square distance of the point determined from the actual position described by the mathematical function. Within the synthetic volume, 36 figures were generated: 25 toruses and 11 spirals. The parameters describing the synthetic figures are summarized in Table 3.

Within each model, a minimum of two central curves were calculated using algorithm 1 for dA = 3 / 2π and *df* = 0.5 (Fig. 9). The obtained curves along with the

Table 3 Summary of the range of parameters used for generation of synthetic models in the process of validating the accuracy of the center curves

Model	D_{min}	D_{max}	$R1_{min}$	$R1_{max}$	$R2_{min}$	$R2_{max}$
Torus	0.6	2	5	25	Nd	Nd
Spiral	1	2	5	30	8	100

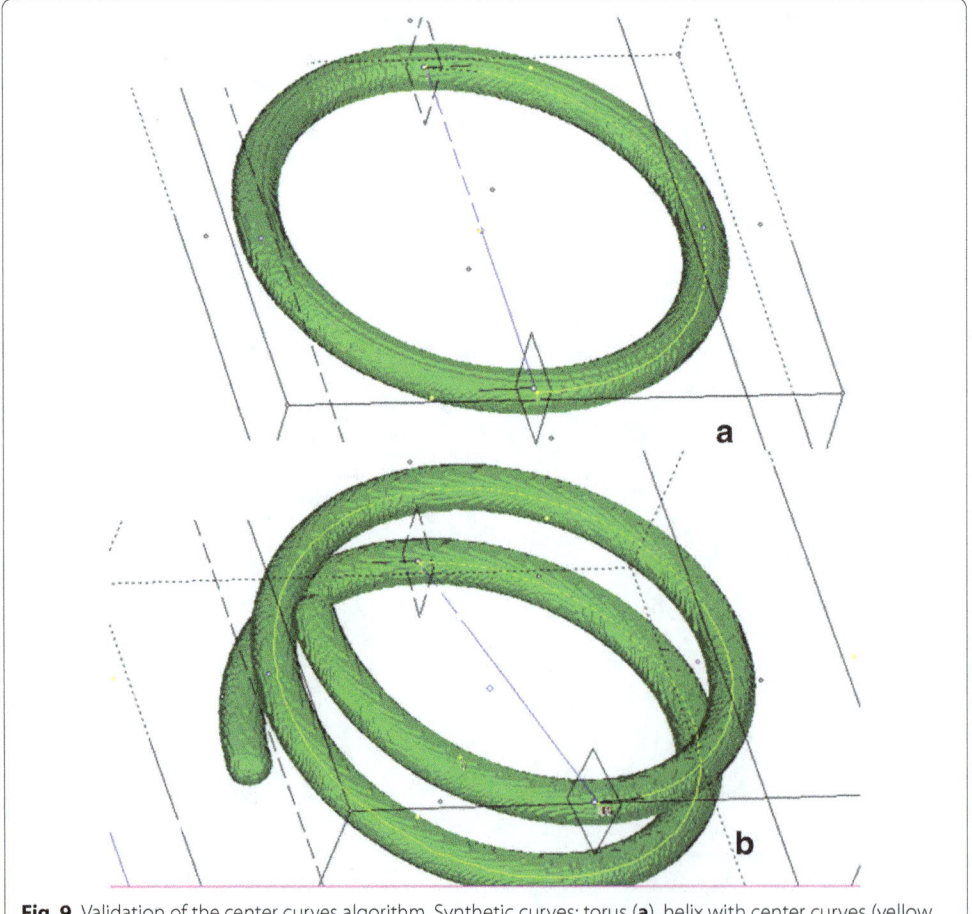

Fig. 9 Validation of the center curves algorithm. Synthetic curves: torus (**a**), helix with center curves (yellow lines) between the measuring planes (**b**)

Table 4 Calculated RMS of precision of the points of the centerline positioning, separately for helix and torus models

Group	N	RMS_{mean}	RMS_{median}	RMS_{min}	RMS_{max}	RMS_{SD}
Helix	11	0.063	0.062	0.059	0.069	0.0034
Torus	25	0.061	0.062	0.041	0.076	0.0090

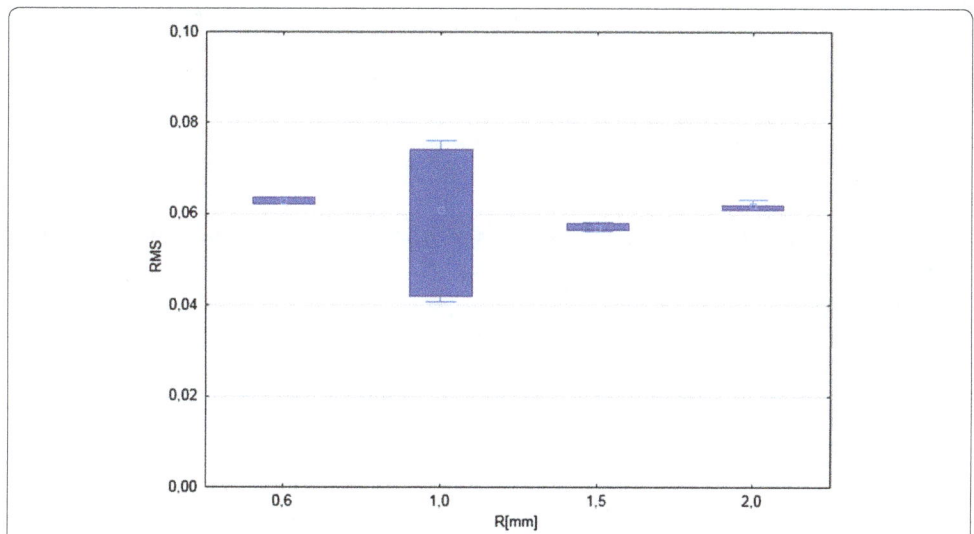

Fig. 10 RMS values of diameter measurements grouped by radius of artificial vessels. Squares represents median values, boxes ranges of 25–75% percentiles and whiskers range of values

model type (HELIX,TORUS) and its parameters are stored in separate files. For each generated central curve, the RMS was calculated in relation to the ideal mathematical model. The results are summarized in Table 4 and Fig. 10.

The obtained data showed very good accuracy of the positioning of the points of the central curves by the developed algorithms. No significant RMS L differences were found for the type and model parameters.

Validation of the vessel bifurcation zone calculation

The process of the validation of the bifurcation zone calculation was conducted utilizing artificial bifurcation zones (ABZ) with known BA, VA_s and C_oI_s. Figure 11 shows artificial bifurcation zone schema and example of artificial model. Ranges of ABZ variables' values are presented in Table 5. The ranges of these values were stated on the basis of the anatomical literature concerning brain vasculature [19, 21–23] and experience of our team gained during more than 8000 digital subtraction angiographies of the brain vasculature.

The results presented as absolute difference between expected and measured value for selected variables are presented in Table 6. Average, absolute difference between real and measured angles were lower than 3.4°. Best results (below 1.9) were observed for dominant branch. When maximal absolute difference were assessed the worst results

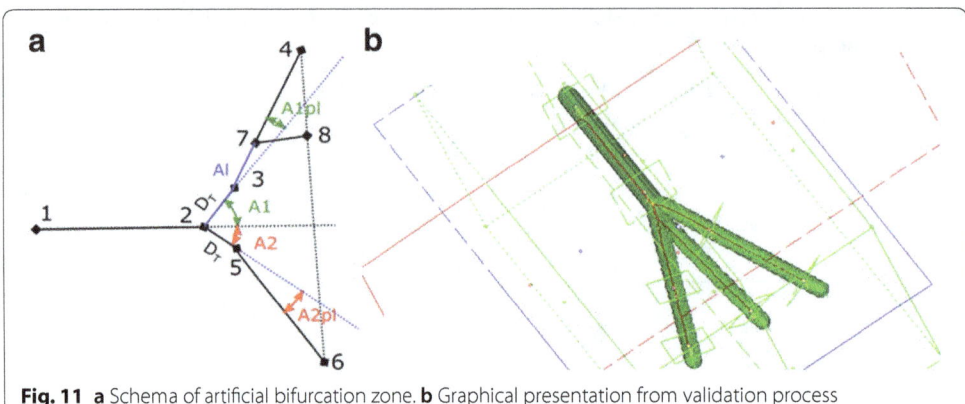

Fig. 11 **a** Schema of artificial bifurcation zone. **b** Graphical presentation from validation process

Table 5 The ranges of the values of variables of ABZ used during algorithms validation process

Parameter	Min	Max
A1	15	80
A2	15	80
$A1_{pl}$	− 15	15
$A2_{pl}$	− 15	15
$_wA_n$	0	0.15
D_T	3.5	4.5
D_1	2.5	4.2
D_2	2.0	3.0
D_{an}	2	6

Table 6 Results of bifurcation zone validation process as average, median, extremes and standard deviation of absolute differences between expected and measured values

Parameter	N	Average Δ	Median Δ	Min Δ	Max Δ	SD Δ
$|\Delta BA|$	70	3.33	2.61	0.010	13.44	2.73
$|\Delta BA|$	70	3.33	2.61	0.010	13.44	2.73
$|\Delta VA_{dom}|$	70	1.83	1.47	0.030	6.58	1.43
$|\Delta VA_{ndom}|$	70	3.27	2.98	0.093	12.93	2.49
$|\Delta C_oI_{dom}|$	70	0.033	0.027	0.0006	0.12	0.026
$|\Delta C_oI_{ndom}|$	70	0.039	0.035	0.0002	0.15	0.033
$|\Delta C_oI_P|$	70	0.0085	0.0068	0.0002	0.054	0.0081

oscillated around 13° for BA and VA of no dominant branch. For dominant branch it was much better—below 7°.

For C_oI average and maximal absolute differences were lowest for parent truck (C_oI_P) slightly over 0.008 and 0.05 respectively comparing to almost 0.04 and 0.15 for both branches.

Table 7 Number and percentage of the main three types of MCA division with divide into male and female and percentage of male and female divisions in each type of division

Division type	Number	Percent	Male	Female
Bifurcation	340	85.0	174 (51.2%)	166 (48.8%)
Multiple trunks	34	6.0	13 (54.2%)	11 (45.8%)
Trifurcation	36	9.0	13 (36.1%)	23 (63.9%)

Table 8 Median values and 25–75% percentile range of analyzed geometrical parameters for bifurcations and multiple trunk vessels divisions with p values for U Mann–Whitney test comparing both groups

Variable	Median values bifurcations	Median values multiple trunks	p (U Mann–Whitney test)
BA (degree)	82.8 (68.0–104.7)	91.5 (72.6–99.0)	> 0.4
VA_{dom} (degree)	46.0 (33.9–59.1)	48.7 (26.2–62.8)	> 0.8
VA_{ndom} (degree)	55.7 (42.1–71.3)	59.9 (51.5–80.0)	> 0.1
Col_{dom}	0.877 (0.785–0.943)	0.886 (0.822–0.974)	> 0.3
Col_{ndom}	0.877 (0.782–0.945)	0.806 (0.764–0.935)	> 0.3
Col_T	0.895 (0.820–0.949)	0.883 (0.803–0.953)	> 0.8
$Kmax_{dom}$ (rad/mm)	0.246 (0.173–0.335)	0.258 (0.159–0.449)	> 0.5
Kav_{dom} (rad/mm)	0.151 (0.115–0.205)	0.187 (0.097–0.268)	> 0.2
Tav_{dom} (rad/mm)	0.321 (0.239–0.454)	0.334 (0.213–0.432)	> 0.5
$Kmax_{ndom}$ (rad/mm)	0.308 (0.185–0.450)	0.381 (0.298–0.640)	< 0.05
Kav_{ndom} (rad/mm)	0.177 (0.125–0.236)	0.216 (0.163–0.318)	< 0.05
Tav_{ndom} (rad/mm)	0.386 (0.281–0.525)	0.440 (0.318–0.549)	> 0.4

The measurements process validation on clinical data

The measurements process validation of clinical data was performed for the middle cerebral artery division. The vessel level-0 was defined as MCA M1 segment (trunk), and level-1 MCA M2 (branches of the first division). Total number of 400 MCA divisions (M = 200) were measured and digitally described. According to Rhoton three main types of the MCA division were identified: bifurcation (BIF), multiple trunks (MT) and trifurcation (TRIF) [19] (Table 7).

For further geometrical analysis only dichotomous divisions: BIF and MT were selected which could be properly, geometrically described by previously defined parameters. TRIFs were rejected because in our opinion they could not be simply described as two combined bifurcations and need much more, also topological parameters which exceeds the scope of the presented paper.

The Shapiro–Wilk tests showed normal distribution only for vessel dimensions. The rest of the analyzed parameters presented with distribution different than normal.

According to the results of normal distribution test observed values of analyzed parameters with p-values of statistical tests comparing BIFs and M9Ts are presented in Tables 8 and 9.

The significant differences between both groups concern $Kmax_{ndom}$ and Kav_{ndom} as well as diameters of non-dominant branches. Both Kmax and Kav presented higher values in MT group comparing to BIFs: 0.381 rad/mm (0.298–0.640) vs. 0.308 rad/

Table 9 Average values and SD of vessels average diameters for bifurcations and multiple trunks with p values for t-test comparing both groups

Variable	Average values bifurcations	Average values multiple trunks	p (t-test)
D_T (mm)	2.9 ± 0.55	2.8 ± 0.51	> 0.5
D_{dom} (mm)	2.4 ± 0.41	2.4 ± 0.56	> 0.6
D_{ndom} (mm)	1.8 ± 0.41	1.5 ± 0.43	*< 0.001*

Statistically significant differences were italicized

Table 10 Spearman's correlations coefficients of analyzed variables with age for divisions of type BIF and MT with specified p values of significance test

Parameter	BIF R	BIF p	MT R	MT p
BA	0.1940	0.0003	0.1836	0.3905
VA_{dom}	0.1378	0.0109	0.0209	0.9228
VA_{ndom}	0.2092	0.0001	0.1866	0.3825
Col_{dom}	− 0.0285	0.6006	0.1806	0.3985
Col_{ndom}	− 0.1023	0.0596	0.1179	0.5832
Col_T	− 0.1604	0.0030	-0.3011	0.1528
$Kmax_{dom}$	0.1132	0.0369	0.4172	0.0425
Kav_{dom}	0.1526	0.0048	0.4333	0.0344
Tav_{dom}	− 0.1489	0.0059	0.1310	0.5419
$Kmax_{ndom}$	0.0979	0.0715	− 0.0191	0.9293
Kav_{ndom}	0.0921	0.0898	0.0379	0.8606
Tav_{ndom}	− 0.2263	0.0000	− 0.0526	0.8070
D_T	0.1713	0.0015	− 0.2001	0.3484
D_{dom}	0.1856	0.0006	− 0.2358	0.2673
D_{ndom}	0.1557	0.0040	− 0.2315	0.2765

Significant (p < 0.05) correlations are italicized

mm (0.185–0.450) and 0.216 rad/mm (0.163–0.318) vs. 0.177 rad/mm (0.125–0.236) respectively.

DT was higher (2.87 mm ± 0.55 vs. 2.79 mm ± 0.51) and Ddom lower (2.37 mm ± 0.41 vs. 2.41 mm ± 0.56) in BIFs comparing to MTs.

D_{ndom} was higher (1.8 mm ± 0.41 vs. 1.5 mm ± 0.43) in BIFs comparing to MTs.

In the next step the parameters were correlated with age. Process of correlation was conducted utilizing Spearman coefficient, separately for each type of division. Additional comparison of significance of differences between both groups were performed. Results are presented in Table 10. For BIF type of division multiple, significant correlations with age were found for variables: BA, VA, Col_T, Kav_{dom}, Tav_{dom} and D. Most of these correlations presented very low p values. For MT type of divisions statistically significant correlations were found for Kmax and Kav values of dominant branch. Form most parameters signs of R coefficients were same for both types of divisions. For BIF type of divisions much lower values of p were observed for lower values of R coefficient comparing to MT. This situation was caused by ten times higher number of BIF type of divisions (n = 340 vs. n = 34) and specificity of p calculation for R coefficients.

Scatter plots of the selected parameters presenting significant correlations with age could be found in Fig. 12. The general impression from presented plots is that for VA and

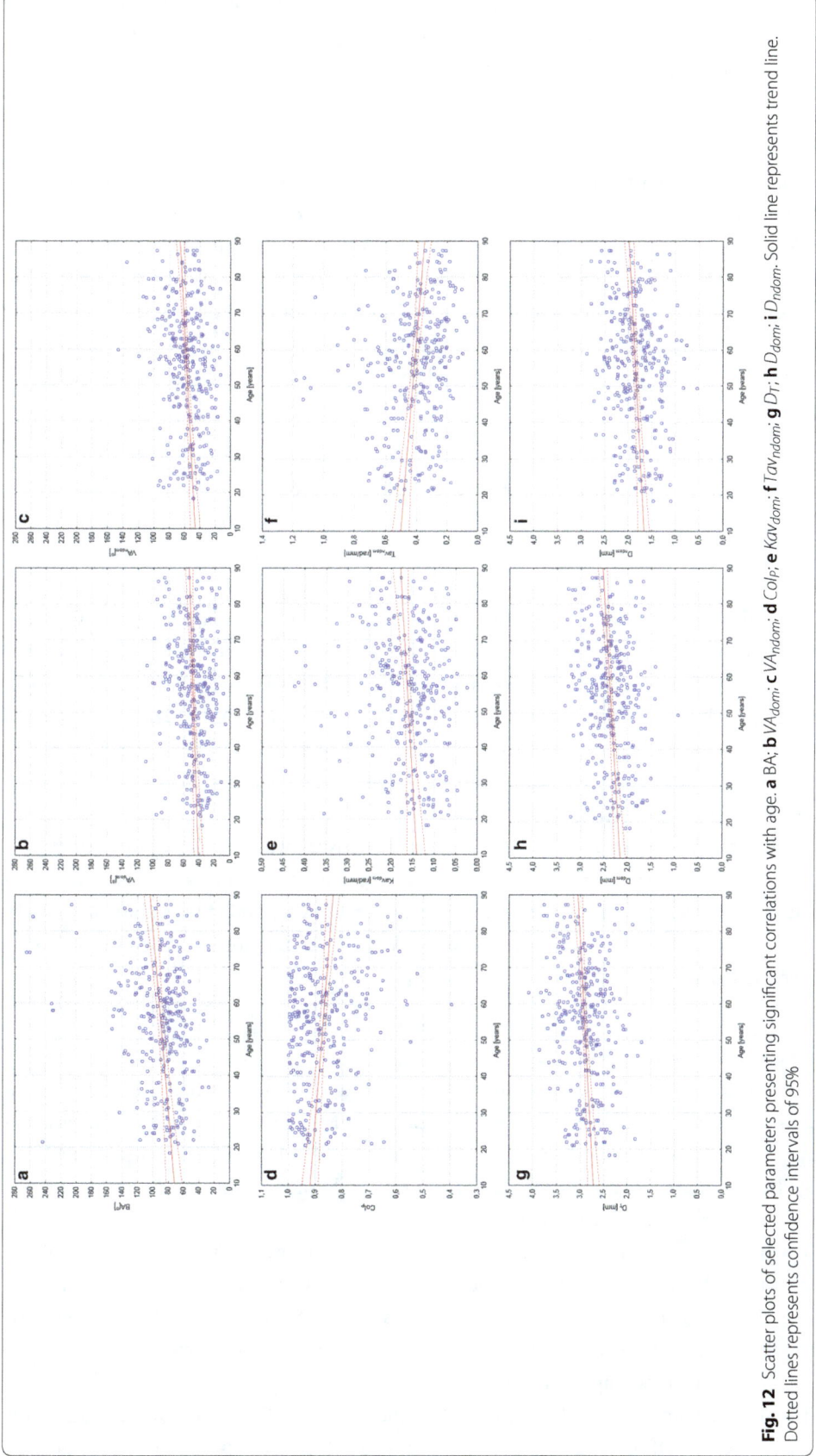

Fig. 12 Scatter plots of selected parameters presenting significant correlations with age. **a** BA; **b** VA_{dom}; **c** VA_{ndom}; **d** CoI_{T}; **e** Kav_{dom}; **f** Tav_{ndom}; **g** D_{T}; **h** D_{dom}; **i** D_{ndom}. Solid line represents trend line. Dotted lines represents confidence intervals of 95%

vessel diameter we can observe much more narrow, and constant value ranges than the rest, especially CoI. In case of BA, Kmax, Tav and CoI despite observed trends we can observe widening of values range with age.

Discussion

Geometrical and mathematical description of vessel geometry was addressed in current literature for a few years. Its linkage to vessel flow was confirmed in many studies [4–7, 24–27]. Most of presented systems, contrary to ours consisted of many separated modules for preprocessing, segmentation, measurement and post-processing of the data. The differences also include different way of vessel models representation and consequently way of centerlines determination. Now we discuss limitations of presented methods, results of algorithm validation on artificial models and finally on native, medical CTA data.

Main features of vessel segmentation and centerline detection

In the presented study, opposite to works of Italian [11–16, 24] and Irish [17] groups vessels models were defined not as 3D meshes but rather group of voxels of the same index values.

The Italian group after segmentation transformed vessel models into 3D meshes and centerlines were determined utilizing Voronoi diagrams of these meshes. In this technique centerline points in vessel cross section are point maximally distant from model surface [11–17, 28, 29]. This idea of maximal distance from model surface is similar to the assumption in the proposed algorithm. Both techniques should also be resistant to errors due to close relationship between two vessels. In this case both generate two local maxima, each in geometrical center of abutting vessels.

Methodology of centerline detection of the Irish group was not clearly described [17]. Only short information, that CLs were lead over geometrical centers of vessel cross sections, is available. This tactic could cause errors in mentioned regions of abutting vessels.

The presented strategy for segmentation is very simple but sufficient as preprocessing for created algorithm for CLs.

Mathematical analysis of the CLs

The basic concept of CLs transformation into mathematical functions was the smoothing and determination of curvature and torsion. This step was realized by transformation into Bezier splines of degree 6 and continuity of C^3 degree [20] was proposed and distribution of control points was dependent on local radius. That was different comparing to the works of Italian and Irish groups.

Transformation into mathematical function allows to avoid indetermination of the curvature and the torsion of CL which could occur in straight vessel segments when using discrete method. The group of OFlynn used transformation into 9th degree polynomials, and Antiga's group utilizes discrete method, as it implies from documentation of the Vascular Modelling Toolkit.

The first advantage of the first approach is that process of approximation additionally cause curve smoothing and allows for easy and continuous calculation of curvature and torsion. The first approach disadvantage is that polynomial could precisely describe only relative small segment of the vessel. Increasing degree of the polynomial increases the length of approximated CL but causes problems with locality approximation. It is also difficult to dynamically change degree of fitting to avoid over and under approximation where vessel changes it diameter.

In the second approach both problems are solved, because separation of CLs points depends on local radius and CL could have any length. There are two disadvantages: before Frénet-Serret frame calculation it is necessary to smooth the curve, which causes known problems with shape preservation, setting proper level of noise filtering [30, 31]. The second, discrete algorithm for Frénet frames calculation could fail on straight portions of vessels: the frame could be indefinable or there could be fast and random fluctuations of its orientation, which causes incorrect calculation of the curvature and torsion.

The approach proposed in this study seems to solve all mentioned problems. Bezier segment degree is chosen to allow construction spline of any number with preservation continuity of third derivative [20]. Approximation within each segment is local. Degree of fitting could be altered by setting control points in distances depending on local vessel diameter.

Vessels diameter measurements

The first limitation was identified during process of measurements validations on artificial models. As presented in Table 2 and in Fig. 8 vessel diameter measurements performs very well down to about 2 * voxel size, which means that measurement of vessels less than this value is not reliable. This issue is a consequence of Nyquist–Shannon sampling theorem. Above this limit presented method reaches subvoxel accuracy of measurements.

Bifurcation zone calculations

Validation of vessel bifurcation zone presented in Table 6, showed very good accuracy of measurements of presented algorithms.

Application of the algorithms to clinical data

The most important step was analysis of clinical data performed on native CTA images of patients. We have observed that the software allows for all measurements to be performed if proper (arterial) or slightly delayed phase of the study was available. In case of mixed arterio-venous phase CLs estimation was possible but required extensive user interactions, so these cases were rejected from calculations.

The percentage representation of various MCA divisions types was similar to presented in anatomical study of Rhoton [19]. The main type of division was BIF consisting of 85% of divisions, the second was TRIF (9%) and the last frequent MT (6%). In the Rhoton's study these percentages are as follows: 78%, 12% and 10% respectively. In each group almost equal numbers of male and female divisions were analyzed (Table 7).

Created application allowed for analysis of 400 MCA divisions. Acquired data and metadata were stored in local database and could be used in the future. In first step, the value distribution analysis, it was showed that almost all parameters had distribution different from normal. In the second step, median and average values of analyzed parameters were calculated for BIF and MT type of division. There were found significant correlations for 9 of analyzed variables with patients age utilizing Spearman correlation coefficients. It was found that BA, VA, Kav_{dom} and D increases their values with patients age. Contrary CoI_P and Tav_{ndom} decreases their values. Further analyzes, especially scatter plots showed that additional widening of Kav, CoI and Tav values range with age. This observation is similar to the widening of CCA bifurcation angle with age reported by Thomas [16].

Conclusions

This article presents a concept of integrated system for measurements and analysis of large amount of vascular and anatomical data. The system implements most precise methods for vascular anatomy description based on centerlines. Use of spline curve consisting of Bezier segment for centerlines approximation is a new concept. Unlike other reported solutions all proposed methodology steps were integrated which allows the analysis of variability of geometrical parameters in a large number of MCA bifurcations using only one application This allows for determination of significant trends in the parameters variability with age and differences between MCA division types.

Splines of 6-th degree and continuity of C3 degree allow unlimited length approximation of the vessels centerlines. Separation of CL spline control points was set on local vessel radius which assures proper vessel geometry approximation without risk of oversampling. Taking into account the obtained results full automation of proposed methodology allows in the future to analyze large amount of medical records and use artificial intelligence for deeper analysis and support for risk stratification for vascular lesions such as aneurysms and atherosclerosis in any vascular territory.

Authors' contributions
Conception and design of the work: JŽ, DS, data analyis: JŽ, medical background, analysis and description, data collection: JŽ, drafting the article: JŽ, GR, DS. All authors read and approved the final manuscript.

Author details
[1] Second Department of Clinical Radiology, Medical University of Warsaw, 1a Banacha, 02-097 Warsaw, Poland. [2] Faculty of Biomedical Engineering, Silesian University of Technology, Roosevelta 40, Zabrze, Poland.

Acknowledgements
Not applicable.

Competing interests
The authors declare that they have no competing interests.

Consent for publication
Not applicable.

Funding
This research was financial supported by the Directional Research at the Second Department of Clinical Radiology at the Medical University of Warsow, Poland.

References

1. Nixon AM, Gunel M, Sumpio BE. The critical role of hemodynamics in the development of cerebral vascular disease. J Neurosurg. 2010;112(6):1240–53.
2. Kleinstreuer C, Hyun S, Buchanan JR, Longest PW, Archie JP, Truskey GA. Hemodynamic parameters and early intimal thickening in branching blood vessels. Crit Rev Biomed Eng. 2017;45(1–6):319–82.
3. Kleinstreuer C, Hyun S, Buchanan JR, Longest PW, Archie JP, Truskey GA. Hemodynamic parameters and early intimal thickening in branching blood vessels. Crit Rev Biomed Eng. 2001;29(1):1–64.
4. Wells DR, Archie JP, Kleinstreuer C. Effect of carotid artery geometry on the magnitude and distribution of wall shear stress gradients. J Vasc Surg. 1996;23(4):667–78.
5. Niu L, Meng L, Xu L, Liu J, Wang Q, Xiao Y, Qian M, Zheng H. Stress phase angle depicts differences in arterial stiffness: phantom and in vivo study. Phys Med Biol. 2015;60(11):4281–94.
6. Niu L, Zhu X, Pan M, Derek A, Xu L, Meng L, Zheng H. Influence of vascular geometry on local hemodynamic parameters: phantom and small rodent study. BioMed Eng Online. 2018;17(1):30.
7. Chiastra C, Iannaccone F, Grundeken MJ, Gijsen FJH, Segers P, De Beule M, Serruys PW, Wykrzykowska JJ, van der Steen AFW, Wentzel JJ. Coronary fractional flow reserve measurements of a stenosed side branch: a computational study investigating the influence of the bifurcation angle. BioMed Eng Online. 2016;15(1):91.
8. Thompson BG, Brown RD, Amin-Hanjani S, Broderick JP, Cockroft KM, Connolly ES, Duckwiler GR, Harris CC, Howard VJ, Johnston SC, Meyers PM, Molyneux A, Ogilvy CS, Ringer J, Torner J. Guidelines for the management of patients with unruptured intracranial aneurysms: a guideline for healthcare professionals from the American Heart Association/American Stroke Association; 2015.
9. Steiner T, Juvela S, Unterberg A, Jung C, Forsting M, Rinkel G. European stroke organization guidelines for the management of intracranial aneurysms and subarachnoid haemorrhage. Cerebrovasc Dis. 2013;35(2):93–112.
10. Wardlaw JM, White PM. The detection and management of unruptured intracranial aneurysms. Brain. 2000;123:205–21.
11. Antiga L, Steinman DA. Robust and objective decomposition and mapping of bifurcating vessels. IEEE Trans Med Imag. 2004;23(6):704–13.
12. Lee SW, Antiga L, Spence JD, Steinman DA. Geometry of the carotid bifurcation predicts its exposure to disturbed flow. Stroke. 2008;39(8):2341–7.
13. Antiga L. Patient-specific modeling of geometry and blood flow in large arteries. Politecnico di Milano; 2002.
14. Piccinelli M, Veneziani A, Steinman DA, Remuzzi A, Antiga L. A framework for geometric analysis of vascular structures: application to cerebral aneurysms. IEEE Trans Med Imag. 2009;28(8):1141–55.
15. Piccinelli M, Bacigaluppi S, Boccardi E, Ene-Iordache B, Remuzzi A, Veneziani A, Antiga L. Geometry of the internal carotid artery and recurrent patterns in location, orientation, and rupture status of lateral aneurysms: an image-based computational study. Neurosurgery. 2011;68(5):1270–851285.
16. Thomas JB, Antiga L, Che SL, Milner JS, Steinman DA, Spence JD, Rutt BK, Steinman DA. Variation in the carotid bifurcation geometry of young versus older adults: implications for geometric risk of atherosclerosis. Stroke. 2005;36(11):2450–6.
17. O'Flynn PM, O'Sullivan G, Pandit AS. Methods for three-dimensional geometric characterization of the arterial vasculature. Ann Biomed Eng. 2007;35(8):1368–81.
18. Rhoton AL. The cerebral veins. Neurosurgery. 2002;51(4 SUPPL.):159–205.
19. Rhoton AL. The supratentorial arteries. Neurosurgery. 2002;51(4 Suppl):53–120.
20. Hagen H. Bezier-curves with curvature and torsion continuity. Rocky Mt J Math. 1986;1986:629–38.
21. Elsharkawy A, Lehecka M, Niemela M, Billon-Grand R, Lehto H, Kivisaari R, Hernesniemi J. A new, more accurate classification of middle cerebral artery aneurysms: computed tomography angiography study of 1,009 consecutive cases with 1,309 middle cerebral artery aneurysms. Neurosurgery. 2013;73(1):94–102102.
22. Rhoton AL. The cerebellar arteries. Neurosurgery. 2000;47(3):29.
23. Rhoton AL. Aneurysms. Neurosurgery. 2002;51(4 SUPPL.):121–58.
24. Passerini T, Sangalli LM, Vantini S, Piccinelli M, Bacigaluppi S, Antiga L, Boccardi E, Secchi P, Veneziani A. An integrated statistical investigation of internal carotid arteries of patients affected by cerebral aneurysms. Cardiovasc Eng Technol. 2012;3(1):26–40.
25. Fan J, Wang Y, Liu J, Jing L, Wang C, Li C, Yang X, Zhang Y. Morphological-hemodynamic characteristics of intracranial bifurcation mirror aneurysms. World Neurosurg. 2015;84(1):114–20.
26. Xu J, Yu Y, Wu X, Wu Y, Jiang C, Wang S, Huang Q, Liu J. Morphological and hemodynamic analysis of mirror posterior communicating artery aneurysms. PLoS ONE. 2013;8(1):55413.
27. Liu J, Xiang J, Zhang Y, Wang Y, Li H, Meng H, Yang X. Morphologic and hemodynamic analysis of paraclinoid aneurysms: ruptured versus unruptured. J Neurointerv Surg. 2014;6(9):658–63.
28. Hassan T, Timofeev EV, Saito T, Shimizu H, Ezura M, Matsumoto Y, Takayama K, Tominaga T, Takahashi A. A proposed parent vessel geometry-based categorization of saccular intracranial aneurysms: computational flow dynamics analysis of the risk factors for lesion rupture. J Neurosurg. 2005;103(4):662–80.
29. Ramachandran M, Retarekar R, Harbaugh RE, Hasan D, Policeni B, Rosenwasser R, Ogilvy C, Raghavan ML. Sensitivity of quantified intracranial aneurysm geometry to imaging modality. Cardiovasc Eng Technol. 2013;4(1):75–86.
30. Pastva TA. Bezier curve fitting. Ph.D. Thesis; 1998.
31. Mansouryar M, Hedayati A. Smoothing via iterative averaging (sia) a basic technique for line smoothing. Int J Comput Elec Eng. 2012;4(3):307–11.

12

New remote centre of motion mechanism for robot-assisted minimally invasive surgery

Xiaoqin Zhou[1], Haijun Zhang[1], Mei Feng[1]* , Ji Zhao[1] and Yili Fu[2]

*Correspondence:
fengmei@jlu.edu.cn
[1] Jilin University, Nan Guan District, Changchun, China
Full list of author information is available at the end of the article

Abstract

Background: Robot-assisted minimally invasive surgery (RMIS) is promising for improving surgical accuracy and dexterity. As the end effector of the robotic arm, the remote centre of motion mechanism is one of the requisite terms for guaranteeing patient safety. The existing remote centre of motion mechanisms are complex and large in volume, as well as high assembly requirement and unsatisfactory precise. This paper aimed to present a new remote centre of motion mechanism for solving these problems.

Methods: A new mechanism based on the RMIS requirements is proposed for holding the laparoscope and generating a remote centre of motion for the laparoscope. The mechanism kinematics is then analysed from the perspective of the structural function, and its inverse kinematics is determined with a small number of calculations. Finally, the position deviation of the laparoscope rotational point is chosen as the index to evaluate the mechanism performance. The experiments are performed to test the deviation.

Results: The position deviations of the laparoscope rotational point do not exceed 2 mm, which is lower than that of the existing remote centre of motion mechanism. The 2 mm positioning error of the laparoscope won't affect surgeon observation of the surgical field, and the pressure caused by the positioning error was acceptable for the skin elasticity. The proposed mechanism meets the RMIS requirement.

Conclusions: The proposed mechanism can achieve the remote centre of motion for the laparoscope. Its simple and compact structure is beneficial to avoid the collision of robotic arms, and it can be applied on other robots for providing the instrument necessary motion in minimally invasive surgery.

Keywords: Robot-assisted minimally invasive surgery, Remote centre of motion mechanism, Motion error

Introduction

Robot-assisted minimally invasive surgery (RMIS) has had a revolutionary impact on surgery, having the ability to satisfy the requirements of higher precision and dexterity for surgery operations. In traditional minimally invasive surgery (MIS), a surgeon holds instruments to perform a surgical operation [1]. In RMIS, the instrument is operated by a robot manipulator to penetrate the patient's body and perform surgical operations (e.g., cutting, tying, and suturing) [2, 3]. Under the constraint imposed by the 'minimally invasive' incision, two tangential motions of the instrument must be confined at the incision port to ensure patient safety [4]. Hence, the instrument has four degrees of freedom (DOFs), namely, pitch, translation, roll, and yaw [5]. For convenience, the concept of the remote centre of motion (RCM) has been devised to describe the pitching and yawing movements around the incision port [6], and RCM generation is one of the requisite terms for an MIS robot that is directly related to patient safety.

Many researchers and institutes have conducted studies on the generation of RCM-based motion. There are two representative techniques: a control method for creating a virtual RCM [7–14], and a mechanical method for creating a physically constrained RCM [15].

The control method employs software algorithms to generate a virtual RCM. This approach can be applied regardless of the robot structure and has the great advantage of robot design simplification. However, it is difficult to guarantee safety in the case of electronic-component or power malfunction; thus, this method is rarely used in commercialised models [5].

The mechanical method employs special mechanisms to provide RCM-based motion for surgical instruments. This approach is more reliable than the previous method, as the injury risk from unexpected control failures is inherently minimised by the structure [16]. The mechanisms to perform RCM-based motion are collectively called 'RCM mechanisms' [17, 18] and utilise parallelograms, spherical linkages, gear trains, etc. The characteristics of each RCM mechanism are described in the following.

The parallelogram mechanism is commonly adopted in MIS robots having high rigidity and, such as the Neurobot [19], BlueDRAGON [20], and the famous da Vinci Surgical System (Intuitive Surgical Inc.) [21]. In addition, many manipulators have been developed based on this architecture, which have diverse structural forms [5, 20, 22–32]. However, a conflict exists between the mechanism movement range and structure. Because the distance between the two transverse bars of the parallelogram becomes shorter when the mechanism attains an extreme angle, the two transverse bars should be mounted far from each other to prevent overlapping. This requirement yields a large-volume structure that does not effectively prevent multi-robotic arm collision. In addition, the RCM position is affected by the relative positions of the upright and transverse bars.

Using the geometric features of circles and spheres, researchers have developed both circular arc [33–38] and spherical [39–46] mechanisms. The installed instrument's axis passes through the mechanism arc or sphere centre, which is set to coincide with the incision port during surgery, thus the instrument can be steered to rotate around the incision port. These types of mechanisms have a small structure; however, this structure is specialised and, thus, high processing and assembly precision are required. In addition, these mechanisms are mainly used to hold lightweight instruments, because of their slightly low rigidity [46, 47].

Lehman et al. [48] have assembled several bevel gears in combination, with the gear axes passing through the incision port. Thus, the driving gear rotation causes the instrument to rotate around the incision port. However, gear clearance exists for gear drive, which has an influence on the control precise of the instrument movement. In addition, the gears require lubrication, which has a negative effect on sterilisation [49].

Li et al. [50, 51] have used three identical CRRR structures to construct a mechanism, where C denotes a cylindrical joint and R a revolute joint. The mechanism employs linear actuators and benefits high rigidity and load capacity from the parallel structure. This parallel mechanism has provided new perspectives for achievement of RCM-based motion. However, the parallel mechanism has a large volume, rendering this design unsuited to surgical operation requiring multi-robotic arm cooperation.

To sum up, the software method has lower security than the mechanical method to achieve RCM-based motion in the case of electronic-component or power malfunction, however, the existing RCM mechanisms are complex and large in volume, as well as high assembly and machining requirement. In view of the importance of safety for MIS, this paper reports realisation of RCM-based motion using the mechanical approach. Considering the contradiction between stiffness and volume in current RCM mechanisms, a planar symmetrical-rod structure in the parallel form is proposed to generate RCM-based motion. Combined with a simple axis-driving joint in series, a new 2-DOF RCM mechanism is presented that merges the advantages of parallel architecture, decoupled motion, and a simple design. Its simple structure is convenient as regards machining and assembly. Moreover, the RCM location is independent of the joint relative position; therefore, repeat adjustment of the joint initial posture for calibration of the RCM position is unnecessary. Because of its compact and small structure, the proposed mechanism can be applied to multi-robotic arms as the end effector to provide the necessary instrument motion and can effectively prevent collision between the robotic arms.

Materials and methods

Mechanical structure design

In a traditional minimally invasive surgery, two assistants always help the surgeon accomplish a surgical operation: one holds the laparoscope to provide the surgeon with a visual display of the surgical site, while the other holds surgical instruments to perform some auxiliary work. For example, when the surgeon needs to cut a particular part of the body, the assistant uses an instrument to elevate the part to facilitate the cutting operation, because the tissues and the organs have a soft texture. The assistant may become tired, especially in some major surgeries, where instruments must be held for long durations. Consequently, the movements performed by hand may not provide sufficient stability, thereby affecting the surgical visual display or the tissue boundary dissociation.

We custom designed a robot that could perform the assistant's work to overcome this problem. As shown in Fig. 1, the robot had three arms: the middle arm and the left and right arms. The middle arm was used to hold the laparoscope and provide a visual display of the surgical site. This arm consisted of one linear joint, three positioning joints, and an RCM mechanism. The linear joint adjusted the height of the robotic arm relative to the operating bed, while the positioning joints enabled the laparoscope to reach the incision port. The RCM mechanism, which served as the end effector of the middle arm, held the laparoscope and facilitated its position change without causing pain at the incision port. The left and right arms were used to hold surgical instruments to accomplish auxiliary work. Unlike the laparoscope, the surgical instrument was designed to

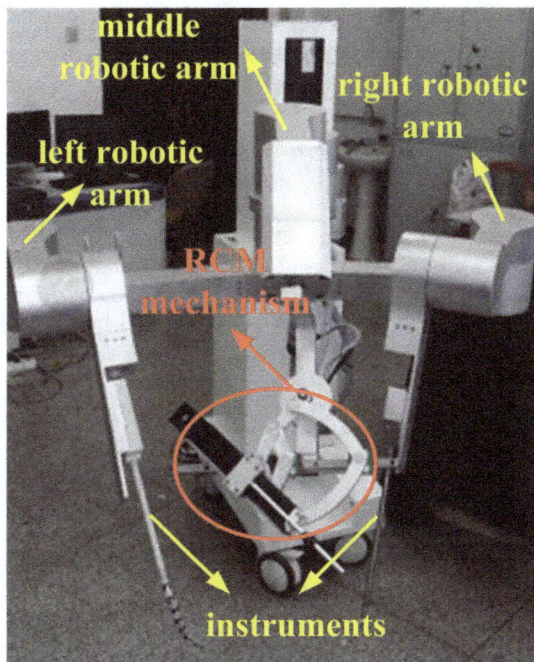

Fig. 1 Custom-designed robot

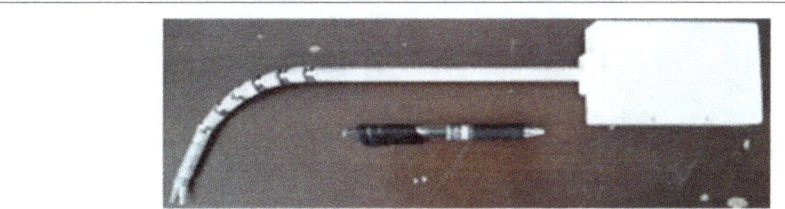

Fig. 2 Flexible surgical instrument

be flexible. As shown in Fig. 2, the instrument had a snake-like configuration with wire actuation, which allowed the instrument to achieve a position change in the patient's body. Thus, the left and right arms did not need to have a RCM mechanism. Therefore, the key points of the robot development lay in the flexible design of the surgical instrument and the RCM mechanism. This study focused on the RCM mechanism design.

The RCM mechanism was used herein to hold the laparoscope and provide its motion for the surgical visual display. As mentioned earlier, the movements along the incision tangential direction were constrained to ensure patient safety. The visual display does not need to be rotated during surgery; thus, the motions of the laparoscope were left as shown in Fig. 3 (i.e., two rotations (pitching and yawing) around the incision port and one insertion motion along the incision port). Therefore, the RCM mechanism was designed to achieve the pitching and yawing motions and the insertion movement of the laparoscope.

A new RCM mechanism was proposed according to the abovementioned requirements. The mechanism consisted of a rotating joint, a symmetrical-rod joint, and a

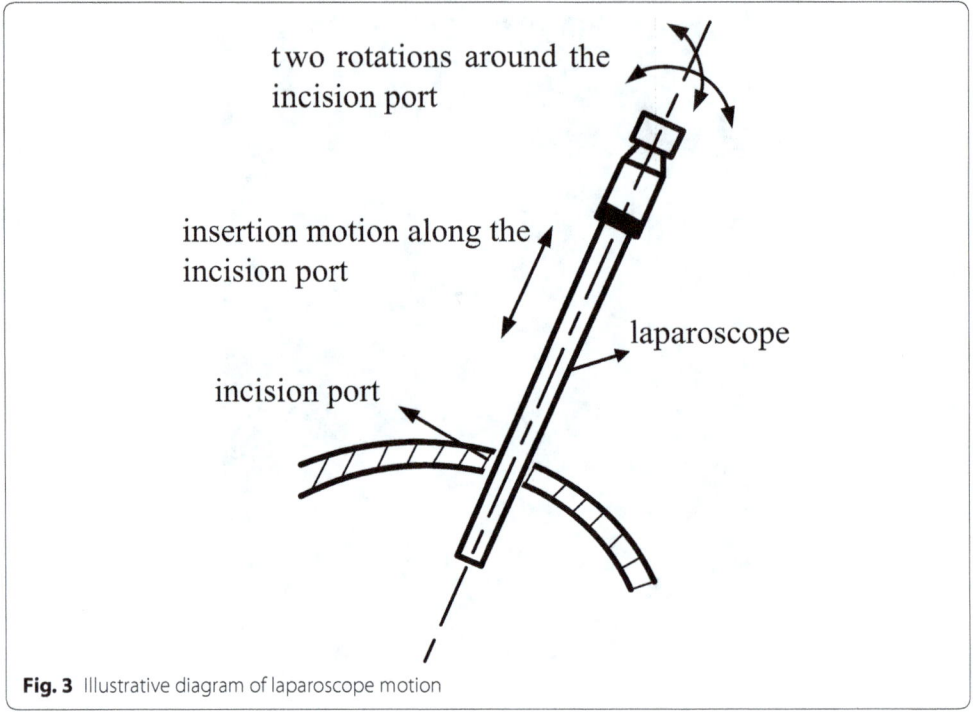

Fig. 3 Illustrative diagram of laparoscope motion

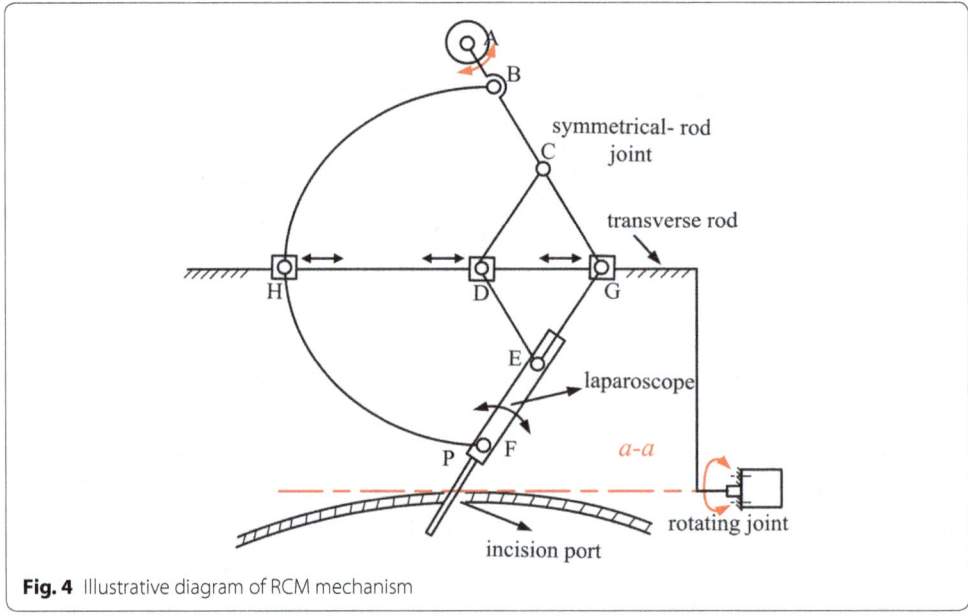

Fig. 4 Illustrative diagram of RCM mechanism

linear joint (Fig. 4). The three joints were connected in series: the symmetrical-rod joint was connected to the end of the rotating joint; the linear joint was connected to the symmetrical-rod joint; and the laparoscope was mounted on the linear joint. The rotating joint was an axis-driving joint (Fig. 4). Its $a–a$ axis passed through the incision port; thus, its rotation moved the laparoscope in a yawing motion around the incision port with no pressure.

The symmetrical-rod joint had a symmetrical structure (Fig. 4) and consisted of rods AC, CG, CD, DE, EG, EF, BH, and HF. The rods were hinged together and were symmetrical to the transverse rod, along which hinge points D, G, and H could slide. Rod AC was connected to the rotating joint by hinge point A and actuated by a motor. Hinge point A was located on the $a-a$ axis and could be treated as a fixed point relative to moving rods. The velocity values of hinge points E and F were equal to those of hinge points C and B, respectively, because of the symmetrical structure. Therefore, the laparoscope maintained the same motion with rod AC. Point P, which was the intersection of the $a-a$ axis and the laparoscope, was symmetrical to and had the same velocity as hinge point A. Thus, the velocity of point P was zero. It could also be seen as a fixed point. During surgery, point P was set to coincide with the incision port; hence, the rotation (pitching) of the laparoscope was around the incision port, and no pressure could be applied on the incision port. To summarise, the proposed mechanism can output a fixed rotational centre with no pressure on the incision site. Moreover, the location of point P depends on the hinge point A's position according to the restriction provided by the structure's symmetrical relationship, which is not affected by the initially assembled mechanism posture. Note that linear joint responsible for the laparoscope-insertion movement was a ball–screw pair, which is omitted from Fig. 4 as it has the 'as-known' structure.

In this paper, the RCM mechanism was used to hold laparoscope for providing the surgery visual display, different from surgical instrument, laparoscope had no force interaction with surgical site, so the loads for this RCM mechanism were the laparoscope gravity and the friction force from trocar port. In this condition, the structural strength was analyzed with finite element method as shown in Fig. 5, which demonstrated the RCM mechanism had good static stability.

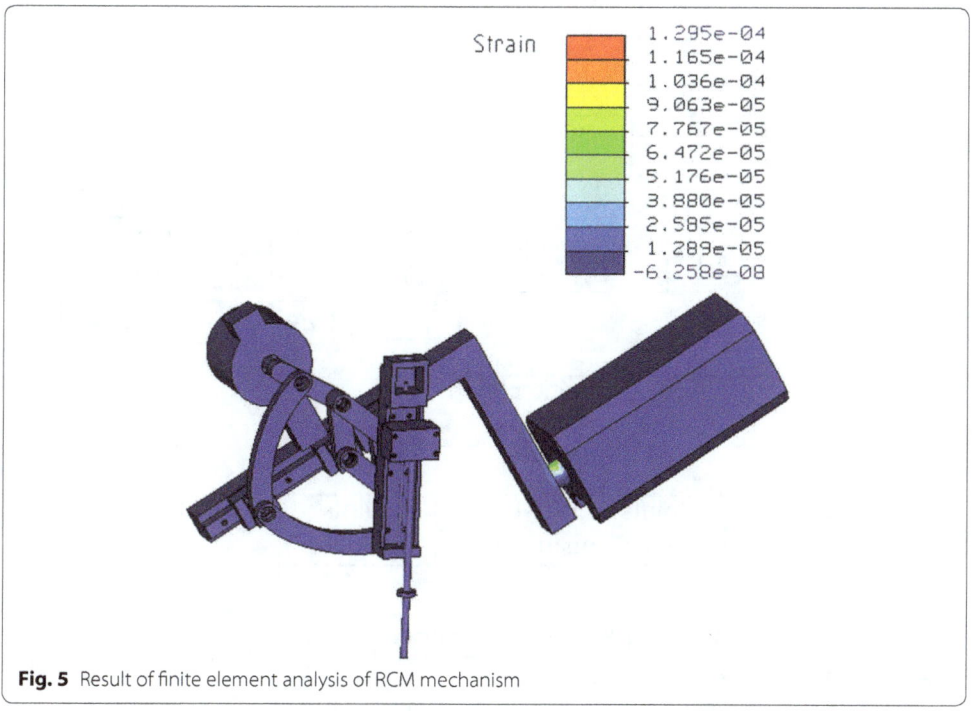

Fig. 5 Result of finite element analysis of RCM mechanism

According to the RMIS characteristics, the proposed RCM mechanism has two working modes: passive and active. The motion of the mechanism in the passive mode is driven by the dragging motion of the surgeon to adjust the laparoscope posture, which is helpful in saving the adjustment time required for configuring the preoperative settings. Meanwhile, in the active mode, the movement of the RCM mechanism is actuated by motors, and the rotation angles of the motors are acquired through calculation of the inverse kinematics using the movement information pertaining to the end of the laparoscope. The kinematics of the RCM mechanism was discussed in the following section.

Kinematic analysis of RCM mechanism

In RMIS, the surgeon uses a pair of master manipulators to control the robot movement. The robot makes the laparoscope (or the instruments) move synchronistically with the master manipulators. In this manner, the operation of the surgeon's hands is reproduced by the robot. Therefore, the information pertaining to the operation of the surgeon's hands should be extracted to control the robot movement by collecting the movement information of the master manipulators. In this study, position control of the laparoscope only is needed, because the proposed RCM mechanism is used to hold the laparoscope. In the active mode, the movement of the master manipulators described by dx, dy, and dz in Fig. 6 is mapped to the laparoscope as the laparoscope position W in every sampling cycle, considering the fact that the movement of the laparoscope is provided by the RCM mechanism, to reproduce the operation of the surgeon's hands. Thus, the inverse kinematics should be solved to control the laparoscope movement and determine the range of the joint movement of the RCM mechanism. Figure 6 shows the control strategy, where k is the scaling factor used to scale the movement of the surgeon's hands. For pre-operative setting, to save the adjustment time, the mechanism worked in passive mode, the electromagnetic clutch released, the surgeon dragged the RCM mechanism to make the laparoscope locate above the surgical site, in this process the joints

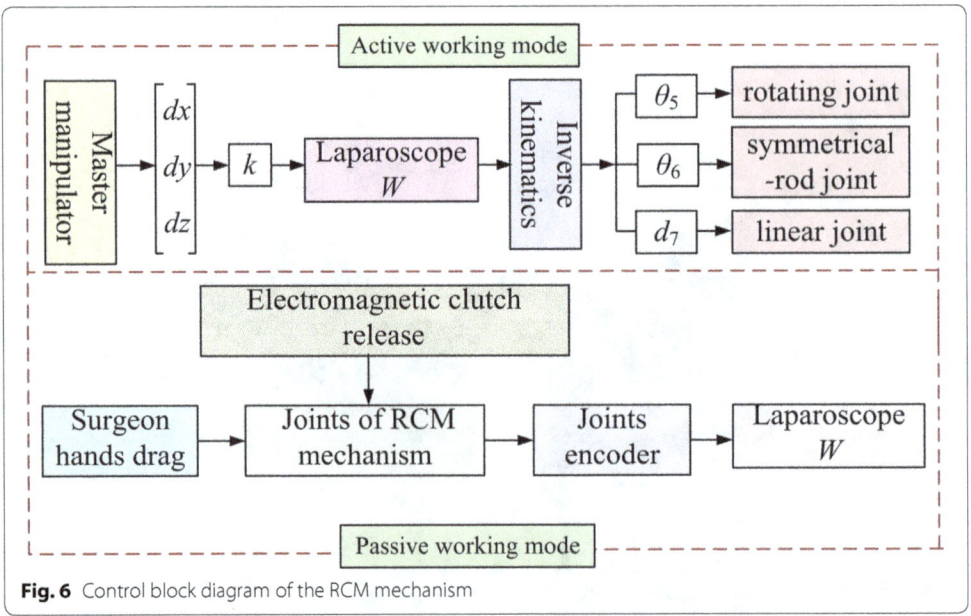

Fig. 6 Control block diagram of the RCM mechanism

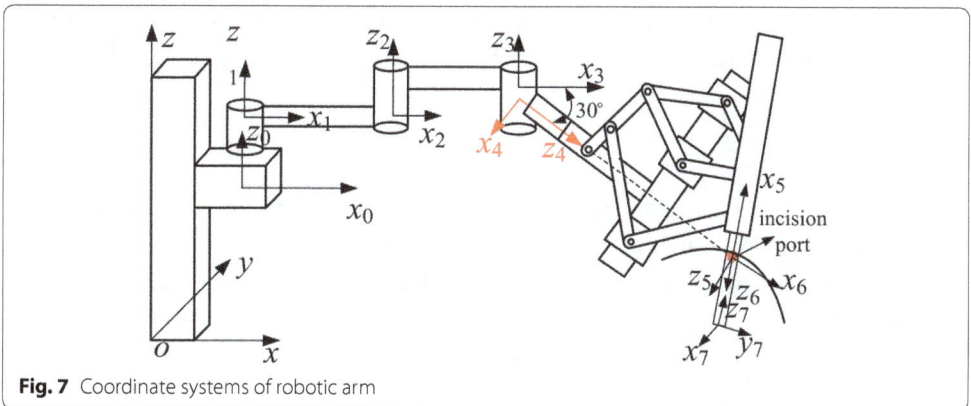

Fig. 7 Coordinate systems of robotic arm

Table 1 D–H parameters of RCM mechanism

Joint	a (mm)	α (°)	d (mm)	θ (°)
1	0	0	100	0
2	270	0	0	θ_2
3	150	0	0	θ_3
4	0	$-(90+30)$	0	θ_4
5	0	-90	620	θ_5
6	0	90	0	θ_6
7	0	0	d_7	0

encoders recorded the joints rotation angles, which were used for the calculation of the laparoscope position W by using the following forward kinematics.

The laparoscope was held by the RCM mechanism; hence, coordinate systems of the middle arm joints were built using the Denavit–Hartenberg (D–H) method to describe the laparoscope position in the robot coordinate system. The function of the RCM mechanism was to provide the pitching and yawing movements of the laparoscope around the incision port as well as the insertion movement along the incision port. Thus, to simplify the establishment of the forward kinematics model, we built the coordinate systems of the RCM mechanism joints from the perspective of a structural function, such that the joints could be treated as two rational joints and one linear joint located at the incision port (Fig. 7). Table 1 presents the kinematic parameters of the middle arm joints.

As mentioned earlier, once the laparoscope reaches the incision port, the positioning joints of the robotic arm are locked during surgery. The homogeneous coordinate transformation matrix $^{o}T_{o4}$ of the coordinate system $o_4 - x_4y_4z_4$ of the rotating joint relative to the base coordinate system $o-xyz$ is a constant, which can be obtained using the feedback from the positioning joint encoders. Thus, to facilitate the calculation, the laparoscope position W in the coordinate system $o-xyz$ can be converted to the position W' in the coordinate system $o_4 - x_4y_4z_4$, as follows:

$$W' = (^{o}T_{o4})^{-1}W \tag{1}$$

Considering that the changes in the laparoscope position are provided by the movements of the RCM mechanism joints, the relationship between the joint movements and the laparoscope position can be described as follows:

$$^5T_4 \cdot {}^6T_5 \cdot {}^7T_6 = W' \tag{2}$$

where ^jT_i is the homogeneous coordinate transformation matrix of the coordinate system i of joint i relative to the coordinate system j of joint j.

The inverse kinematics solution of the RCM mechanism joints 5–7 is obtained as follows to solve Eq. (2):

$$
\begin{cases}
\theta_5 = \arctan\left(\dfrac{W'_y}{W'_x}\right) \\[2mm]
\theta_6 = \arctan\left(\dfrac{W'_y}{\sin(\theta_5)\cdot(W'_z-620)}\right) \\[2mm]
d_7 = \dfrac{W'_z-620}{\cos(\theta_6)}
\end{cases} \tag{3}
$$

where θ_5, θ_6, and d_7 are the ranges of motion of the rotating, symmetrical-rod, and linear joints, respectively. There are no coupling motion of the three joints.

Experiments

As mentioned earlier, the RCM mechanism is used to hold the laparoscope and provide its pitching and yawing movements around a point. This point is set to coincide with the incision port to alleviate stress and ensure patient safety, and should be a fixed point to keep the coincidence with the incision port when the laparoscope moves. Thus, the "fixed" point is the testing point and its the position deviation is used herein to evaluate the performance of the proposed mechanism and verify the effectiveness of the proposed mechanism for achieving RCM-based motion. For convenience, its initial position is referred to as P_0 (x_0, y_0, z_0), and the collected position during the mechanism movement is referred to as P_i (x_i, y_i, z_i). Therefore, the deviation between P_0 and P_i is as follows:

$$e_i = |P_i - P_0| = \sqrt{(x_0 - x_i)^2 + \left(y_0 - y_i\right)^2 + (z_0 - z_i)^2} \quad i = 1, 2, 3, \ldots \text{n} \tag{4}$$

where the deviation e_i represents the position change between the collected position P_i and the initial position P_0, and subscript i represents the serial number of the collected position.

We replaced the laparoscope with a tube having the same diameter to test the position change of the point around which the laparoscope rotates. The movement of the tube was divided into steps, and we used two cameras to test the position change e_i in every step. Figure 8 illustrates the measuring principle. First, the rotating and symmetrical-rod joints were set to their initial positions to mark P_o (Fig. 8). Two coordinate sheets were then placed vertically, and two cameras were set at the location, where the lines between the camera lenses and P_o were vertical to the coordinate sheets. Thus, the e_i of the collected position P_i in every step could be acquired by its projections (R_{ix}, R_{iy}) on the coordinate sheets. The distances between the cameras and P_i and those between the cameras and the projections on

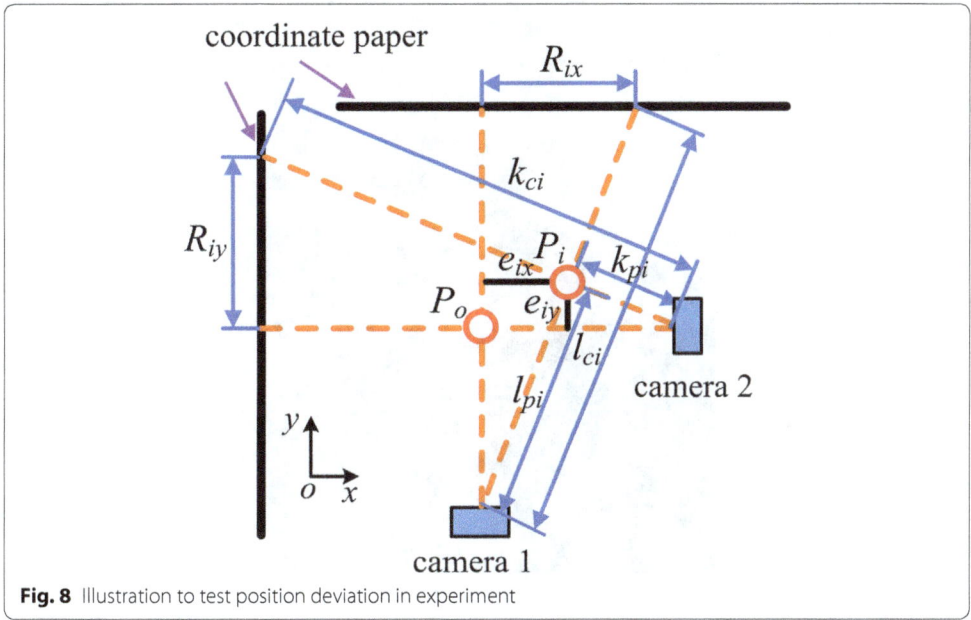

Fig. 8 Illustration to test position deviation in experiment

the coordinate sheets in every step were measured. The components of e_i can be calculated as follows:

$$\begin{cases} e_{ix} = R_{ix} \cdot \dfrac{l_{pi}}{l_{ci}} \\ e_{iy} = R_{iy} \cdot \dfrac{k_{pi}}{k_{ci}} \end{cases} \tag{5}$$

where e_{ix} and e_{iy} are the components of e_i along the x- and y-axes, respectively; R_{ix} and R_{iy} are the projections of e_i on the coordinate sheets; l_{pi} and l_{ci} are the distance between camera 1 and P_i and that between camera 1 and the projection on the coordinate sheet, respectively; and k_{pi} and k_{ci} are the distance between camera 2 and P_i and that between camera 2 and the projection on the coordinate sheet, respectively.

Thus, the e_i of the P_i in every step can be obtained as follows:

$$e_i = \sqrt{e_{ix}^2 + e_{iy}^2} \tag{6}$$

Three groups of experiments were performed to test e_i. Groups 1 and 2 were used to test e_i when the rotating and symmetrical-rod joints moved separately, and group 3 was used to test e_i when the laparoscope moved in a defined trajectory achieved by coordinated motion of the rotating and symmetrical-rod joints. Figure 9 shows the experimental setup. We measured the distances l_{pi}, l_{ci}, k_{pi}, and k_{ci} and used two cell phones to take photographs in every step. We read the projections R_{ix} and R_{iy} from the coordinate sheets in the photographs.

The movement ranges of the rotating and symmetrical-rod joints in the group 1 and 2 experiments were both $-70°$ to $+70°$, and the position of the testing point

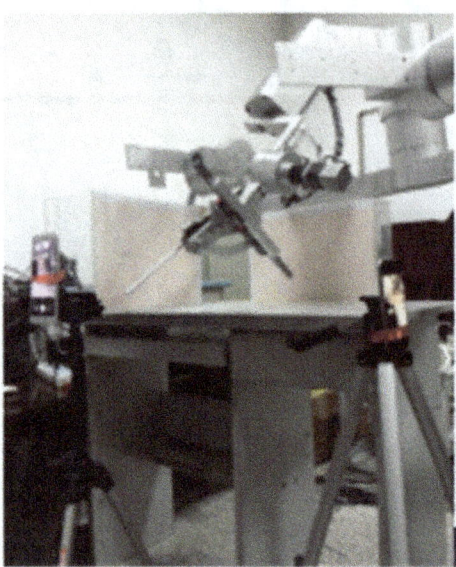

Fig. 9 Experimental setup for evaluation of the RCM mechanism performance

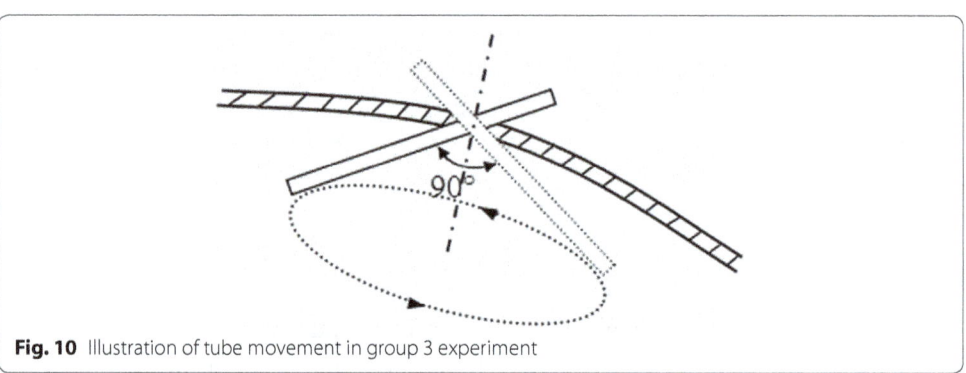

Fig. 10 Illustration of tube movement in group 3 experiment

was collected for every 5° rotation of the joints. Figure 10 illustrates the results of e_i between the collected and initial positions.

According to surgeons, in surgery, the laparoscope usually rotates in a range where the angle formed by the laparoscope axis and the normal of the incision port is less than 45°. Thus, in the group 3 experiment, we defined the tube movement such that the tube moved around the incision port, and the spatial angle between the tube and the normal of the incision port was 45°. Accordingly, the traces of the tube formed a cone, and the trajectory of the tube tip was a circle (Fig. 10). This tube movement was achieved through coordinated motion of the rotating and symmetrical-rod joints. Hence, according to the trajectory of the tube tip, the ranges of the joint movement were calculated using the above mentioned inverse kinematics solution and the joint movements were divided into 52 steps.

As mentioned above, during surgery, to alleviate patient skin stress, the fixed point achieved by the mechanical structure of RCM mechanism is set to keep the coincidence with the incision port, to evaluate the mechanism "certring" performance under the

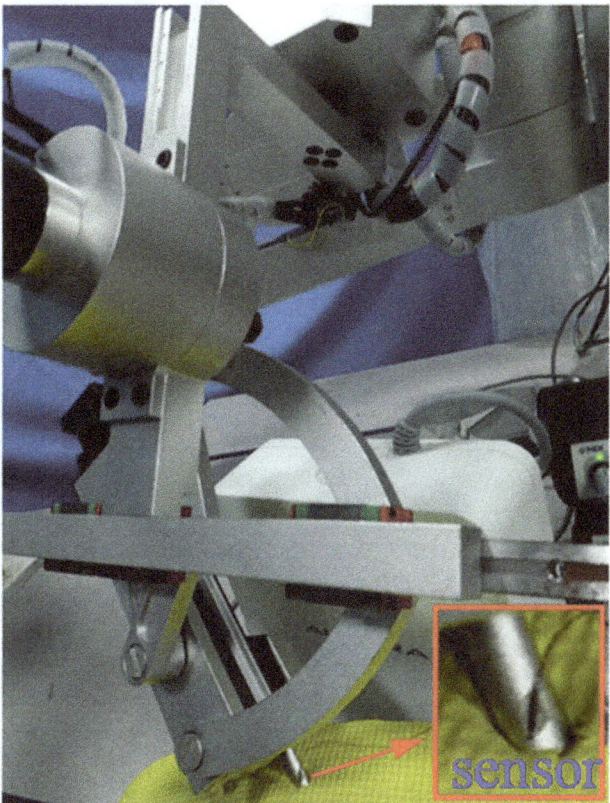

Fig. 11 The experimental picture of the imitative experiment

action of incision port, an imitative experiment was carried out by measuring the position deviation of the fixed point. As shown in Fig. 11, an elastic cushion was adopted to represent the skin, the tube stood for laparoscope was inserted through the cushion, NDI Auraro testing system was employed and its sensor was stuck on the cylindrical surface of the fixed point, thus the position of the point stuck on sensor can be captured by the electromagnetic tracking system even under obstructions. Due to the restriction of the Auraro magnetic field, the position deviations were tested when the rotating and symmetrical-rod joints moved separately. The sensor collecting positions were the positions of the point on the cylindrical surface centred on fixed point, to observe the deviations of the fixed point position, they were converted to the positions of the fixed point.

Results

The experimental results in Fig. 12 showed that the position deviations of the testing point in the group 1 and 2 experiments were not greater than 1.5 mm while the joints moved from the zero position to the left and right limited positions, and Fig. 13 showed the position deviation and its components in group 3 in every step.

With the NDI Auraro testing system, after the calculation of the position deviations, the results were shown in Fig. 14. The position deviations of the fixed point were no more than 1.3 mm, which was smaller than that measured by cameras images. That because the elastic cushion provided a constraint for the movement backlash of the

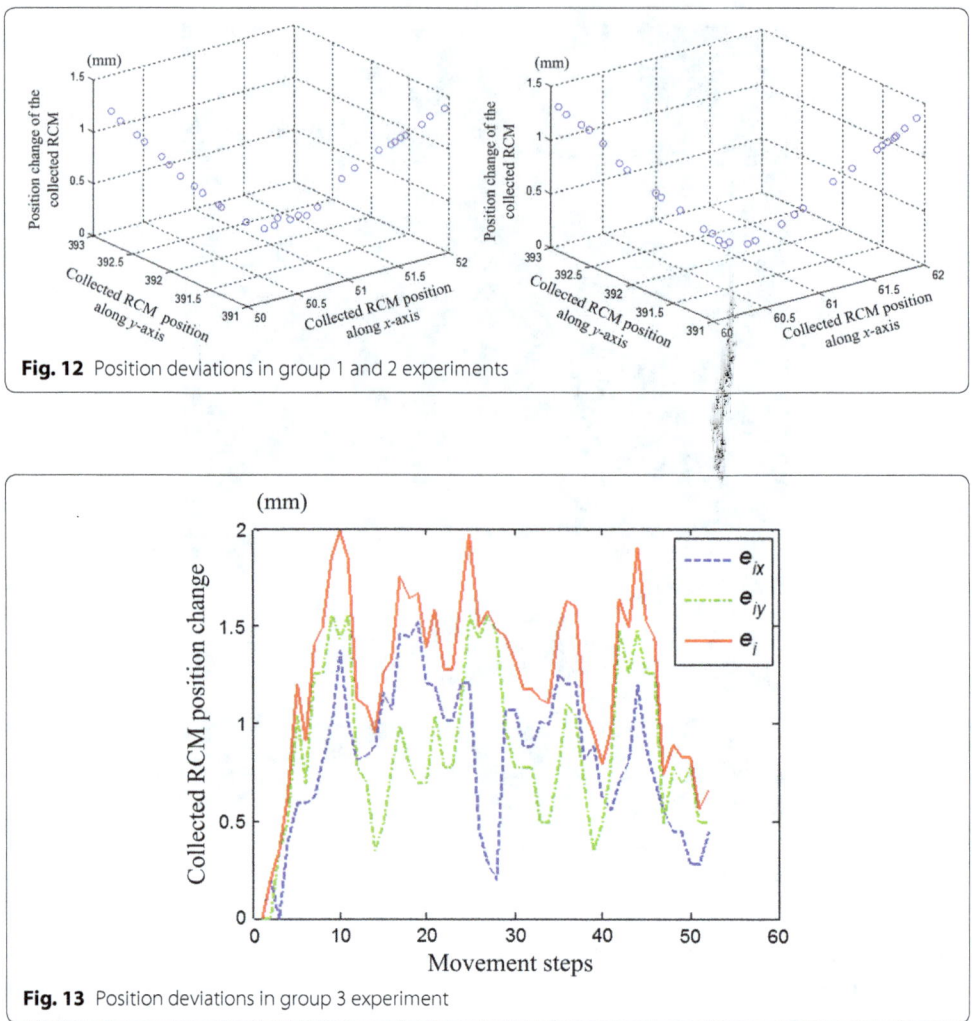

Fig. 12 Position deviations in group 1 and 2 experiments

Fig. 13 Position deviations in group 3 experiment

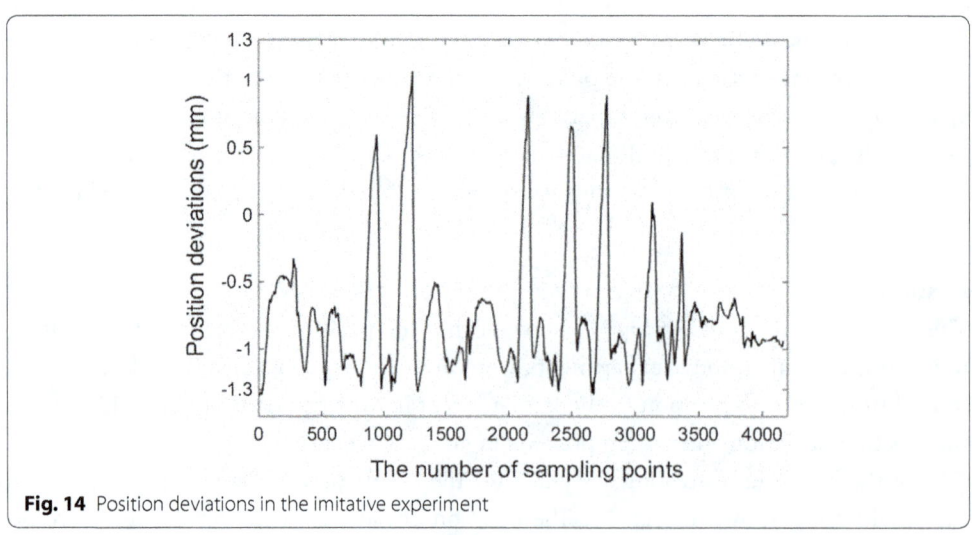

Fig. 14 Position deviations in the imitative experiment

mechanism, which avoided the error caused by the joints backlash; in addition, the distance measurement and image processing involved in the cameras testing method may introduce errors. The test results of the two test methods are relatively close, which can reflect the performance of the proposed mechanism in achieving RCM-based motions.

According to results of the experiments validation, the position deviations fluctuated for approximately 1 mm, but did not exceed 2 mm. The proposed RCM mechanism was used herein to hold the laparoscop for providing surgeon visual display, for the wild field of view in RMIS, the 2 mm error of laparoscope position won't affect surgeon observation. Meanwhile, the pressure on the incision port caused by the laparoscope position deviation was acceptable considering that the body surface tissue was sufficiently elastic to bear a 2-mm deviation, thus the proposed RCM mechanism met the RMIS requirement. In addition, the proposed RCM mechanism was a proof-of-principle prototype. The hinge points suffered from backlash because of the machining and assembly errors, and the motion error was mainly caused by the lack of machining accuracy. These problems could be overcome by improving the structural accuracy of the mechanism.

Conclusion

In this paper, a new RCM mechanism for holding a laparoscope in RMIS was proposed, which consists of a symmetrical-rod joint, an axis-driving joint, and a linear joint for the laparoscope to achieve two RCM-based motions and insertion, respectively. The symmetrical-rod joint was designed using the characteristics of the symmetrical structure; thus, the RCM point is symmetrical to a fixed point so as to achieve 'fixed' RCM performance to alleviate the pressure caused by the surgical instrument. Thus, the laparoscope can pitch and yaw around the incision port. The proposed mechanism has high rigidity, as well as having a small volume due to the planar assembly mode. The entire mechanism is primarily composed of straight rods, which are easily machined. Moreover, because of its compact structure, the proposed mechanism can be applied to multi-robotic arms, effectively preventing collisions between the arms.

To simplify the kinematics calculation, the inverse kinematics was calculated from the perspective of the structural function, but not the mechanical structure; this greatly reduced the number of calculations and considerably aided control. As the function of the RCM mechanism is to provide instrument rotation around the RCM point, the RCM position deviation was selected as an index to verify the performance of the proposed mechanism. The deviations were tested by examining the position projection with two cameras and by using NDI Auraro testing system, respectively. The testing results show that the position deviations did not exceed 2 mm, which is lower than the 7 mm "certring" error of the RCM mechanism used in da Vinci surgical robot system [52]. The proposed RCM mechanism can achieve the RCM-based motion and meet the RMIS requirements, and its compact and simple structure can help to prevent collision between the robotic arms.

Authors' contributions
XQ Z and HJ Z designed the mechanism structure, M F and J Z conducted the experiments to test the movement errors of each joint and the joints compound motion, M F was also a major contributor in writing the manuscript, Yili Fu analyzed the experiments data. All authors read and approved the final manuscript.

Author details

[1] Jilin University, Nan Guan District, Changchun, China. [2] Harbin Institute of Technology, Nan Gang District, Harbin, China.

Acknowledgements

Not applicable.

Competing interests

The authors declare that they have no competing interests.

Consent for publication

Not applicable.

Funding

The National Natural Science Foundation of China under Grant No. 61305102 provided the fund for the remote centre of motion mechanism design and machining, the Open Foundation of the State Key Laboratory of Robotics and System under Grant No. SKLRS-2013-MS-02 helped the experiments data collection and analysis, and the fifty-fourth batch of China Postdoctoral Science Fund under Grant No. 2013M540247 supports the manuscript writing and polishing.

References

1. White Alan D, Giles Oscar, Sutherland Rebekah J, et al. Minimally invasive surgery training using multiple port sites to improve performance. Surg Endosc. 2014;28(4):1188–93.
2. Palep JH. Robotic assisted minimally invasive surgery. J Minim Access Surgery. 2009;5(1):1–7.
3. Konstantinova J, Jiang A, Althoefer K, et al. Implementation of tactile sensing for palpation in robot-assisted minimally invasive surgery: a review. IEEE Sens J. 2014;14(8):2490–501.
4. Zong G, Pei X, Yu J, et al. Classification and type synthesis of 1-DOF remote center of motion mechanisms. Mech Mach Theory. 2008;43(12):1585–95.
5. Nisar S, Endo T, Matsuno F. Design and kinematic optimization of a 2 degrees-of-freedom planar remote center of motion mechanism for minimally invasive surgery manipulators. J Mech Robot. 2017;9(3):031013.
6. Taylor RH, Funda J, Grossman DD, et al. Improved remote center-of-motion robot for surgery. European Patent No. EP0595291. 1994.
7. Dahroug B, Tamadazte B, Andreff N. 3D Path Following with Remote Center of Motion Constraints. In: Proceeding of the international conference on informatics in control, automation and robotics. 2017.
8. Dombre E, Michelin M, Pierrot F, et al. MARGE Project: design, modeling and control of assistive devices for minimally invasive surgery. In: Proceedings of medical image computing and computer-assisted intervention. 2004; p. 1–8.
9. Konietschke R, Ortmaier T, Hagn U, et al. Kinematic design optimization of an actuated carrier for the DLR multi-arm surgical system. In: Proceedings of IEEE/RSJ international conference on intelligent robots and systems. 2006; p. 4381–7.
10. Yang D, Wang L, Li Y. Kinematic analysis and simulation of a MISR system using bimanual manipulator. In: Proceedings of IEEE international conference on robotics and biomimetics. 2017; p. 271–6.
11. Seibold U, Kübler B, Hirzinger G. Prototype of instrument for minimally invasive surgery with 6-axis force sensing capability. In: Proceedings of IEEE international conference on robotics and automation. 2005; p. 496–501.
12. Li M, Kapoor A, Taylor RH. Telerobotic control by virtual fixtures for surgical applications. Adv Telerobot. 2007;31:381–401.
13. Marinho M M, Harada K, Mitsuishi M. Comparison of Remote Center-of-Motion Generation Algorithms. In: Proceedings of IEEE/SICE international symposium on system integration. 2017.
14. Yang D, Wang L, Xie Y, et al. Optimization-based inverse kinematic analysis of an experimental minimally invasive robotic surgery system. In: Proceedings of IEEE international conference on robotics and biomimetics. 2015; p. 1427–32.
15. Kuo CH, Dai JS, Dasgupta P. Kinematic design considerations for minimally invasive surgical robots: an overview. Int J Med Robot Comput Assist Surg. 2012;8(2):127–45.
16. Zhang N, Huang P, Li Q, et al. Modeling, design and experiment of a remote-center-of-motion parallel manipulator for needle insertion. Robot Comput Integr Manuf. 2017;50:193–202.
17. Gandhi P, Bobade R, Chen C. On the novel compliant remote center mechanism. In: proceedings of the 1st international and 16th national conference on machines and mechanisms. 2013; p. 576–81.
18. Li J, Xing Y, Liang K, et al. Kinematic Design of a novel spatial remote center-of-motion mechanism for minimally invasive surgical robot. J Med Devices. 2015;9(1):011003.

19. Davies B, Starkie S, Harris SJ, et al. Neurobot: a special-purpose robot for neurosurgery. Proc IEEE Int Conf Robot Autom. 2000;4:4103–8.

20. Rosen J, Brown JD, Chang L, et al. The BlueDRAGON—a system for measuring the kinematics and the dynamics of minimally invasive surgical tools in vivo. In: Proceedings of IEEE international conference on robotics and automation. 2002; p. 1876–81.

21. Sun L W, Meer F V, Yan B, et al. Design and development of a da vinci surgical system simulator. In: proceedings of the IEEE international conference on mechatronics and automation. 2007; p. 1050–5.

22. Schurr MO, Arezzo A, Neisius B, et al. Trocar and instrument positioning system TISKA: an assist device for endoscopic solo surgery. Surg Endosc. 1999;13(5):528–31.

23. Eldridge B, Gruben K, LaRose D, et al. A remote center of motion robotic arm for computer assisted surgery. Robotica. 1996;14(1):103–9.

24. Baumann R, Maeder W, Glauser D, Clavel R. The PantoScope: a spherical remote-center-of-motion parallel manipulator for force reflection. In: Proceedings of IEEE international conference on robotics and automation. 1997; p. 718–23.

25. Davies B, Starkie S, Harris SJ, et al. Neurobot: a special-purpose robot for neurosurgery. In: Proceedings of IEEE international conference on robotics and automation. 2000; p. 4103–8.

26. Salcudean SE, Zhu WH, Abolmaesumi P, et al. A robot system for medical ultrasound. In: Proceedings of the 9th international symposium of robotics research (ISRR'99). 1999; p. 195–202.

27. Stoianovici D, Whitcomb LL, Anderson JH, et al. A modular surgical robotic system for image guided percutaneous procedures. In: Proceedings of medical image computing and computer-assisted intervention—MICCAI. 1998; p. 404–10.

28. Stoianovici D, Cleary K, Patriciu A, et al. AcuBot: a robot for radiological interventions. IEEE Trans Robot Automat. 2003;19(5):927–30.

29. Shahriari N, Heerink W, Van KT, et al. Computed tomography (CT)-compatible remote center of motion needle steering robot: fusing CT images and electromagnetic sensor data. Med Eng Phy. 2017;45:71–7.

30. Sun J, Yan Z, Du Z. Optimal design of a novel remote center-of-motion mechanism for minimally invasive surgical robot. In: IOP conference series: earth and environmental science 69. 2017.

31. Stoianovici D, Jun C, Lim S, et al. Multi-imager compatible, mr safe, remote center of motion needle-guide robot. IEEE Trans Bio-Med Eng. 2017;99:1–1.

32. Trochimczuk R. Analysis of parallelogram mechanism used to preserve remote center of motion for surgical telemanipulator. Int J Appl Mech Eng. 2017;22(1):229–40.

33. Davies BL, Hibberd RD, Ng WS, Wickham JEA. A surgeon robot for prostatectomies. Proc Int Conf Adv Robot. 1991;1:871–5.

34. Buess GF, Arezzo A, Schurr MO, et al. A new remote-controlled endoscope positioning system for endoscopic solo surgery. Surg Endosc. 2000;14(4):395–9.

35. Wang S, Li Q, Ding J, Zhang Z. Kinematic design for robotassisted laryngeal surgery systems. In: Proceedings of IEEE/RSJ international conference on intelligent robots and systems. 2006; p. 2864–9.

36. Yamauchi Y, Ohta Y, Dohi A, et al. Needle insertion manipulator for ct-guided stereotactic neurosurgery. J Life Supp Eng. 2010;5(4):91–8.

37. Berkelman P, Ma J. A compact modular teleoperated robotic system for laparoscopic surgery. Int J Robot Res. 2009;28(9):1198–215.

38. Harris SJ, Arambulacosio F, Mei Q, et al. The Probot—an active robot for prostate resection. Proc Inst Mech Eng [H]. 1997;211(4):317–25.

39. Farz A, Payandeh S. A robotic case study: optimal design for laparoscopic positioning stands. Int J Robot Res. 1998;17(9):1553–60.

40. Ouerfelli M, Kumar V. Optimization of spherical five-bar parallel drive linkage. ASME J Mech Design. 1994;116:166–73.

41. Li T, Payandeh S. Design of spherical parallel mechanisms for application to laparoscopic surgery. Robotica. 2002;20(2):133–8.

42. Lum MJH, Rosen J, Sinanan MN, Hannaford B. Optimization of a spherical mechanism for a minimally invasive surgical robot: theoretical and experimental approaches. IEEE Trans Biomed Eng. 2006;53(7):1440–5.

43. Zemiti N, Ortmaier T, Morel G. A new robot for force control in minimally invasive surgery. In: Proceedings of IEEE/RSJ international conference on intelligent robots and systems. 2004; p. 3643–8.

44. Zemiti N, Morel G, Ortmaier T, Bonnet N. Mechatronic design of a new robot for force control in minimally invasive surgery. IEEE/ASME Trans Mech. 2007;12(2):143–53.

45. Guojun Niu, Bo Pan, Fuhai Zhang, et al. Multi-optimization of a spherical mechanism for minimally invasive surgery. J Central South Univ. 2017;24(6):1406–17.

46. Essomba T, Vu LN. Kinematic analysis of a new five-bar spherical decoupled mechanism with two-degrees of freedom remote center of motion. Mech Mach Theory. 2018;119:184–97.

47. Hong J, Dohi T, Hashizume M, et al. An ultrasound-driven needle-insertion robot for percutaneous cholecystostomy. Phys Med Biol. 2004;49(3):441–55.

48. Lehman AC, Tiwari MM, Shah BC, et al. Recent advances in the CoBRASurge robotic manipulator and dexterous miniature in vivo robotics for minimally invasive surgery. Proc Inst Mech Eng C. 2015;224(224):1487–94.

49. Pio GD, Pennestrì E, Valentini PP. Kinematic and power-flow analysis of bevel gears planetary gear trains with gyroscopic complexity. Mech Mach Theory. 2013;70(6):523–37.

50. Li Q, Herve JM, Peng Huang. Type synthesis of a special family of remote center-of-motion parallel manipulators with fixed linear actuators for minimally invasive surgery. J Mech Robot. 2017;9:031012.

51. Zhang N, Huang P, Li Q, et al. Modeling, design and experiment of a remote-center-of-motion parallel manipulator for needle insertion. Robot Comput Integr Manufact. 2017;50:193–202.

52. Wilson JT, Tsao TC, Hubschman JP, et al. Evaluating remote centers of motion for minimally invasive surgical robots by computer vision. In: Proceedings of IEEE/ASME international conference on advanced intelligent mechatronics. 2011; p. 1413–18.

13

Comparison of named entity recognition methodologies in biomedical documents

Hye-Jeong Song[1,2], Byeong-Cheol Jo[1,2], Chan-Young Park[1,2], Jong-Dae Kim[1,2] and Yu-Seop Kim[1,2]*

From International Conference on Biomedical Engineering Innovation (ICBEI) 2016 Taichung, Taiwan. 28 October–1 November 2016

*Correspondence:
yskim01@hallym.ac.kr
[2] Bio-IT Research Center,
Hallym University,
Chuncheon, South Korea
Full list of author information
is available at the end of the
article

Abstract

Background: Biomedical named entity recognition (Bio-NER) is a fundamental task in handling biomedical text terms, such as RNA, protein, cell type, cell line, and DNA. Bio-NER is one of the most elementary and core tasks in biomedical knowledge discovery from texts. The system described here is developed by using the BioNLP/NLPBA 2004 shared task. Experiments are conducted on a training and evaluation set provided by the task organizers.

Results: Our results show that, compared with a baseline having a 70.09% F1 score, the RNN Jordan- and Elman-type algorithms have F1 scores of approximately 60.53% and 58.80%, respectively. When we use CRF as a machine learning algorithm, CCA, GloVe, and Word2Vec have F1 scores of 72.73%, 72.74%, and 72.82%, respectively.

Conclusions: By using the word embedding constructed through the unsupervised learning, the time and cost required to construct the learning data can be saved.

Keywords: Biomedical named entity recognition (Bio NER), Recurrent neural network (RNN), Conditional random fields (CRFs), Word embedding

Background

Named entity recognition (NER) assigns a named entity tag to a designated word by using rules and heuristics. The named entity, which presents a human, location, and an organization, should be recognized [1]. Named entity recognition is a task that extracts nominal and numeric information from a document and classifies the word into a person, an organization, or a date category [2]. NER classifies all words in the document into existing categories and "none-of-the-above".

Biomedical named entity recognition is very important in language processing of biomedical texts, especially in extracting information of proteins and genes such as RNA or DNA from documents. Finding named entities of genes from texts is a very important and difficult task [3]. Finding a gene name in texts corresponds to finding a company name or a human name in newspapers. Recognizing biomedical named entities seems to be more difficult than recognizing normal named entities [4]. Numerous research studies have recognized named entities by using supervised learning algorithms based on many rules [5].

Supervised learning approaches have used Hidden Markov Models (HMMs) [6], decision trees [7], support vector machines (SVMs) [8], and conditional random fields (CRFs) [9, 10]. Supervised learning methods normally train with data of many features based on various linguistic rules, and evaluate the performance with test data that could not be found in the training data.

In this paper, we compare the performances of recurrent neural networks of deep learning with conditional random fields. A recurrent neural network (RNN) uses a Jordan-type algorithm and an Elman-type algorithm. We also measure the performance of conditional random fields using word embedding as their features. Word embedding has increased performance in natural language processing, machine translation, voice recognition, and so on [11]. Word embedding has been used as features in natural language processing and is mapped from a word in the higher-dimensional space into a real-numbered vector in the lower-dimensional space. Word2Vec, canonical correlation analysis (CCA), and global vector (GloVe) are used as word embedding methodologies in this paper. We compared two RNN algorithms and CRFs using three word embedding methods for named entity recognition in biomedical literature.

In the rest part of "Background", we explain named entity recognition, particularly for biomedical texts. We introduce detailed methodologies and basic features used in this paper in "Methods". "Results and discussion" shows the experimental results and evaluations, and "Conclusion" is our conclusion.

Biomedical named entity recognition

Named entity recognition (NER) classifies all unregistered words appearing in texts and is a subtask of information extraction. Normally, NER uses eight categories—location, person, organization, date, time, percentage, monetary value, and "none-of-the-above" [12, 13]. NER first finds named entities in sentences and declares the category of the entity. In the sentence:

> *"Apple [***organization***] CEO Tim Cook [***Person***] Introduces 2 New, Lager iPhones, Smart Watch at Cupertino [***Location***] Flint Center [***Organization***] Event [14]."*

"Apple" is recognized as an organization name instead of a fruit name in terms of its context. The words "Tim" and "Cook" are altogether recognized as a single word having a meaning of CEO of the Apple Company and a person's name. "Cupertino" is a city name in California and is recognized as a location name, and "Flint" and "Center" are considered as a single name and recognized as an organization name.

Named entity recognition has three approaches—dictionary based, rule based, and machine learning based. A dictionary-based approach stores as many named entities as possible in a list called a gazetteer. This approach seems to be very simple, but at the same time has limitations. The NER is difficult because the target words are mainly proper nouns or unregistered words. In addition, new words can be generated frequently, and even the same word stream could be recognized as diverse named entities in terms of their current context [15, 16]. The second approach of the NER is a rule-based approach [17]. This approach ordinarily depends on the rules and patterns of named entities appearing in real sentences. Although rule-based approaches can use context to solve the problem of multiple named entities, every rule should be written before it is actually used. The third approach, the machine learning-based approach, tags the named entities to words even when the words are not listed in the dictionary and the context is not described in the rule set. For these approaches, support vector machines (SVMs) [18], Hidden Markov Models (HMMs) [6, 19], Maximum Entropy Markov Models (MEMMs) [20], and conditional random fields (CRFs) [9, 10] are mainly utilized.

Natural language processing researchers have been interested in the information extraction of genes, cancer, and protein from biomedical literature [21–24]. Biomedical named entity recognition, which is essential to biomedical information extraction, has been treated as the first stage of text mining in biomedical texts. For years, recognizing technical terms in the biomedical area has been one of the most challenging tasks in natural language processing related to biomedical research [25]. In this paper, we use five categories (protein, DNA, RNA, cell type, and cell line) instead of the categories used in the ordinary NER process. An example of the NER tagged sentence is as follows:

"IL-2 [B-protein] responsiveness requires three distinct elements [B-DNA] within the enhancer [B-DNA]."

Biomedical NER faces difficulties for five reasons. First, because of current researches, the number of new technical terms is rapidly increasing. It is very difficult to build a gazetteer that includes all of the new terms. Second, the same words or expressions could be classified as differently named entities in terms of their context. Third, the length of an entity is quite long, and the entity could include control characters such as hyphens (e.g., "12-*o*-tetradecanoylphorbol 13-acetate"). Fourth, abbreviation expressions are frequently used in the biomedical area, and they experience sense ambiguity. For example, "TCF" could refer to "T cell factor" or to "Tissue Culture Fluid" [26, 27]. Finally, in biomedical terms, normal terms or functional terms are combined, which is why a biomedical term can become too long. For example, "HTLV-I-infected" and "HTLV-I-transformed" include the normal terms "I", "infected", and "transformed". It is difficult for biomedical NER to segment the sentence with named entities. Spelling changes also create a problem [28]. In addition, the named entity of one category could subsume another named entity of another category [29].

Methods

We perform named entity recognition for words in a sentence by using CRFs and RNN, and compare the performance of each method. We use a BioNLP/NLPBA 2004 corpus [30, 31] of 22,402 sentences. We use 18,546 sentences as a training data set, and 3856 sentences as a test data set. The corpus are tagged with "protein", "DNA", "RNA", "cell line", and "cell type" categories. The next section describes CRFs and word embedding, and the rest explains RNN.

Conditional random fields

A CRF is a statistical sequence modeling framework first introduced in [32]. CRFs are a class of statistical modeling methods often applied in pattern recognition and machine learning, where they are used for structure prediction. Whereas an ordinary classifier

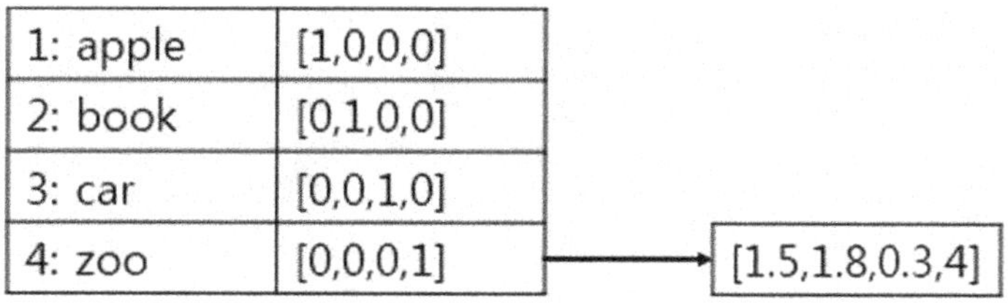

1: apple	[1,0,0,0]
2: book	[0,1,0,0]
3: car	[0,0,1,0]
4: zoo	[0,0,0,1]

$\longrightarrow$ [1.5,1.8,0.3,4]

Fig. 1 One-hot vs. word embedding. The left vector inside a table is one-hot representation, and the right vector inside a small rectangle is word embedding representation

predicts a label for a single sample without regard to "neighboring" samples, a CRF can take context into account [33]. The reason why CRFs are more effective than HMMs is that CRFs use the conditional probability property instead of the independence assumption mainly used in HMMs. CRFs also avoid label bias problems and avoid the weaknesses of other Markov models derived from MEMMs and graphic models. CRFs show better performance than MEMMs and HMMs in bioinformatics, computational linguistics, and voice recognition. CRFs are also used for the prediction and analysis of labels for data in natural language writing. Features can be chosen randomly, and they are to be normalized to obtain solution [32, 34].

In this model, $X = \{x_1, x_2, x_3, \ldots x_T\}$ are the input data in which components are connected in sequence, and $Y = \{y_1, y_2, y_3, \ldots y_T\}$ are the labels for each component of the input data. In other words, when a new x is given, a y value is predicted using the following model:

$$p(y|\mathbf{x}) = \frac{1}{z(\boldsymbol{x})} \prod_{t=1}^{T} \exp\left\{ \sum_{k=1}^{k} \omega_k f_k\left(y_t, y_{t-1}, \boldsymbol{x}_t\right) \right\} \tag{1}$$

$$z(\mathbf{x}) = \sum_y \exp\left(\sum_k \omega_k f_k(y, \boldsymbol{x}) \right), \tag{2}$$

where z(x) standardizes the probability value, and f_k is a feature function, which is a characteristic function on feature k. This function returns 1 when the given input $y_t, y_{t-1}, \boldsymbol{x}_t$ includes a feature k, and returns 0 otherwise. ω_k is the weight of the feature. In this study, a CRF suite [35] was used to make predictions by using the average perceptron generated by the CRF algorithm.

Word embedding

Word embedding is also called word representation or distributed representation. It learns vector representation for every word appearing in the corpus. Previous research studies represented a word as a one-hot representation. The one-hot representation uses a vocabulary-sized vector, and takes a 1 when the word appears in the document and 0 when it does not [36]. Word embedding reduces the dimensions and sparseness of the original vector and fills the vector with real numbers. Figure 1 shows the difference between one-hot representation and word embedding.

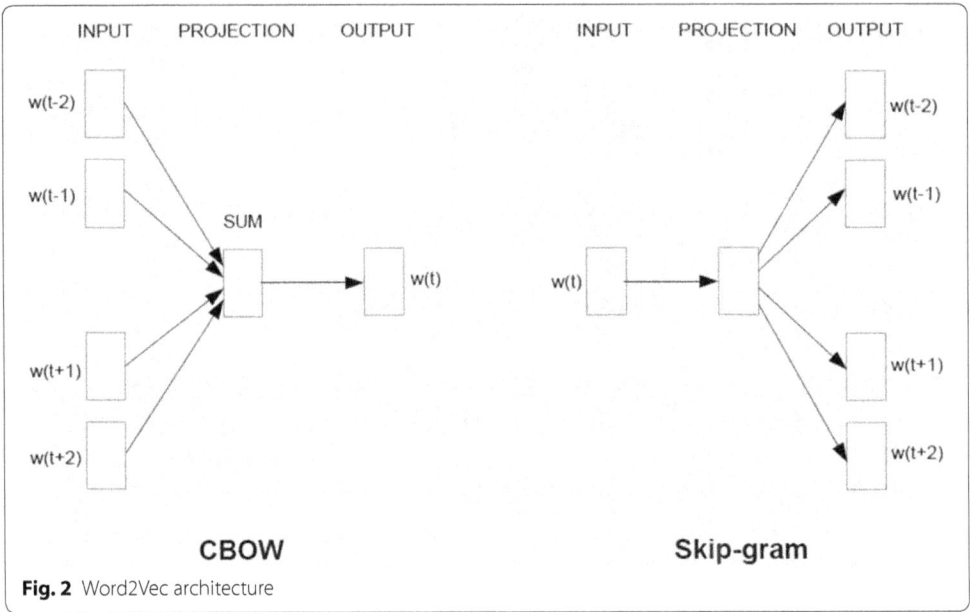

Fig. 2 Word2Vec architecture

Word2Vec

Word2Vec assumes that the words sharing the same context could have similar meanings. It classifies words near to the given word into related words and learns the words using artificial neural networks. Word2Vec has two structures: Continuous Bag of Words (CBOW) and skip gram architectures. Figure 2 shows the Word2Vec architecture [37].

The CBOW side has surrounding words $w(t-2)$, $w(t-1)$, $w(t+1)$, and $w(t+2)$ as input, and predicts $w(t)$ as output. The skip-gram side uses $w(t)$ as its input and predicts $w(t-2)$, $w(t-1)$, $w(t+1)$, and $w(t+2)$ as output.

Global vector (GloVe)

GloVe is an unsupervised learning algorithm for obtaining vector representations for words. Training is performed on aggregated global word-word co-occurrence statistics from a corpus, and the resulting representations showcase interesting linear substructures of the word vector space [38]. GloVe considers the global context as well as the local context [39].

$$\sum_{i,j=1}^{v} f\left(x_{ij}\right)(e_i^T \widetilde{e}_j + y_i + \widetilde{y}_j - \log X_{ij})^2, \tag{3}$$

where X is a word co-occurring in a matrix, X_{ij} is the frequency of the co-occurrence of word i and word j, and $X_i = \sum_{k}^{v} X_{ik}$ is the total number of occurrences of word i in the corpus. The occurrence probability of a word j in the context of a word i is $X_{ij} = P(j|i) = X_{ij}/x_i$. e is word embedding, and $\widetilde{e}$ is a separate-context word embedding. $f(X_{ij})$ indicates the weight and has three conditions. First, $f(0) = 0$. Second, $f(x)$ does not decrease not to give weights to very rarely co-occurring words. Third, $f(x)$ should be relatively smaller than the large value of x so it does not give weight to frequently co-occurred words.

Canonical correlation analysis (CCA)

Canonical correlation analysis (CCA) was introduced by Hotelling [40]. CCA is a statistical method to investigate the relationship between two variable sets, and it can concurrently examine the correlation of variables belonging to different sets. CCA finds correlations between two variable sets (X, Y), and also finds parameters that maximize the correlation coefficients [41]. CCA can be calculated directly from the data set, and can also be calculated after transforming the data sets into covariance matrices. These two methods are represented based on singular value decomposition. In [42, 43], if CCA is used to predict labels in data, string theory guarantees the correspondence to lower-dimensional embedding. CCA tries to find two projection vectors to maximize the correlation. Using random variables $(X, Y \in R)$, where X is a word representation and Y is its related context representation, CCA tries to find k-dimensional projection vectors that maximize the correlation between two variables [44].

Assuming that we have two variables $x \in C^{d_1}$, $\quad y \in C^{d_2}$, CCA can be defined as a problem to maximize the correlation between two variables on X and Y vectors. With a pair of vectors $\mathrm{x} = \hat{w}_x^T x$, $\quad \mathrm{y} = \hat{w}_y^T y$, we can use the following correlation expression:

$$\mathrm{p} = \frac{E[\mathrm{xy}]}{\sqrt{[x^2] E[y^2]}} = \frac{w_x^T C_{xy} w_y}{\sqrt{w_x^T C_{xx} w_x w_y^2 C_{yy} w_y}} \tag{4}$$

where $C_{xy} = E[xy^T]$, $C_{xx} = E[xx^T]$, and $C_{yy} = E[yy^T]$. The first eigenvectors $\hat{w}_{x_1}, \hat{w}_{y_1}$ can be the first correlation P_1, and the second eigenvectors can be the second correlation P_2 [45].

Recurrent neural network

In machine learning and cognitive science, artificial neural networks (ANNs) are a family of models inspired by biological neural networks that are used to estimate or approximate functions that can depend on a large number of inputs and are generally unknown [46]. ANNs work well in nonlinear functions and pattern recognition. Many researchers working in data mining, artificial intelligence, and bioinformatics have been interested in ANNs for its diverse applications [47].

Figure 3 [48] shows a simple ANN structure. ANNs use an activation function with a combination function of input variables and input values. The input layer takes input values for its training, and the hidden layer is located between the input layer and the output layer. Training is performed mainly in the hidden layer and tries to find the optimum weight value set labeled on each edge. A sigmoid function is used on each node to calculate each node's output after summing its inputs.

A recurrent neural network (RNN) has connections between nodes to form a directed cycle. Unlike normal feedforward networks, RNN can also use feedback systems [49]. RNN has shown outstanding performance in various natural-language processing tasks. The basic idea of RNN comes from the mechanism of sequential labeling. Normal ANNs

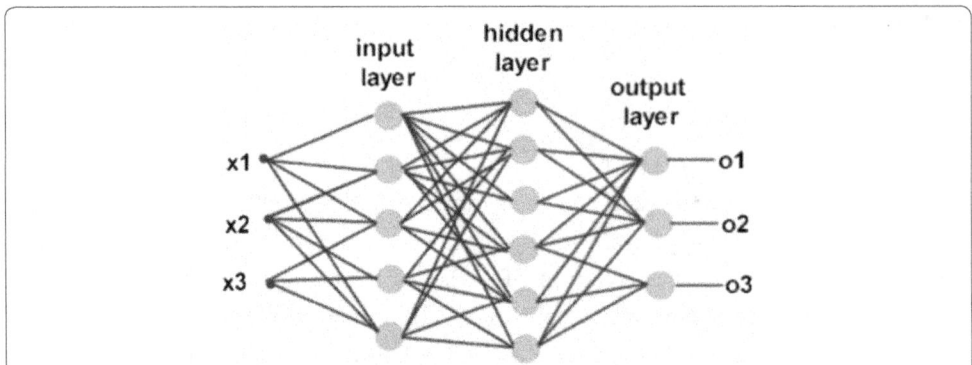

Fig. 3 Simple neural network. This network takes *x* values as its inputs and makes *o* values as its outputs. This network is composed of an input layer, a hidden layer, and an output layer

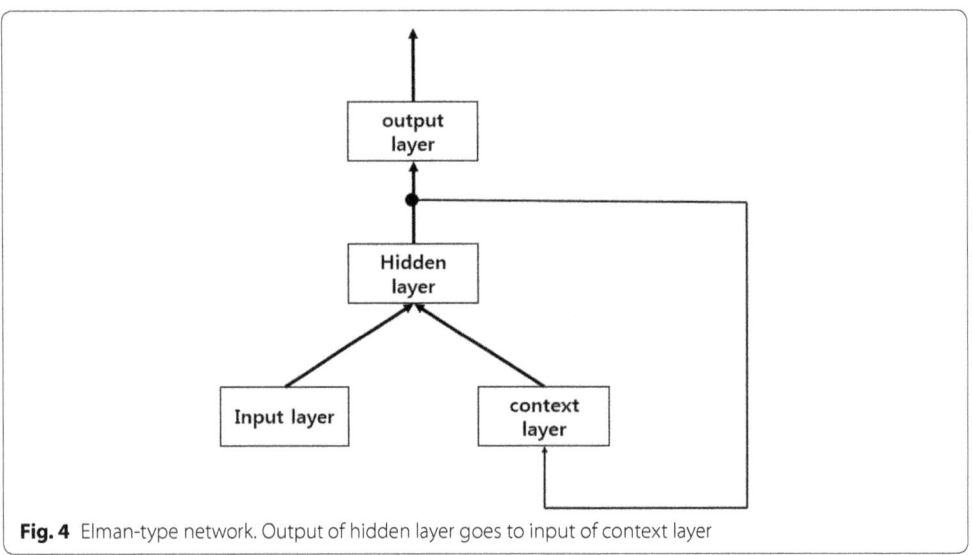

Fig. 4 Elman-type network. Output of hidden layer goes to input of context layer

assume independence between their inputs and outputs. RNN applies the same tasks to every component of the sequence, and the output is affected by the previous calculation results. In other words, the network is designed so that input x_t of time t and the previous hidden layer's output of time $t-1$ can contribute to the hidden layer's output of time t. Although RNN can be applied to any sequence length, shorter sequences show better performance [50].

We apply an RNN algorithm by using an RNN tutorial [51]. RNN has two types: the Elman-type network [52] and the Jordan-type network [53]. The Elman-type network adds a context layer to the normal RNN and feeds back the hidden layer's output to the context layer's input. This network feeds back the output value to the hidden layer rather than the input layer. The hidden layer of this network plays the same role as the input layer of a normal RNN. Figure 4 shows the basic structure of the Elman-type RNN. The output of the hidden layer, a sigmoid function of each node, and the output value of this network are explained below:

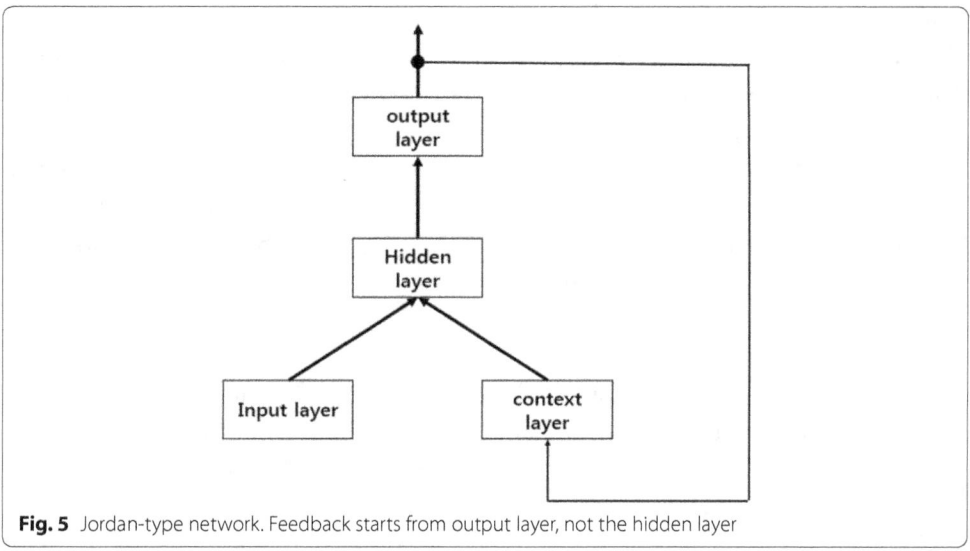

Fig. 5 Jordan-type network. Feedback starts from output layer, not the hidden layer

$$h(t) = f(Ux(t) + Vh(t-1)) \tag{5}$$

$$f(x) = \frac{1}{1 + e^{-x}} \tag{6}$$

$$y_t = g(Wh(t)) \tag{7}$$

In (5), which shows the output of the hidden node, U is a matrix of raw input values and current hidden nodes, and V is a matrix of the context node and the previous hidden node. Expression (6) shows a sigmoid function, and (7) shows the output value.

The Jordan-type network shown in Fig. 5 is very similar to the Elman-type network, except that the feedback is coming from the output layer rather than the hidden layer. The hidden layer's output is calculated by the following expression:

$$h(t) = f\left(Ux(t) + Vy(t-1)\right) \tag{8}$$

We use negative log-likelihood as a loss function. The gradient descent uses the mini-batch gradient descent method. This method does not apply the gradient descent method to each data, but calculates the gradient by batch and reflects it to the next learning. We apply a mini-batch gradient descent to one batch in one sentence. Because the length of the sentences in the corpus are all different, this method works well.

Table 1 N-Gram description

Feature	Description
Unigram	$w_{i-2}, w_{i-1}, w_i, w_{i+1}, w_{i+2}$
Bigram	$w_{i-2}|w_{i-1}, w_{i-1}|w_i, w_i|w_{i+1}, w_{i+1}|w_{i+2}$
Trigram	$w_{i-2}|w_{i-1}|w_i, w_{i-1}|w_i|w_{i+1}, w_i|w_{i+1}|w_{i+2}$

Feature

This study uses n-Gram features for a baseline experiment of the conditional random fields. Recurrent neural networks use raw word sequences for their inputs. Table 1 lists the unigrams, bigrams, and trigrams used in this study.

For the sentence, "Tumor and serum beta-2-microglobulin expression in women with breast cancer", let us assume that w_i is "breast". Then, w_{i-2}, w_{i-1}, w_{i+1} and w_{i+2} are "women", "with", "cancer" and ".", respectively. $w_{i-1}|w_i$ of the bigram is "with|breast", and $w_{i-2}|w_{i-1}|w_i$ is "with|breast|cancer".

Results and discussion

We use a BioNLP/NLPBA 2004 shared corpus for the experiment. In this experiment, we compare the performance of RNN and CRFs with word embedding. For the baseline, only n-Gram (unigram, bigram, trigram) features of CRFs are utilized. The Jordan-type RNN and Elman-type RNN are compared, and at the same time, Word2Vec, GloVe, and CCA of the CRFs are also compared. For performance evaluation, we set the word embedding dimension to 100, the window size to 5, the number of hidden units to 100, and the number of hidden layers to 1.

We use the F1 score as the performance measurement. The F1 score is calculated by the following expression:

$$F1 \text{ score} = \frac{2 * \text{precision*recall}}{\text{precision} + \text{recall}}, \tag{9}$$

where the precision is a ratio of true positives from the positive side, and recall is a ratio of true positives from the true side.

In this experiment, the Jordan-type RNN shows an F1 value of 60.75%, and the Elman-type RNN has an F1 value of 58.80%. For the CRFs' performance measurement, we apply various dimensions of word embedding (10, 30, 50, 80, 100), window sizes (3, 5, 7, 9, 11), and the minimum frequency (3).

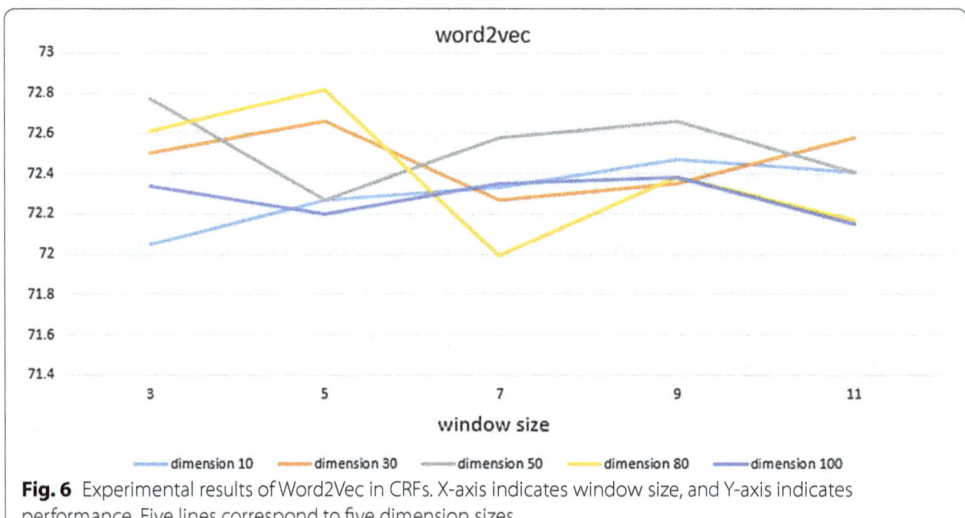

Fig. 6 Experimental results of Word2Vec in CRFs. X-axis indicates window size, and Y-axis indicates performance. Five lines correspond to five dimension sizes

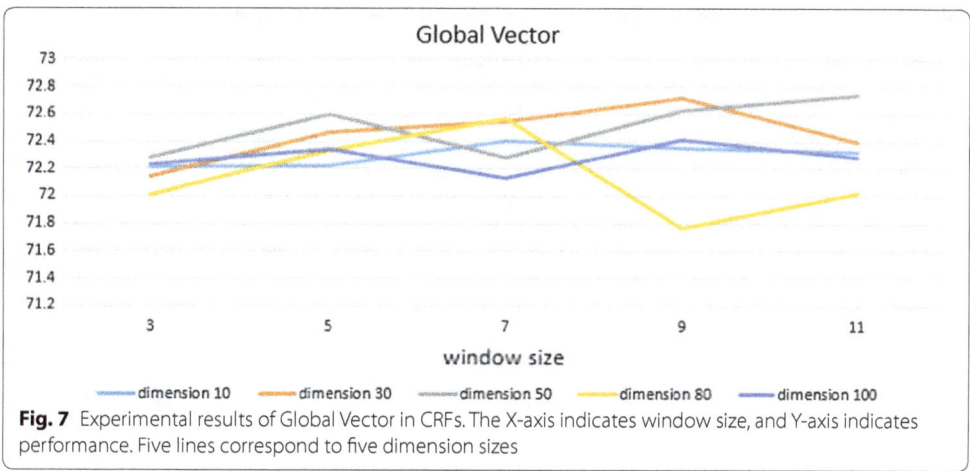

Fig. 7 Experimental results of Global Vector in CRFs. The X-axis indicates window size, and Y-axis indicates performance. Five lines correspond to five dimension sizes

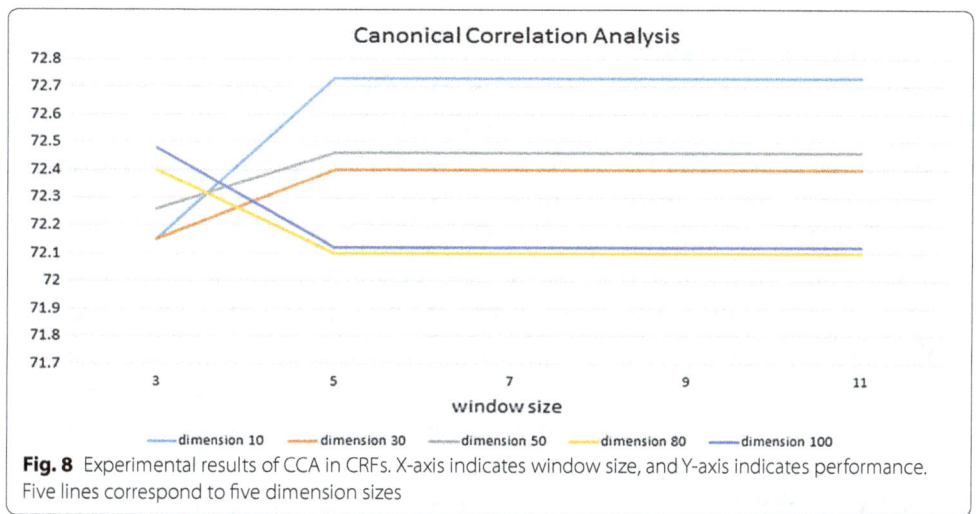

Fig. 8 Experimental results of CCA in CRFs. X-axis indicates window size, and Y-axis indicates performance. Five lines correspond to five dimension sizes

Figures 6, 7 and 8 show the experimental results of each word embedding method with various dimensions and window sizes.

Word2Vec shows the highest performance when the dimension size is 80 and the window size is 5. However, the same line shows the lowest performance when the window size is changed to 7. In Fig. 6, the line with dimension size of 50 shows a relatively stable and high performance for all window sizes. Word2Vec does not seem to need high-dimensional representation, and lower-dimensional representations show an increase in performance proportional to the window size. Higher-dimensional representations do not exhibit particular characteristics in Word2Vec.

In GloVe, a representation of 50 dimensions and 11 window sizes shows the highest performance. Like Word2Vec, GloVe also shows relatively stable and high performance when its size of dimension is 50. Of course, the 30-dimensional representation also shows a good result.

Figure 8 shows that lower-dimensional cases have relatively higher performance than higher-dimensional cases when CCA is used for word embedding.

Table 2 Performance comparison by using BioNLP/NLPBA 2004 corpus

System	Methodology	F1 score (%)
Our system	CRF	
	Base line	71.09
	Word2vec	72.82
	Glove	72.74
	CCA	72.73
	RNN	
	Jordan	60.75
	Elman	58.80
Zhou and Su [54]	HMM, SVM	72.55
Song et al. [9]	SVM, CRF	66.28
Ponomareva et al. [55]	HMM	65.7
Saha et al. [29]	Maximum entropy	67.41

Table 2 lists the results compared with well-known former research results. Our system shows an F1 score of up to 72.82%, which is the highest of all the results in Table 2, when CRFs of Word2Vec are used. Zhou et al. [54] used HMM and SVM to achieve 72.55% for the BioNLP/NLPBA shared task 2004, and their achievement has been the highest until now. At the same competition, Song et al. achieved 66.28% by using HMM and CRF. Ponomareva et al. [55] used HMM and achieved 65.7%, and Saha et al. [29] used Maximum Entropy to obtain an F1 score of 67.41%. Findel et al. [56], Settles [57], and Tsai et al. [58] reported scores of 69.8% to 70.2%, which could not overcome the results from Zhou and Su. Our system shows a maximum score of 72.82%, which is approximately 0.3% points higher than Zhou and Su's scores when using Word2Vec-based CRFs. Word embedding is also advantageous in that it is automatically constructed through the unsupervised learning, while the existing methodology uses data that is directly labeled by a person. Our approach does not require any domain knowledge, a dictionary, or other outside resources, but we were able to show the highest performance of all tested methods.

Conclusion

Bio-NER has more difficulties than normal NER because technical terms in biomedical texts have unusual characteristics. We compared various machine-learning approaches based on CRFs and RNN. In this research, RNN exhibited a lower performance than CRFs. The disadvantage of RNN is that it does not remember old information. Also, since we did not find the optimal activation function and initialization method, RNN has lower performance than CRFs. We use a single hidden layer. However, RNN could be a very useful method in Bio-NER because of its unsupervised learning property. From an experiment, our method shows the highest performance of all the other experiments.

For the future study, our research will proceed in three directions. First, we will design a more optimized deep artificial neural network structure for the Bio-NER. Because we had limited knowledge and experience in deep artificial neural network, this study used a relatively simple model. Therefore, we will develop deep artificial neural network specialized on this problem based on accumulated knowledge and technology. Second, we would like to develop unsupervised learning methods for the Bio-NER. The lack of an annotated corpus is a barrier to new research. Although it has unsupervised learning properties, RNN requires an annotated corpus. We should develop fully- or semi-supervised learning methods for Bio-NER. Third, various linguistic resources for domain knowledge should be built for performance development. Gazetteers, word embedding methods, and other resources should be developed.

Abbreviation
NER: named entity recognition; RNN: recurrent neural network; CRF: conditional random fields; CCA: canonical correlation analysis; GloVe: global vector; SVM: support vector machines; HMM: Hidden Markov Models; MEMM: Maximum Entropy Markov Models; NLP: natural language processing.

Declarations
Authors' contributions
Song took parts in deep learning design and implementation. Jo prepared for the whole experiment, including data and programming. Park advised Jo in implementing CRFs and CNN. J Kim expertized in biomedical IT convergence research. He analyzed input and output data. Y Kim is a corresponding author. He design this research and directed this research team. All authors read and approved the final manuscript.

Author details
[1] School of Software, Hallym University, Chuncheon, South Korea. [2] Bio-IT Research Center, Hallym University, Chuncheon, South Korea.

Acknowledgements
Not applicable.

Competing interests
The authors declare that they have no competing interests.

Consent for publication
Not applicable.

Funding
This research was supported by Basic Science Research Program through the National Research Foundation of Korea (NRF) funded by the Ministry of Science, ICT and future Planning (2015R1A2A2A01007333), and (NRF-2017M3C4A7068188). The publication charge for this article was funded by the first grant and the second grant supported a part of this research.

References
1. Sang EFTK, Meulder FD. Introduction to the CoNLL-2003 shared task: language-independent named entity recognition. In: Proceedings of the seventh conference on natural language learning at HLT-NAACL, vol. 4. 2003. p. 142–7.
2. Isozaki H, Kazawa H. Efficient support vector classifiers for named entity recognition. In: Proceedings of the 19th international conference on computational linguistics. Association for Computational Linguistics, vol. 1. 2002. p. 1–7.
3. Leaman R, Gonzalez G. BANNER: an executable survey of advances in biomedical named entity recognition. Pac Symp Biocomput. 2008;13:652–63.
4. Wilbur J, Smith L, Tanaben L. Biocreative 2 gene mention task. In: Proceedings of second BioCreative challenge evaluation workshop. 2007.

5.   Rau LF. Extracting company names from text. In: Proceedings of the conference on artificial intelligence applications of IEEE, vol. 1. 1991. p. 29–32.

6.   Zhao S. Named entity recognition in biomedical texts using an HMM model. In: Proceedings of the international joint workshop on natural language processing in biomedicine and its applications. Association for Computational Linguistics. 2004. p. 84–7.

7.   Sekine SN. Description of the Japanese NE system used for Met-2. In: Proceedings of the message understanding conference. 1998. p. 1314–9.

8.   Lee KJ, Hwang YS, Rim HC. Two phase biomedical NE recognition based on SVMs. In: Proceedings of the ACL 2003 workshop on natural language processing in biomedicine. Association for Computational Linguistics, vol. 13. 2003. p. 33–40.

9.   Song Y, Kim E, Lee GG, Yi B. POSBIOTM-NER in the shared task of BioNLP/NLPBA 2004. In: Proceedings of the international joint workshop on natural language processing in biomedicine and its applications. Association for Computational Linguistics. 2004. p. 100–3.

10. McCallum A, Li W. Early results for named entity recognition with conditional random fields, features induction and web-enhanced lexicons. In: Proceedings of the seventh conference on natural language learning at HLT-NAACL, vol. 4. Association for Computational Linguistics. 2003. p. 188–91.

11. Qiu L, Cao Y, Nie Z, Yu Y, Rui Y. Learning word representation considering proximity and ambiguity. In: Twenty-eighth AAAI conference on artificial intelligence. 2014. p. 1572–8.

12. Borthwick A. A maximum entropy approach to named entity recognition. Doctoral dissertation. New York: New York University; 1999.

13. Santos CN, Milidiú RL. Entropy guided transformation learning: algorithms and applications. New York: Springer; 2012.

14. Adam W. Named entity recognition at RAVN-part 1. https://www.ravn.co.uk/named-entity-recognition-ravn-part-1/. Accessed Apr 2016.

15. Cohen KB, Hunter L. Natural language processing and systems biology. Artificial intelligence and systems biology. New York: Springer; 2005. p. 145–73.

16. Liu H, Hu Z, Torii M, Wu C, Friedman C. Quantitative assessment of dictionary-based protein named entity tagging. J Am Med Inform Assoc. 2006;13:497–507.

17. Fukuda K, Tsunoda T, Tamura A, Takagi T. Toward information extraction: identifying protein names from biological papers. Pac Symop Biocomput. 1998;707:707–18.

18. Kazama J, Makino T, Ohta Y, Tsujii J. Tuning support vector machines for biomedical named entity recognition. In: ACL-02 workshop on natural language processing in biomedical applications, vol. 3. 2002. p. 1–8.

19. Cutting D, Kupiec J, Pedersen J. Sibun P. A practical part-of-speech tagger. In: Proceedings of the third conference on applied natural language processing. 1992. p. 133–40.

20. McCallum A, Freitag D, Pereira FC. Maximum entropy Markov models for information extraction and segmentation. ICML. 2000;17:591–8.

21. Aronson AR, Rindflesch TC, Browne AC. Exploiting a large thesaurus for information retrieval. Proc RIAO. 1994;94:197–216.

22. Baker PG, Goble CA, Bechhofer S, Paton NW, Stevens R, Brass A. An ontology for bioinformatics applications. Bioinformatics. 1999;15:510–20.

23. Blaschke C, Miguel AA, Ouzounis C, Valencia A. Automatic extraction of biological information from scientific text: protein–protein interactions. ISMB. 1999;7:60–7.

24. Craven M, Kumlien J. Constructing biological knowledge bases by extracting information from text sources. ISMB. 1999;7:77–86.

25. Krauthammer M, Nenadic G. Term identification in the biomedical literature. J Biomed Inform. 2004;37:512–26.

26. Wang H, Zhao T, Tan H, Zhang S. Biomedical named entity recognition based on classifiers ensemble. IJCSA. 2008;5:1–11.

27. Campos D, Matos S, Oliveira JL. Biomedical named entity recognition: a survey of machine learning tools. New York: INTECH Open Access Publisher; 2012.

28. Tsai RTH, Sung C, Dai H, Hung H, Sung T, Hsu W. NERBio: using selected word conjunctions, term normalization, and global patterns to improve biomedical named entity recognition. BMC Bioinform. 2006;7:11.

29. Saha SK, Sarkar S, Mitra P. Feature selection techniques for maximum entropy based biomedical named entity recognition. J Biomed Inform. 2009;42:905–11.

30. Jin-Dong K. Report on Bio-Entity Recognition Task at BioNLP/NLPBA 2004. http://www.nactem.ac.uk/tsujii/GENIA/ERtask/report.html. Accessed Apr 2016.

31. Kim JD, Ohta T, Tsuruoka Y, Tateisi Y. Introduction to the bio-entity recognition task at JNLPBA. In: Proceedings of the international joint workshop on natural language processing in biomedicine and its applications. Association for Computational Linguistics. 2004. p. 70–75.

32. Lafferty J, McCallum A, Pereira F. Conditional random fields: probabilistic models for segmenting and labeling sequence data. In: Proceedings of the eighteenth international conference on machine learning, vol. 1, ICML. 2001. p. 282–289.

33. Wikipedia. https://en.wikipedia.org/wiki/Conditional_random_field. Accessed Apr 2016.

34. Mahmoud A, Pattar A, Hamdulla A. Uyghur stemming using conditional random fields. Int J Sign Process Image Process Pattern Recognit. 2015;8:43–50.

35. Naoaki Okazaki. CRFsuite. http://www.chokkan.org/software/crfsuite/. Accessed April 2016.

36. Collobert R, Weston J, Bottou L, Karlen M, Kavukcuogle K, Kuksa P. Natural language processing (almost) from scratch. J Mach Learn Res. 2011;12:2493–537.

37. Mikolov T, Chen K, Corrado G, Dean J. Efficient estimation of word representations in vector space. In: ICLR. 2013.

38. Pennington J, Socher R, Manning CD. Glove: global vectors for word representation. https://nlp.stanford.edu/projects/glove/. Accessed Apr 2016.

39. Pennington J, Socher R, Manning CD. Glove: Global vectors for word representation. In: Proceedings of the empirical methods in natural language processing. EMNLP. 2014. p. 1532–43.
40. Hotelling H. Relations between two sets of variates. Biometrika. 1936;28:321–77.
41. Härdle W, Simar L. Canonical correlation analysis. In: Applied multivariate statistical analysis. 2007. p. 321–30.
42. Kakade SM, Foster DP. Multi-view regression via canonical correlation analysis. Learning theory. Berlin: Springer; 2007. p. 82–96.
43. Sridharan K, Kakade SM. An information theoretic framework for multi-view learning. In: Conference on learning theory. COLT. 2008. p. 403–14.
44. Stratos K, Collins M, Hsu D. Model-based word embeddings from decompositions of count matrices. In: Proceedings of the annual meeting of the association for computational linguistics. 2015. p. 1282–91.
45. Johansson B, Borga M, Knutsson H. Learning corner orientation using canonical correlation. 2001.
46. Wikipedia. https://en.wikipedia.org/wiki/Artificial_neural_network. Accessed Apr 2016.
47. Hagan MT, Demuth HB, Beale MH, Jesus OD. Neural network design. Boston: PWS Publishing Company; 1996.
48. Data Mining Server (DMS). Neural Networks http://dms.irb.hr/tutorial/tut_nnets_short.php. Accessed Apr 2016.
49. Nielsen MA. Neural networks and deep learning. http://neuralnetworksanddeeplearning.com. Accessed Apr 2016.
50. Mesnil G, He X, Deng L, Bengio Y. Investigation of recurrent-neural-network architectures and learning methods for language understanding. Graz: INTERSPEECH; 2013. p. 3771–5.
51. Gregoire M. Recurrent neural networks with word embeddings. http://deeplearning.net/tutorial/rnnslu.html. Accessed Mar 2016.
52. Elman JL. Finding structure in time. Cognit Sci. 1990;14:179–211.
53. Jordan MI. Serial order: a parallel distributed processing approach. Adv Psychol. 1997;121:471–95.
54. Zhou GD, Su J. Exploring deep knowledge resources in biomedical name recognition. In: Proceedings of the international joint workshop on natural language processing in biomedicine and its applications. Association for Computational Linguistics. 2004. p. 96–9.
55. Ponomareva N, Pla FA, Molina A, Rosso P. Biomedical named entity recognition: a poor knowledge HMM-based approach. New York: LNCS; 2007. p. 382–7.
56. Finkel J, Dingare S, Nguyen H, Nissim M, Manning C, Sinclair G. Exploiting context for biomedical entity recognition: From syntax to the Web. In: Proceedings of the joint workshop on natural language processing in biomedicine and its applications. 2004. p. 88–91.
57. Settles B. Biomedical named entity recognition using conditional random fields and rich feature sets. In: Proceedings of joint workshop on natural language processing in biomedicine and its applications. 2004. p. 104–7.
58. Tsai T, et al. Integrating linguistic knowledge into a conditional random field framework to identify biomedical named entities. Expert Syst Appl. 2006;30:117–28.

Permissions

All chapters in this book were first published in BMEO, by BioMed Central; hereby published with permission under the Creative Commons Attribution License or equivalent. Every chapter published in this book has been scrutinized by our experts. Their significance has been extensively debated. The topics covered herein carry significant findings which will fuel the growth of the discipline. They may even be implemented as practical applications or may be referred to as a beginning point for another development.

The contributors of this book come from diverse backgrounds, making this book a truly international effort. This book will bring forth new frontiers with its revolutionizing research information and detailed analysis of the nascent developments around the world.

We would like to thank all the contributing authors for lending their expertise to make the book truly unique. They have played a crucial role in the development of this book. Without their invaluable contributions this book wouldn't have been possible. They have made vital efforts to compile up to date information on the varied aspects of this subject to make this book a valuable addition to the collection of many professionals and students.

This book was conceptualized with the vision of imparting up-to-date information and advanced data in this field. To ensure the same, a matchless editorial board was set up. Every individual on the board went through rigorous rounds of assessment to prove their worth. After which they invested a large part of their time researching and compiling the most relevant data for our readers.

The editorial board has been involved in producing this book since its inception. They have spent rigorous hours researching and exploring the diverse topics which have resulted in the successful publishing of this book. They have passed on their knowledge of decades through this book. To expedite this challenging task, the publisher supported the team at every step. A small team of assistant editors was also appointed to further simplify the editing procedure and attain best results for the readers.

Apart from the editorial board, the designing team has also invested a significant amount of their time in understanding the subject and creating the most relevant covers. They scrutinized every image to scout for the most suitable representation of the subject and create an appropriate cover for the book.

The publishing team has been an ardent support to the editorial, designing and production team. Their endless efforts to recruit the best for this project, has resulted in the accomplishment of this book. They are a veteran in the field of academics and their pool of knowledge is as vast as their experience in printing. Their expertise and guidance has proved useful at every step. Their uncompromising quality standards have made this book an exceptional effort. Their encouragement from time to time has been an inspiration for everyone.

The publisher and the editorial board hope that this book will prove to be a valuable piece of knowledge for researchers, students, practitioners and scholars across the globe.

List of Contributors

Chang Liu
Department of Information Technology and Engineering, Chengdu University, Chengdu 610106, China
The Clinical Hospital of Chengdu Brain Science Institute, MOE Key Lab for Neuroinformation, Center for Information in Medicine, University of Electronic Science and Technology of China, Chengdu 610054, China
School of Life Science and Technology, University of Electronic Science and Technology of China, Chengdu 610054, China
Key Laboratory of Pattern Recognition and Intelligent Information Processing in Sichuan, Chengdu 610106, China

Xi Yu
Department of Information Technology and Engineering, Chengdu University, Chengdu 610106, China
Key Laboratory of Pattern Recognition and Intelligent Information Processing in Sichuan, Chengdu 610106, China

Xi Wu and JiLiu Zhou
Department of Computer Science, Chengdu University of Information Technology, Chengdu 610225, China

Yuan Yan Tang
Faculty of Science and Technology, University of Macau, Macau, China.

Jian Zhang
School of Physics and Electronic Engineering, Sichuan Normal University, Chengdu, China

Feng Huang
College of Metrology & Measurement Engineering, China Jiliang University, Xueyuan Road 258, Hangzhou, China
State Key Laboratory of Fluid Power and Mechatronic Systems, Zhejiang University, Hangzhou, China

Zhe Gou and Xiaodong Ruan
State Key Laboratory of Fluid Power and Mechatronic Systems, Zhejiang University, Hangzhou, China

Yang Fu
School of Mechanical and Automotive Engineering, Zhejiang University of Science and Technology, Hangzhou, China

Chuan He, Rui Xu, Shenglong Jiang, Feng He and Dong Ming
Lab of Neural Engineering & Rehabilitation, Department of Biomedical Engineering, College of Precision Instruments and Optoelectronics Engineering, Tianjin University, Tianjin, China
Tianjin International Joint Research Center for Neural Engineering, Academy of Medical Engineering and Translational Medicine, Tianjin University, Tianjin, China

Meidan Zhao and Yongming Guo
College of Acupuncture and Massage, Tianjin University of Traditional Chinese Medicine, Tianjin, China

Yi-Horng Lai
Department of Health Care Administration, Oriental Institute of Technology, No. 58, Sec. 2, Sichuan Rd., Banqiao Dist., New Taipei City 22061, Taiwan

Fang-Ming Hsu and Kuo-Hui Yeh
Department of Information Management, National Dong Hwa University, Hualien 97401, Taiwan, ROC

Cheng-Fa Chiang
Department of Information Management, National Dong Hwa University, Hualien 97401, Taiwan, ROC
Physical Education Center, National Dong Hwa University, Hualien 97401, Taiwan, ROC

Pritika Dasgupta
Department of Biomedical Informatics, School of Medicine, University of Pittsburgh, Pittsburgh, PA, USA

Jessie VanSwearingen
Department of Physical Therapy, School of Health and Rehabilitation Sciences, University of Pittsburgh, Pittsburgh, PA, USA

Ervin Sejdic
Department of Electrical and Computer Engineering, Swanson School of Engineering, University of Pittsburgh, Pittsburgh, PA, USA

Rana Sadeghi Chegani and Carlo Menon
School of Mechatronic Systems Engineering and
Engineering Science, Simon Fraser University,
250-13450-102 Avenue, Surrey, BC V3T 0A3, Canada

**Cheung-Hwa Hsu, Chao-Hui Ou, Wei-Lun Hong
and Yu-Han Gao**
Department of Mold and Die Engineering, National
Kaohsiung University of Science and Technology,
Kaohsiung, Taiwan, ROC

Ying Xin, Aili Zhang and Lisa X. Xu
School of Biomedical Engineering, 400 Med-X
Research Institute, Shanghai Jiao Tong University,
1954 Huashan Rd, Shanghai, China

J. Brian Fowlkes
Department of Radiology, University of Michigan
Health System, 3226C Medical Sciences Building I,
1301 Catherine Street, Ann Arbor, MI, USA

Chunlan Yang, Min Lu and Yanhua Duan
College of Life Science and Bioengineering, Beijing
University of Technology, Beijing 100124, China

Bing Liu
Department of Ophthalmology, Hospital of Beijing
University of Technology, Beijing 100124, China

Jarosław Żyłkowski and Grzegorz Rosiak
Second Department of Clinical Radiology, Medical
University of Warsaw, 1a Banacha, 02-097 Warsaw,
Poland

Dominik Spinczyk
Faculty of Biomedical Engineering, Silesian
University of Technology, Roosevelta 40, Zabrze,
Poland

**Xiaoqin Zhou, Haijun Zhang, Mei Feng and Ji
Zhao**
Jilin University, Nan Guan District, Changchun,
China

Yili Fu
Harbin Institute of Technology, Nan Gang District,
Harbin, China

**Hye-Jeong Song, Byeong-Cheol Jo, Chan-Young
Park, Jong-Dae Kim and Yu-Seop Kim**
School of Software, Hallym University, Chuncheon,
South Korea
Bio-IT Research Center, Hallym University,
Chuncheon, South Korea

Index